Mathematical Methods and Economic Theory

Mathematical Methods and Economic Theory

Anjan Mukherji

Subrata Guha

OXFORD

UNIVERSITY PRESS

OXFORD

UNIVERSITY PRESS

YMCA Library Building, Jai Singh Road, New Delhi 110 001

Oxford University Press is a department of the University of Oxford. It furthers the
University's objective of excellence in research, scholarship, and education
by publishing worldwide in

Oxford New York

Auckland Cape Town Dar es Salaam Hong Kong Karachi Kuala Lumpur
Madrid Melbourne Mexico City Nairobi New Delhi Shanghai Taipei Toronto

With offices in

Argentina Austria Brazil Chile Czech Republic France Greece Guatemala
Hungary Italy Japan Poland Portugal Singapore South Korea Switzerland
Thailand Turkey Ukraine Vietnam

Oxford is a registered trademark of Oxford University Press
in the UK and in certain other countries

Published in India
by Oxford University Press, New Delhi

ISBN-13: 978-0-19-806997-3
ISBN-10: 0-19-806997-9

Typeset in CMR 10/13
By Dash Publishing Solutions, Noida, UP

Contents

Preface

According to one of the most well-known textbooks in basic economics, there are three fundamental questions for any economy: 'what should be produced?', 'how to produce?', and 'for whom to produce?'. We need to address similar questions too: 'what does the book contain?', 'how does the book address these questions?', and 'who is the book addressed to?'. It would be convenient to begin with the last since that will have a bearing on the answers to the other two.

Briefly the first question may be answered thus: this book is about economic theory in general and decision-making in particular. It discusses how some fundamental aspects of related theory may be developed and deduced. This axiomatic development is crucial to our approach and the book tries its best to lay bare this method. The book is meant for first year graduate (masters level) students in economics pursuing a course on mathematical methods in economics and for their teachers. This course is so standard and commonplace that its existence does not require any explanation. It will thus be assumed that students have done some basic calculus, real analysis, and linear algebra at the undergraduate level. Thus the tools chapters, which contain the mathematics section, are not as comprehensive as in comparable mathematics books. In principle, however, no prerequisite mathematics is required: we provide definitions and the theorems which will be invoked; and also provide some exercises to check the reader's understanding of the concepts. However, exposure to some undergraduate level mathematics will surely be helpful. We have generally assumed that as a prerequisite, students should have passed mathematics in the Class XII examinations. We begin by presenting proofs in some detail; as the results become more complicated, we state them without proof and try to provide an intuitive explanation wherever possible; whenever the proofs are presented, they are of great importance, since these methods will teach us how to proceed when we encounter similar situations in the realm of economics.

Economic theory is supposed to provide us with a book of rules or a manual of what to expect when we encounter some economic problem. It is very important to follow and understand these rules carefully because in these rules lies the secret of their success. If we do not pay attention, then we may observe that what the rules predict does not take place and, consequently, we may discard theory as being useless. This belief that something may be alright in theory but not sensible in practice is perhaps the worst conclusion that we may arrive at. Sometimes by following rules mechanically, without any understanding, we may arrive at a solution to some problem or the other; most often, though, this may lead us nowhere. And to, therefore, argue that the rules are pointless would be gravely erroneous.

Given the complexity of economic institutions and their interaction with decision-makers, existing theory often needs very strict preconditions before it can be applied. So a proper understanding of these rules will help us identify what these conditions are. Sometimes existing theory is not applicable because of some peculiarities and in such cases, it needs to be modified. Clearly, this is possible only if there is a deep understanding of the background. Note too that these deductions must be made correctly and carefully, with attention to the smallest details; otherwise there is little point in following through with such a scheme.

Often people do not pay attention to rigour and tend to think that the emphasis on rigour is misplaced. And further, people often are seen to argue that there is a trade-off between relevance and rigour. What is the point of such mathematical arguments people may argue? Obviously, there is no point of attempting any argument that is not correct; nor is there any point in wasting time with arguments which are analytically weak. It is not as if there is a choice in the matter. If you wish to have relevant economic theory, then you must be very careful and as rigorous as possible in deriving it: else the theory will be useless. Rigour and relevance must, therefore, necessarily move hand in hand. Now an argument may be easily checked for rigour if the argument is mathematical. Herein lies the great attraction of a mathematical argument: assumptions are clearly stated and steps follow one after another. It is easy to check whether there are any flaws. In literary arguments, rigour is equally desirable; unfortunately it is a lot more difficult to spot flaws. Hence our interest in mathematical arguments stems entirely from this facility in maintaining rigour.

One should be aware, however, that the use of mathematics in formulating models leads to some abstraction, that is, emphasizing certain things and neglecting others. Clearly one hopes that nothing important gets left out. And critics have generally used this aspect to shower abuses on this approach. No less a person than Bertrand Russell (1931) argued: '...many people have a passionate hatred of abstraction, chiefly, I think, because of its intellectual difficulty; but as they do not wish to give this reason, they invent all sorts of others that sound grand. They say that all abstraction is falsification...'

It is by now well established that even to understand and follow developments in basic economic theory, a fairly diverse bag of mathematical tools is required. Further, a lot of economic theory is based on certain fundamental mathematical propositions. The wide acceptance of this fact is also strengthened by the availability of a wide variety of texts with titles such as 'Mathematical Methods in Economics'. We have found that while most of these books are useful, they are not adequate as they develop mathematical tools, without providing convincing or revealing applications to problems in economics. Economics is unlike the natural sciences in that there is a major problem of measurement: functional forms are not known, nor are the values of coefficients known with any degree of exactitude. Consequently learning some mathematics merely to apply to numerical problems which are unlikely to be encountered does not seem to be the right approach. Our approach is an attempt to set matters right so far as this aspect is concerned. What we try to do is to provide an understanding of the mathematical propositions necessary for the fundamentals of economic theory. Consequently the applications of these mathematical propositions have to be suited to the needs of economics.

As a brief resume, the book primarily deals with problems of decision-making under various situations. Beginning with the simplest possible individual choice problem, we discuss the problem of aggregation of individual choice. This naturally leads to a discussion of two distinct types of decisions: demand and supply (Section I). We also discuss problems of interactive decision-making when one person's decision is affected by the decision taken by others (game theoretic issues); we provide a discussion of decision-making under uncertainty as well. All these problems may be classified as problems of static decision-making: where *time* does not play a role (Section II). We make up for this omission when we consider some types of decision-making over time and the kinds of problems which need to be addressed in such situations (Section III). At the end of each section, we provide a brief set of further readings, including material from both mathematics and economics. The mathematics references are to the material that we have used ourselves; other equivalent sources can also be used. For example, a good reference to mathematical analysis will have different answers and we have used the ones that we found the most convenient. The economics references are all to books, articles, and surveys which we found the most instructive and relevant for the task at hand. Naturally these references are not exhaustive but we do hope that our treatment will encourage some inquisitive minds to look at them. Finally, the exercises in the mathematics sections are to test the reader's understanding of definitions and the constructs. Some of them have been collected from standard references and from problem sets that our colleagues have used in the past.

To reiterate, our objective is to analyse some basic problems in economic theory that is, the problems associated with decision-making under various alternative settings. We set these problems out and develop the mathematics that we have found necessary for this purpose. We are sure that it is only with the help of such tools that all steps will be clear. There is a beautiful cartoon by Sidney Harris, available on the internet, titled 'I-think-you-should-be-more-explicit-here-in-step-two' which is worth considering to drive home this point.[1] Our attempt in this book has been to follow the Professor's advice as well as we could.

The book began as a project of the Reserve Bank of India Unit at the Centre for Economic Studies and Planning of the Jawaharlal Nehru University, New Delhi. The authors were both members of the Unit and without implicating the RBI in any way,

---

[1]Step two of an argument, in the cartoon, is a wish: we need a miracle here. The Professor asks that this be made more explicit.

the support received from them is gratefully acknowledged. The book was developed over a fairly long period and within a year of its development, the RBI Unit had shrunk, with Subrata Guha having moved to an open position at the Centre of Economic Studies and Planning while Anjan Mukherji remained with the Unit as the RBI Chair Professor of Economic Theory. The work proceeded with Subrata Guha being primarily responsible for the material in Chapters 2 and 14 and sections 15.3 and 15.4 of Chapter 15, while Anjan Mukherji was responsible for all the remaining chapters. We are, of course, indebted to the many sources that we have used to put this text together and these have been mentioned. Several versions were circulated and comments were received from Amitava Bose, Satish Jain, Dipankar Dasgupta, Soumyen Sikdar, Krishnendu Ghosh Dastidar, Uday Bhanu Sinha, Mousumi Das, Subir Chakrabarti, Rajendra Prasad Kundu, Tridip Ray and, unless already named, a couple of reviewers of the Oxford University Press (OUP). We are grateful to them all. We have tried to accommodate as many of the recommendations as we could. Some of our students also helped and special mention must be made of Amarjyoti Mahanta who cheerfully helped us gather exercises and check various matters which arose from time to time. Nitasha Devasar of OUP prompted us from time to time and kept us on course. Surely without these interventions the book could not have been completed. The final version of the book benefitted greatly from stylistic suggestions from the editorial team at OUP. We are grateful to all of them for their support. Finally, we are indebted to our families for their contribution to this project: Anjan Mukherji thanks his wife Shormila and son Arnab for having made the project possible. Arnab also helped whenever the programme LaTex, in which the manuscript was prepared, did not perform the desired task, and this was often. Subrata Guha thanks his wife Ranja for support throughout this endeavour. The book would not have been completed without the support we received from this quarter.

SECTION ONE

- Introduction
- Basic Mathematical Logic
- Set Theory
- Functions of a Single Variable
- Economic Applications I: Choice, Utility, and Aggregation

CHAPTER ONE

Introduction

1.1 The Objective

We shall be concerned in Section I with the problem of choice. In its starkest form, a decision-maker has to choose from among a collection of alternatives. A choice can be made on the basis of a criterion using which the decision-maker compares and ranks alternatives. Such an exercise is a prerequisite for any problem of choice. The first question we shall analyse is what kinds of criteria permit choice.

For example, consider that you have to choose one person from a class of 30 students for a prize. To begin with, you need to know what the prize is for. Say it is to be given for the highest score in a recently conducted examination. So you need to look at the scores of these students and choosing the winner reduces to the simple matter of looking at those who have the highest scores. Thus, we need to know the class of students (the alternatives) and their scores in the examination (the criterion being the highest score in that examination); the choice then is easy and straightforward.

Note that the criterion leads us to the scores which in turn enable us to rank the students in the class from those who have the highest scores to the ones who have the lowest scores. Thus, the criterion provides the ranking among the alternatives and we shall use the words criterion and ranking interchangeably. Our first task is to analyse the nature of rankings over alternatives which permit a choice to be made, that is, to select a 'best' according to this ranking.

A second, related question, concerns the nature of ranking itself. To continue with our class of 30 students, if the criterion, instead of being scores in an examination, had been qualitative, how would we have proceeded? Suppose that you are interested in choosing the healthiest in the class. Now one way to proceed is to construct an index of health and then use this index to arrive at a number associated with each student, which would reflect the health of that student. Once again we are in the familiar territory of ranking students according to a numerical score. Thus we need to ask whether all criteria or rankings can be provided with such a numerical representation.

A third and final problem for Section I would be to reconsider the second question. It should be clear that attaching a numerical score to being healthy, for instance, is likely to be arbitrary. So is it possible that we get a committee (of at least two experts) to provide their ranking of healthy students and then use their conclusions to arrive at an aggregation of their ranking?

The first is the problem of choice; the second is the problem of representability whereas the third is the problem of aggregation. These are the target problems for Section I, in a sense, and constitutes the very basic set of problems in economic theory; not only are they of interest in themselves but are also crucial to our discussions in later sections. We shall set up and define our targets for each of the sections.

1.2 The Tools for Section I

We begin with a chapter in mathematical logic which will help us construct claims and their proofs. We proceed next to a discussion of sets and relations and operations with sets. In particular we discuss the set of real numbers and their properties. We go on to discuss the notion of a function of a single variable, its continuity and differentiability. As we have mentioned earlier, these are in the nature of review sections, where we present all that is essential for our purpose. We also present some exercises to check whether these concepts and results have been properly grasped.

KEY TERMS

Aggregation
Choice
Criterion
Ranking
Representability

CHAPTER TWO

Basic Mathematical Logic

2.1 Introduction

Any economic theory begins by assuming the truth of certain assertions (about the objects that it studies) known as the assumptions of the theory. The theory then claims that, given the truth of these assumptions, we can conclude that certain other assertions, known as the implications of the theory, must also be true. If this claim is to carry conviction the theory must be able to demonstrate that, according to some generally accepted rules of correct reasoning, whenever the assumptions of the theory are true any given implication of the theory is also true.

Whether or not a clearly specified set of rules can be shown to underpin a specific process of reasoning is important because there are significant benefits associated with rule-based reasoning. Rules eliminate inconsistencies in the process of reasoning. This also ensures that if we criticize the kind of reasoning used in a theory whose conclusions we are predisposed to reject, we cannot accept the same kind of reasoning in a theory whose conclusions we are predisposed to accept. Therefore, proponents of alternative theories, despite their differences, can agree to adhere to certain standards of argument. Moreover, rule-based reasoning ensures that the underlying bases for a process of reasoning are transparent and with the resulting absence of ambiguity, the chain of reasoning can be independently verified by persons other than those advancing the theory. Finally, since rules of reasoning are applied to derive the truth of a theory's implications from the truth of its assumptions, all underlying assumptions of the theory must be clearly stated. This helps to make a discussion regarding the theory more fruitful because it reduces the possibility of disagreement arising from a misunderstanding about the theory's assumptions.

The relative success of the experimental sciences in predicting the truth of exactly verifiable assertions, which are the implications of its theories, suggests that the method of reasoning used in these sciences provides a possible basis for developing generally acceptable rules of reasoning. The exact quantitative predictions in these sciences are derived using the language and the results of mathematics. Therefore, one way to develop generally acceptable rules of correct reasoning is to study the nature of reasoning used in mathematics and to frame a corresponding set of rules of reasoning which can be used to obtain most, if not all, mathematical results. This is the objective of the discipline of mathematical logic and the rules of reasoning developed within the discipline may be referred to as 'the rules of mathematical logic' or, more simply, as 'the rules of logic'. Note that the rules of mathematical logic provide us, in turn, with sets of rules which can support any given process of mathematical reasoning used in economic theory.

In the context of our discussion, a rule of reasoning could naturally be taken to mean a kind of rule which states that we can conclude certain things about the truth or falsehood of certain assertions, if certain conditions as regards the truth or falsehood of certain other assertions are satisfied. There are other kinds of rules as well in the discipline of mathematical logic and we will, therefore, refer to rules of the above kind as truth rules of logic.

Our shared judgements about the correctness of the process of reasoning used in a theory cannot be based on our beliefs about the truth or falsehood of the theory's assumptions. Supporters and detractors of a theory might disagree about the truth of its assumptions. Frequently, both will agree that most of the theory's assumptions representing a simplified picture of reality are false (not exactly true), although they might disagree about the degree to which these assumptions 'approximate reality', that is, the degree to which they are false. However, the basic test which any theory has to pass is whether we can conclude, using the truth rules of logic, that its implications would be exactly true *if* its assumptions were exactly true. This suggests that truth rules of logic should be based on connections between the truth or falsehood of different assertions revealed by the *form* or *structure* of these assertions and must be independent of beliefs about the truth or falsehood of these assertions, which would depend on their content.

Consider the following pair of assertions: 'It is not true that economic growth reduces poverty' and 'It is not true that redistribution of wealth reduces economic growth'. Different persons may have different combinations of beliefs about the truth of these assertions. However, note that both the assertions have the form 'It is not true that A' where A can represent either of the two assertions 'Economic growth reduces poverty' and 'Redistribution of wealth reduces economic growth'. Whatever assertion A stands for, two persons with different beliefs about the truth of the assertions can still agree that an assertion of the form 'It is not true that A' is true if the assertion A is false and is false if the assertion A is true. Note, however, that in the English language the same assertion can take a multiple number of forms. Depending on the context in which the assertion is made 'It is not true that $2 + 2$ is equal to 5' can alternatively be expressed as '$2 + 2$ can never be equal to 5' or 'If $2 + 2$ is equal to 5 then pigs can fly!'. If truth rules of logic had to be defined for all these alternative forms, the set of rules would become unmanageably large. Therefore, before setting out any set of truth rules of logic we need to outline a system of notation by symbols which allows us to transform alternative forms of the same assertion in the English language into the same form in terms of symbols. The truth rules of logic are then defined for the symbolic forms of assertions obtained using this system of notation.

In this chapter, we will discuss the set of truth rules of logic belonging to an elementary branch of the subject known as **sentential logic**. We will introduce the system of notation used for this set of rules, define the rules, and then discuss how using a decision procedure we can check in simple cases whether a process of reasoning is correct according to these truth rules. Broadly speaking, a process of reasoning which is correct according to some set of truth rules is described in logic as being *logically valid* according to those rules. The main objective of this chapter is to use the simple truth rules of sentential logic to introduce to the reader the notion of logical validity and the related notions of logical consistency and logical independence.

For any economic theory, besides the question of whether the truth of the implications of a theory follows by reasoning from the truth of its assumptions, two other questions are also important. The implications of a theory are supposed to be true if *all* the assumptions of a theory are true. However, a theory is of no use if we can, simply by applying rules of reasoning, rule out the possibility of a situation in which all the assumptions of the theory are simultaneously true. In other words, it is necessary to ensure that the assumptions of a theory are *consistent*. The assumptions of a theory may also be viewed as providing us with a set of features of the world which give rise to the phenomena described by the implications of the theory. However, if the truth of some of these assumptions follows, simply by applying rules of reasoning, from the truth of the other assumptions, the set of conditions which are sufficient for the implications of the theory to be true may actually be smaller than made out by the theory. It is, therefore, often important to ensure that the assumptions of a theory are *independent*. This chapter, therefore, also discusses the meaning of **consistency** and **independence** of a set of assertions and how we can establish or question them according to the set of truth rules of sentential logic.

We end the chapter by introducing a system of notation and a set of truth rules which are extensions of those in sentential logic. These are widely used in subsequent chapters and form the basis of the branch of logic known as **predicate logic**.

2.2 Sentential Logic

2.2.1 Sentences, truth values, and notations

From the discussion in the previous section it is clear that the main objects dealt with by rules of logic are assertions or, more precisely, assertive sentences. The characteristic of assertive sentences which sets them apart from other types of sentences, as far as the rules of logic are concerned, is that one can meaningfully attach the qualities of truth or falsehood to these sentences. For example, 'All firms sell an identical product' is an assertive sentence. Both the claims 'It is true that all firms sell an identical product' and 'It is false (not true) that all firms sell an identical product' are meaningful. On the other hand, it is meaningless to claim that the sentences 'Are not the products of all the firms identical?' or 'How identical indeed are the products of all the firms!' are either true or false. In discussing the rules of logic we shall refer to assertive sentences or assertions as **sentences**.[1]

Before proceeding further we state our first and most basic truth rule of logic:

Principle of Bivalence: For any sentence exactly one of the following assertions is true:

(i) The sentence is true,

(ii) The sentence is false.

The Principle of Bivalence rules out the possibility that sentences can have degrees of truth and are, to some degree, both true and false (not true). The Principle also implies that when we have considered the possibility that a sentence is true and the possibility that the sentence is false (not true) we have considered all possible states that a sentence can possess as regards the quality of truth. In mathematical logic, this is usually expressed by saying that a sentence must have one and only one of two truth values. If a sentence is true we shall say it has the truth value T. If it is false we shall say that it has the truth value F.[2]

In sentential logic, sentences are denoted either by **sentential letters**, which are upper case letters from the English alphabet with or without numerical subscripts, for example, A, P_1, A_{23}, or are denoted using a combination of sentential letters, parentheses, and logical symbols called **sentential connectives**. There are five sentential connectives, namely, $\sim$, $\wedge$, $\vee$, $\rightarrow$, and $\leftrightarrow$ which are used in the following way:

(i) A sentence may be denoted by '$\sim (P)$' just in the case that there is a sentence P such that the sentence is false if P is true and the sentence is true if P is false. The sentence $\sim (P)$ is read as 'not P' and is known as 'the **negation** of P'. If $P = $ ('is identical with') 'The price of rice will rise' then the sentence 'The price of rice will not rise' may be denoted by '$\sim (P)$'.

(ii) A sentence may be denoted by '$(P \wedge Q)$' just in the case that there is a pair of sentences P and Q such that the sentence is true if both P and Q are true and the sentence is false if at least one of P and Q is false. The sentence $(P \wedge Q)$ is read as 'P and Q' and is referred to as 'the **conjunction** of P and Q'. If $P = $ 'The price of rice will rise' and $Q = $ 'The price of wheat will fall' then the sentence 'The price of rice will rise but the price of wheat will fall' may be denoted by '$(P \wedge Q)$'.

(iii) A sentence may be denoted by '$(P \vee Q)$' just in the case that there is a pair of sentences P and Q such that the sentence is true if at least one of

---

[1] Many texts in logic use the term 'proposition' instead of 'sentence'.

[2] Sometimes the symbols $\top$ and $\bot$ (inverted $\top$) are used instead of T and F, respectively.

P and Q is true and the sentence is false if both P and Q are false. The sentence $(P \vee Q)$ is read as 'P or Q' and is referred to as 'the **disjunction** of P and Q'. If $P =$ 'The government can increase employment by increasing expenditure' and $Q =$ 'The government can raise employment by cutting taxes' then the sentence 'The government can raise employment by increasing expenditure or cutting taxes' may be denoted by '$(P \vee Q)$'. However, if $P =$ 'The price of rice will rise' and $Q =$ 'The price of wheat will fall' then the sentence 'Either the price of rice will rise or the price of wheat will fall' cannot be denoted by '$(P \vee Q)$' because the sentence is not true if both P and Q are true.

(iv) A sentence may be denoted by '$(P \rightarrow Q)$' just in the case that there is a pair of sentences P and Q such that the sentence is true if either P is false or both P and Q are true and the sentence is false if P is true and Q is false. The sentence $(P \rightarrow Q)$ is read as 'If P then Q', 'P only if Q', 'P is a sufficient condition for Q', or 'Q is a necessary condition for P'. The sentence $(P \rightarrow Q)$ is referred to as 'the **conditional sentence** with **antecedent** P and **consequent** Q'. The word '**implication**' is often used instead of the expression 'conditional sentence'. Moreover, the expression 'antecedent P and consequent Q' may be replaced with the expression '**left member** P and **right member** Q'. If $P =$ 'The price of rice will rise' and $Q =$ 'The demand for rice will fall' then the sentence 'If the price of rice increases the demand for rice will fall' may be denoted by '$(P \rightarrow Q)$'. However, if $P =$ 'India is a free market economy' and $Q =$ 'India is an industrialized nation', the sentence 'If India was a free market economy then India would be an industrialized nation' cannot be denoted by '$(P \rightarrow Q)$' because it is possible to consider the given sentence to be false even though P is false.

(v) A sentence may be denoted by '$(P \leftrightarrow Q)$' just in the case that there is a pair of sentences P and Q such that the sentence is true if either both P and Q are true or both P and Q are false and the sentence is false if one of the sentences P and Q is false and the other is true. The sentence $(P \leftrightarrow Q)$ is read as 'P if and only if (iff) Q' or 'P is a necessary and sufficient condition for Q'. The sentence $(P \leftrightarrow Q)$ is referred to as 'the **biconditional sentence** with left member P and right member Q'. The word '**equivalence**' is often used instead of the expression 'biconditional sentence'. If $P =$ 'The price of rice will rise' and $Q =$ 'The demand for rice will fall' then the sentence 'The demand for rice will fall just in the case that the price of rice rises' may be denoted by '$(P \leftrightarrow Q)$'.

A number of symbols written down one after the other without any spaces in between, the symbols being either sentential letters, sentential connectives, or parentheses will be called an **SL string**, for example, $(DS \wedge \vee S) \rightarrow A)$, $(\sim (P \rightarrow Q) \wedge Q)$. We will denote a string by the letters p, q, and r, with or without numerical subscripts.

An SL string will be called a **well-formed formula in sentential logic (SLwff)** iff its being so can be justified solely by one or more of the following rules:

1. Any sentential letter is an SLwff,
2. For any string p, if p is an SLwff then $\sim(p)$ is an SLwff,
3. For any two strings p and q, if p and q are SLwffs then $(p \wedge q)$, $(p \vee q)$, $(p \to q)$, and $(p \leftrightarrow q)$ are all SLwffs.

Note that among the two examples of SL strings given above, only the second one is an SLwff according to the above definition. By the first rule 'P' and 'Q' are SLwffs. Therefore, by the third rule, '$(P \to Q)$' is an SLwff. Therefore, by the second rule, '$\sim(P \to Q)$' is an SLwff. Moreover, because 'Q' is also an SLwff, by the third rule '$(\sim(P \to Q) \wedge Q)$' is an SLwff.

Consider an SLwff which has at least one sentential connective. Suppose we are allowed to add only a single sentential connective and (if necessary) parentheses to some SLwff or pair of SLwffs in order to obtain the given SLwff. The main connective of the given SLwff is the sentential connective which must then be added to obtain the given SLwff. Thus, $\wedge$ is the main connective of '$(\sim(P \to Q) \wedge Q)$', $\sim$ is the main connective of '$\sim(P \to Q)$' and $\to$ is the main connective of '$(P \to Q)$'. We will refer to the largest SLwffs to the left and right of the main connective of an SLwff as the left member and right member, respectively of the SLwff. For example, the left and right members of '$(\sim(P \to Q) \wedge Q)$' are '$\sim(P \to Q)$' and 'Q', respectively. An SLwff with $\sim$ as its main connective has no left member.

We use certain conventions for writing SLwffs which reduce the number of parentheses used in SLwffs and make reading SLwffs easier. The conventions are the following:

1. If $\sim$ is not the main connective of an SLwff, the outermost parentheses of the SLwff may be omitted. For example, '$(\sim(P \to Q) \wedge Q)$' can be written as '$\sim(P \to Q) \wedge Q$'.
2. If the main connective of an SLwff is either a '$\to$' or a '$\leftrightarrow$' then the outermost parentheses of the left and right members of the main connective may be omitted provided the main connective of the concerned member is not a '$\sim$','$\to$' or a '$\leftrightarrow$'. For example, '$(P \leftrightarrow Q) \leftrightarrow ((P \to Q) \wedge (Q \to P))$' can be written as '$(P \leftrightarrow Q) \leftrightarrow (P \to Q) \wedge (Q \to P)$'.
3. $\sim(p)$, where p is an SLwff consisting of a single sentential letter, may be written as $\sim p$ in any SLwff. For example, '$(\sim(P \wedge \sim(P)))$' can be written as '$\sim(P \wedge \sim P)$'.
4. Convention 2 applies for any SLwff which is written with its outermost parentheses as part of any other SLwff. For example, $((P \wedge Q) \to P) \to P \vee \sim P$ may be written as $(P \wedge Q \to P) \to P \vee \sim P$.

Sentences may be classified into atomic and compound sentences. A sentence is an **atomic sentence** iff it can only be denoted by an SLwff consisting of a single sentential letter. For example, 'The price of rice will rise' is an atomic sentence. The sentence 'Tea and coffee are substitutes' is an atomic sentence because even though the word 'and' appears in the sentence, the sentence cannot be written as the conjunction of two sentences. For example, the sentence cannot be written as 'Tea is a substitute and coffee is a substitute'. The set of atomic sentences is not empty even though every sentence can be expressed as the negation of some other sentence because we will adopt the convention that a sentence if denoted by P cannot also be denoted as $\sim (\sim P)$. That is, we will distinguish between a sentence such as 'The price of rice will rise' and its double negation 'It is not true that the price of rice will not rise'. If a sentence is not an atomic sentence we will refer to it as a compound sentence. Thus, the sentence 'Neither will prices fall nor will incomes rise' is a **compound sentence** because it can be denoted by the SLwff '$\sim P \wedge \sim Q$' where $P = $ 'Prices will fall' and $Q = $ 'Incomes will rise'.

Note that **if all the sentential letters in any SLwff are taken to denote particular sentences, the SLwff is always taken to denote a sentence.** The clarification is necessary because there are SLwffs which if read by substituting sentences for sentential letters and the expressions 'not', 'and', 'or', 'if ... then', 'if and only if', respectively for $\sim, \wedge, \vee, \rightarrow$, and $\leftrightarrow$, seem to result in expressions which appear meaningless in the English language and, therefore, not countable as sentences. For example, if $P = $ 'The price of rice will rise' and $Q = $ '$2 + 2 = 4$' then the SLwff '$(P \rightarrow Q)$' may be read as 'If the price of rice will rise then $2 + 2 = 4$'. This appears to be a meaningless sentence unless we consider the circumstances permitting the usage of the sentential connective '$\rightarrow$' and note that this actually represents the assertion that 'At least one of the sentences "The price of rice will not rise" and "$2 + 2 = 4$" is true'.

2.2.2 Truth rules and truth tables

The most elementary set of truth rules of logic (the truth rules of sentential logic) are the following:

1. **Principle of Bivalence:** explained in Section 2.2.1.
2. **Truth Rule for Negation:**
   Let p be any SLwff.

   (i) If a sentence denoted by p is true then the sentence denoted by $\sim (p)$ is false and conversely.
   (ii) If a sentence denoted by p is false then the sentence denoted by $\sim (p)$ is true and conversely.

3. **Truth Rule for Conjunction:**
   Let p and q be any two SLwffs.

   (i) If a pair of sentences, denoted by p and q, are true then the sentence denoted by $(p \wedge q)$ is true and conversely.

(ii) If at least one of a pair of sentences, denoted by p and q, is false then the sentence denoted by $(p \wedge q)$ is false and conversely.

4. **Truth Rule for Disjunction:**
   Let p and q be any two SLwffs.

   (i) If at least one of a pair of sentences, denoted by p and q, is true then the sentence denoted by $(p \vee q)$ is true and conversely.

   (ii) If a pair of sentences, denoted by p and q, are false then the sentence denoted by $(p \vee q)$ is false and conversely.

5. **Truth Rule for Implication:**
   Let p and q be any two SLwffs.

   (i) If a sentence denoted by p is false or a sentence denoted by q is true then the sentence denoted by $(p \to q)$ is true and conversely.

   (ii) If a sentence denoted by p is true and a sentence denoted by q is false then the sentence denoted by $(p \to q)$ is false and conversely.

6. **Truth Rule for Equivalence:**
   Let p and q be any two SLwffs.

   (i) If a pair of sentences, denoted by p and q, are both true or both false then the sentence denoted by $(p \leftrightarrow q)$ is true and conversely.

   (ii) If one of a pair of a sentences, denoted by p and q, is true and the other false then the sentence denoted by $(p \leftrightarrow q)$ is false and conversely.

From here onwards, we shall use the notations $\wedge_{i=1}^{n} A_i$ and $\vee_{i=1}^{n} A_i$ to denote respectively the sentences denoted by $((...(((A_1 \wedge A_2) \wedge A_3) \wedge A_4) \wedge ...) \wedge A_n)$ and $((...(((A_1 \vee A_2) \vee A_3) \vee A_4) \vee ...) \vee A_n)$, where n is a positive integer. From the truth rule for conjunction we can conclude that the sentence $\wedge_{i=1}^{n} A_i$ is true whenever all the sentences $A_1, A_2, ..., A_n$ are true and is false when at least one of these sentences is false. In contrast, from the truth rule for disjunction it follows that $\vee_{i=1}^{n} A_i$ is true whenever at least one of the sentences $A_1, A_2, ..., A_n$ is true and is false when each of these sentences is false.

Suppose a compound sentence is denoted by an SLwff with at least one sentential connective and with sentential letters denoting component sentences. The above truth rules of logic can be used to find the truth value of the compound sentence for any arbitrary assignment of truth values to these component sentences. Consider the following example. Suppose there is a sentence denoted by the SLwff '$(\sim Q \wedge P) \vee \sim P$' where '$P$' and '$Q$' denote sentences. Suppose the sentences P and Q are both assigned the truth value 'T', that is, P and Q are both assumed to be true. Then, we know from the truth rule for negation that the sentence $\sim Q$ is false. Applying the truth rule for conjunction it then follows that the sentence $\sim Q \wedge P$ is false. From the truth rule for negation we can conclude that the sentence $\sim P$ is false. Therefore, applying the truth rule for disjunction we can conclude that the sentence $(\sim Q \wedge P) \vee \sim P$ is false.

This constitutes a process of reasoning which begins from an assignment of truth values to component sentences and proceeds in stages. At each stage, the truth value of a sentence is derived through application of a truth rule of logic from truth values assumed or derived for other sentences at previous stages. The process ends with the derivation of the truth value of the given sentence at the final stage. For the given assignment of truth values to the component sentences we can tabulate the sentences considered in this process and their corresponding truth values as follows:

Q	P	$\sim Q$	$\sim Q \wedge P$	$\sim P$	$(\sim Q \wedge P) \vee \sim P$
T	T	F	F	F	F

Repeating this process for every assignment of truth values to the component sentences we can obtain a table showing the derivation (using truth rules of logic) of the truth value of the given sentence for every assignment of truth values to its component sentences. This table, given below, is known as a **truth table** for the sentence $(\sim Q \wedge P) \vee \sim P$.

Q	P	$\sim Q$	$\sim Q \wedge P$	$\sim P$	$(\sim Q \wedge P) \vee \sim P$
T	T	F	F	F	F
T	F	F	F	T	T
F	T	T	T	F	T
F	F	T	F	T	F

Note, in particular, that the entry in each cell of the final column of a truth table for a sentence tells us what we can conclude about the truth or falsehood of any sentence by: (a) assuming the assignment of truth values to component sentences given in the row containing the cell; and (b) applying the truth rules of sentential logic. Note also that all sentences which can be denoted by the same SLwff will have the same truth table. Thus, sentences which share the same form or structure according to the system of notation used in sentential logic have the same truth value if component sentences denoted by the same sentential letter are assigned the same truth value.

2.2.3 Tautologies, contradictions, and contingent sentences

When a sentence is denoted by an SLwff where each sentential letter denotes a component *atomic* sentence, the truth table for that sentence reveals the relation established by the truth rules of logic between the truth values of the sentence and the truth values of its component atomic sentences. Based on the nature of this relation sentences can be usefully classified into three types.

A sentence is a **tautology** or **logical truth** iff we can, by using the truth rules of logic, conclude that the sentence is true irrespective of the truth values of its component atomic sentences. Suppose a truth table for a given sentence is drawn up by assigning truth values to its component atomic sentences. It follows that the sentence is a tautology just in the case that every entry in the final column of this table is 'T'. Note also that, for any assignment of truth values to the component atomic sentences of a given sentence, the only possible truth values for any component sentence are 'T' and 'F'. Therefore, if we find any truth table for a given sentence where every entry in the final column is 'T' we can conclude the sentence is true for every assignment of truth values to component atomic sentences, that is, the sentence is a tautology.

The significance of a tautology is that it represents a sentence which can be adjudged to be true simply by applying the truth rules of logic (without assuming anything about the truth or falsehood of any specific sentences). Thus, a tautology is true due to its form or structure (the way its component sentences are strung together with sentential connectives) and not due to the content or meaning of its various parts.

Given any two sentences A and B, we say that A **tautologically implies** B iff the sentence $A \to B$ is a tautology. Therefore, A tautologically implies B iff, irrespective of the truth values of the atomic sentences in A and B, we can conclude, simply by using the truth rules of logic, that whenever A is true B is also true. Similarly, we say that A is **tautologically equivalent** to B iff the sentence $A \leftrightarrow B$ is a tautology. The tautological equivalence of A and B signifies that simply by applying the truth rules of logic we can conclude that A and B are either both true or both false.

Consider, for example, the sentence 'A fall in the price of wheat is a sufficient condition for the quantity demanded of rice to fall if the price of rice rises, iff a fall in the wheat price and a rise in the price of rice is a sufficient condition for the quantity demanded of rice to fall'. Let W = 'There is a fall in the price of wheat', R = 'There is a rise in the price of rice' and Q = 'The quantity demanded of rice will fall'. The given sentence can then be denoted by the SLwff '$(W \to (R \to Q)) \leftrightarrow (W \wedge R \to Q)$'. Every entry in the final column of the truth table for this SLwff, given below, has the entry 'T'. It follows that the sentence is a tautology. Further, the sentences $W \to (R \to Q)$ and $W \wedge R \to Q$ are tautologically equivalent.

R	Q	W	$R \to Q$	$W \wedge R$	$W \to (R \to Q)$	$W \wedge R \to Q$	$(W \to (R \to Q))$ $\leftrightarrow (W \wedge R \to Q)$
T	T	T	T	T	T	T	T
T	T	F	T	F	T	T	T
T	F	T	F	T	F	F	T
T	F	F	F	F	T	T	T
F	T	T	T	F	T	T	T
F	T	F	T	F	T	T	T
F	F	T	T	F	T	T	T
F	F	F	T	F	T	T	T

A sentence is a **contradiction** or **logical falsehood** iff we can, by using the truth rules of logic, conclude that the sentence is false irrespective of the truth values of its component atomic sentences. For example, given any sentence A, the sentence

$A \wedge \sim A$ is a contradiction. We can check whether a sentence is a contradiction using the same procedure we use to check whether it is a tautology, except that all entries in the final column of the truth table now have to be 'F' rather than 'T'. Given the principles for construction of a truth table, it is obvious that *a sentence is a contradiction iff its negation is a tautology and a sentence is a tautology iff its negation is a contradiction.* A sentence is a **contingent sentence** iff it is neither a tautology nor a contradiction.

2.2.4 Logical consequence and the validity of arguments

A sentence is said to be a **logical consequence** of a non-empty set of sentences (or, of the sentences in the set) iff we can conclude, simply by applying the truth rules of logic, without assuming anything particular about the truth or falsehood of other sentences, that the given sentence is true whenever every sentence in the given set is true. From the discussion in the previous section it follows that a sentence B is a logical consequence of sentences $A_1, A_2, ..., A_n$ iff the sentence $\wedge_{i=1}^{n} A_i$ tautologically implies B. The expression 'is a logical consequence of' is used interchangeably with the expressions 'is a logical implication of', 'is logically implied by', 'can be logically derived from', and 'logically follows from.'

An **argument** is a sequence of sentences which taken together assert that a given sentence (known as the **conclusion** of the argument) is the logical consequence of a set of sentences (known as the **premises** of the argument). The following is an example of an argument: *Agriculture will grow only if there is public investment in agriculture. Either agriculture will grow or the home market will not grow. Firms will not try to reduce competition only if the home market grows. There is no public investment in agriculture. Therefore, firms will try to reduce competition.* The first four sentences in italics are the premises of the argument and the last sentence is its conclusion.

An argument is **logically valid** just in the case that its conclusion logically follows from its premises. Therefore, given an argument with premises $P_1, P_2, ..., P_n$ (n being a positive integer) and conclusion C, the argument is logically valid iff the sentence $\wedge_{i=1}^{n} P_i \rightarrow C$ is a tautology. An argument is **logically sound** just in the case that it is logically valid and all the premises of the argument are true. Thus, a sound argument is always valid but a valid argument is not always sound.

From the definition of logical validity of an argument it follows that one way to check whether an argument with premises $P_1, P_2, ..., P_n$ and conclusion C is logically valid is to construct a truth table for the sentence $\wedge_{i=1}^{n} P_i \rightarrow C$ and consider assignments of truth values to all component atomic sentences. If each entry in the final column of the table is a 'T' the argument is valid, otherwise not. However, if the number of atomic sentences in $\wedge_{i=1}^{n} P_i \rightarrow C$ is very large then constructing the necessary truth table would be extremely tedious. If there are m distinct atomic sentences in the premises and conclusions of the argument then a truth table with 2^m rows will need to be constructed. Therefore, an alternative method is adopted which involves construction of a **proof** (or more correctly, a formal proof) for the argument. We will not consider the precise definition of a proof in sentential logic but content ourselves with presenting the basic idea. Accordingly, we will simply define a proof as a chain of reasoning which begins by assuming that the premises of the argument are true and then uses just these assumptions in conjunction with the truth rules of logic to assert that the conclusion must also be true.

Consider the following argument: *The outcome will be socially optimal only if no houses are built on the flood plain. If at least one house is built on the flood plain then the government will spend on flood control. If developers believe that the government will spend on flood control even if one house is built on the flood plain, developers will build houses on the flood plain. If developers build houses on the flood plain and the government spends on flood control then the developers will be better off. A necessary condition for no houses to be built on the flood plain is that developers will not build houses on the flood plain. Developers believe that the government will spend on flood control even if one house is built on the flood plain. Therefore, the outcome will not be socially optimal but developers will be better off.*

Let O = 'The outcome will be socially optimal.'
H = 'No houses will be built on the flood plain.'
G = 'The government will spend on flood control.'
D = 'Developers believe that the government will spend on flood control even if one house is built on the flood plain.'
B = 'Developers will build houses on the flood plain.'
E = 'Developers will be better off.'

The argument can then be represented as follows with premises separated from the conclusion by the horizontal line:

$$O \rightarrow H$$
$$\sim H \rightarrow G$$
$$D \rightarrow B$$
$$B \wedge G \rightarrow E$$
$$H \rightarrow \sim B$$
$$D$$
$$\overline{}$$
$$\sim O \wedge E$$

The following can then be considered to be a proof of the above argument.

1. Suppose $O \rightarrow H$ is true.
2. Suppose $\sim H \rightarrow G$ is true.
3. Suppose $D \rightarrow B$ is true.
4. Suppose $B \wedge G \rightarrow E$ is true.
5. Suppose $H \rightarrow \sim B$ is true.
6. Suppose D is true.

7. From line (3) and line (6), by the truth rule for implication, B is true.
8. From line (7), by the truth rule for negation, $\sim B$ is false.
9. From line (5) and line (8), by the truth rule for implication, H is false.
10. From line (1) and line (9), by the truth rule for implication, O is false.
11. From line (10), by the truth rule for negation, $\sim O$ is true.
12. From line (9), by the truth rule for negation, $\sim H$ is true.
13. From line (2) and line (9), by the truth rule for implication, G is true.
14. From line (7) and line (13), by the truth rule for conjunction, $B \wedge G$ is true.
15. From line (4) and line (14), by the truth rule for implication, E is true.
16. From line (11) and line (15), by the truth rule for conjunction, $\sim O \wedge E$ is true.

This is the **method of direct proof** of the validity of an argument in which a proof is constructed for the argument itself. In many instances, however, the validity of an argument can be established by constructing a proof for another related argument. We will now discuss the underlying bases for two such alternative methods.

Suppose $P_1, P_2, ..., P_n$ are the premises of a given argument and A and B are sentences such that $A \rightarrow B$ is the conclusion of the argument. Also, suppose the argument with premises $P_1, P_2, ..., P_n, A$ and conclusion B is valid. Then we know that $\wedge_{i=1}^n P_i \wedge A \rightarrow B$ is a tautology. We will now establish that $\wedge_{i=1}^n P_i \rightarrow (A \rightarrow B)$ must be a tautology. Suppose $\wedge_{i=1}^n P_i$ is true. Then, it follows from the truth rule for conjunction that $\wedge_{i=1}^n P_i \wedge A$ is true whenever A is true. However, because $\wedge_{i=1}^n P_i \wedge A \rightarrow B$ is a tautology, it follows from the truth rule for implication that B must be true whenever A is true. Therefore, using the truth rule for implication, it follows that $A \rightarrow B$ must be true. Therefore, using simply the truth rules of logic we have established that $A \rightarrow B$ is true whenever $\wedge_{i=1}^n P_i$ is true. In other words, $\wedge_{i=1}^n P_i \rightarrow (A \rightarrow B)$ is a tautology. Thus, the validity of the argument with premises $P_1, P_2, ..., P_n, A$ and conclusion B is sufficient to establish the validity of the given argument. This is the underlying basis for the **method of conditional proof** which can be used to establish the validity of any argument that has a conditional sentence as its conclusion. The method involves constructing a proof for a related argument where the premises of the new argument must consist of the premises of the given argument and the antecedent of its conclusion and the conclusion of the new argument must be the consequent of the conclusion of the given argument.

Suppose $P_1, P_2, ..., P_n$ are the premises of a given argument and C is its conclusion. Suppose there exists a sentence Q such that Q is a contradiction and the argument with premises $P_1, P_2, ..., P_n$ and $\sim C$ and the conclusion Q is valid. For the given argument to be valid, $\wedge_{i=1}^n P_i \rightarrow C$ must be a tautology. Since Q is a contradiction and $\wedge_{i=1}^n P_i \wedge \sim C \rightarrow Q$ is a tautology, using the truth rule for implication it follows that $\wedge_{i=1}^n P_i \wedge \sim C$ must also be a contradiction. From the truth rule for conjunction it then follows that $\sim C$ must be false whenever $\wedge_{i=1}^n P_i$ is true. Then, by the truth rule for negation, C must be true whenever $\wedge_{i=1}^{n} P_i$ is true. Thus, $\wedge_{i=1}^{n} P_i \rightarrow C$ is a tautology and the given argument is valid if the related argument with premises $P_1, P_2, ..., P_n, \sim C$ and conclusion Q is valid. This is the justification for the **method of proof by contradiction** wherein a given argument can be proved to be valid by constructing a proof for a related argument where the premises of the new argument consist of the premises of the given argument and the negation of its conclusion and the conclusion of the new argument is a contradiction.

Till now we have discussed how to establish the validity of an argument. How do we establish the invalidity of an argument when the number of atomic sentences in the argument is large? Note that a necessary and sufficient condition for a sentence $A \rightarrow B$ *not* to be a tautology is that A must be true and B must be false for at least one assignment of truth values to the atomic sentences in A and B. Thus, an argument with premises $P_1, P_2, ..., P_n$ and conclusion C will be invalid if we can find at least one assignment of truth values to the atomic sentences in $P_1, P_2, ..., P_n$ and C such that each of the premises of the given argument is true and its conclusion is false.

To illustrate, consider the following argument.

If the money supply increases and Firm 1 adjusts its price then the increase in Firm 2's optimal price will be large if the prices of Firm 1 and Firm 2 are strategic complements. If the money supply increases and Firm 1 does not adjust its price then the increase in Firm 2's optimal price will not be large. If the cost to Firm 2 of changing its price is positive then Firm 2 will adjust its price iff the increase in Firm 2's optimal price is large. Therefore, if the money supply increases and the cost to Firm 2 of changing its price is positive then both Firm 1 and Firm 2 will adjust their prices or both Firm 1 and Firm 2 will not adjust their prices.

> Let $M =$ 'The money supply will increase.'
> $A =$ 'Firm 1 will adjust its price.'
> $L =$ 'The increase in Firm 2's optimal price will be large.'
> $S =$ 'Prices of Firm 1 and Firm 2 are strategic complements.'
> $C =$ 'The cost to Firm 2 of changing its price is positive.'
> $B =$ 'Firm 2 will adjust its price.'

Then, the above argument may be represented as follows:

$$M \wedge A \rightarrow (S \rightarrow L)$$
$$M \wedge \sim A \rightarrow \sim L$$
$$C \rightarrow (B \leftrightarrow L)$$

$$\overline{}$$

$$M \wedge C \rightarrow (A \wedge B) \vee (\sim A \wedge \sim B)$$

To check whether the above argument is invalid we need to check whether it is possible to have an assignment of truth values to the sentences A, B, C, L, M, and S such that T is the truth value of the three premises and F is the truth value of the conclusion of the argument. Note that the truth value of the conclusion $M \wedge C \rightarrow$

$(A \wedge B) \vee (\sim A \wedge \sim B)$ will be F iff the truth values of both M and C are equal to T and the truth value of $(A \wedge B) \vee (\sim A \wedge \sim B)$ is F. The truth value of $(A \wedge B) \vee (\sim A \wedge \sim B)$ will be F iff A and B have different truth values.

Suppose the truth values of M, C and A are all equal to T and the truth value of B is F. Then, because the truth values of both M and A are equal to T, the truth value of the first premise $M \wedge A \rightarrow (S \rightarrow L)$ will be T iff the truth value of $(S \rightarrow L)$ is T. The truth value of $(S \rightarrow L)$ will be T iff the truth value of S is F or the truth value of L is T.

If the truth values of M and A are equal to T then the truth value of the second premise $M \wedge \sim A \rightarrow \sim L$ is also T. Moreover, if the truth value of C is T then the truth value of the premise $C \rightarrow (B \leftrightarrow L)$ is T iff the truth value of $B \leftrightarrow L$ is T. The truth value of $B \leftrightarrow L$ is T iff the truth values of B and L are the same. Therefore, if the truth value of B is F, the third premise is true if the truth value of L is F.

Therefore, if the sentences A, C, and M are assigned the truth value T and the sentences B, L, and S are assigned the truth value F then all the premises of the argument will be true and the conclusion of the argument will be false. This then establishes the invalidity of the given argument.

2.2.5 Logical consistency and independence

A set of sentences is **logically consistent** iff the conjunction of the sentences is not a contradiction. Therefore, if a set of sentences is logically consistent then there exists an assignment of truth values to the atomic sentences within these sentences for which, using the truth rules of logic, we can conclude that each of these sentences is true. Suppose n is a positive integer and $P_1, P_2, ..., P_n$ are n arbitrarily given sentences. Then, it follows that $P_1, P_2, ..., P_n$ are logically consistent iff at least one row of a truth table constructed for $(P_1 \wedge P_2 \wedge ... \wedge P_n)$ by assigning truth values to the atomic sentences in $P_1, P_2, ..., P_n$ has the entry 'T' in the final column.

In the previous section we considered an example of an invalid argument. Do the premises and the conclusion of this argument form a logically consistent set of sentences? We saw that '$M \wedge A \rightarrow (S \rightarrow L)$', '$M \wedge \sim A \rightarrow \sim L$', '$C \rightarrow (B \leftrightarrow L)$', and '$M \wedge C \rightarrow (A \wedge B) \vee (\sim A \wedge \sim B)$' can be used to denote respectively the three premises and the conclusion of the argument if the sentential letters 'A', 'B', 'C', 'L', 'M', and 'S' are used to denote appropriate atomic sentences. If we assign the truth value 'F' to the sentences C and M, whatever the truth values we assign to A, B, L, and S, the truth values of all the four compound sentences are equal to T. Thus, the premises and conclusion of this invalid argument are logically consistent.

From the definition of logical consistency it also follows that the sentences P_1, $P_2, ..., P_n$ are **logically inconsistent** iff $\wedge_{i=1}^{n} P_i$ is a contradiction. Note that if $\wedge_{i=1}^{n} P_i$ is a contradiction then, given any sentence Q, $\wedge_{i=1}^{n} P_i \rightarrow Q$ is a tautology. Therefore, if $P_1, P_2, P_3, ..., P_n$ are logically inconsistent then any sentence Q can be logically derived from this set of sentences. Note also that if Q is a contradiction and $\wedge_{i=1}^{n} P_i \rightarrow Q$ is a tautology then $\wedge_{i=1}^{n} P_i$ must be a contradiction. Therefore, in order to establish that $P_1, P_2, ..., P_n$ are logically inconsistent it is sufficient to demonstrate that there exists a contradiction which is a logical consequence of $P_1, P_2, ..., P_n$.

A set of sentences is **logically independent** just in the case that none of the sentences in the set is a logical consequence of the other sentences in the set. That is, the n sentences $P_1, P_2, ..., P_n$ are logically independent iff every argument, which has one of these sentences as its conclusion and the remaining sentences as its premises, is invalid. In order to prove that $P_1, P_2, ..., P_n$ are logically independent it is, therefore, necessary and sufficient to demonstrate the invalidity of n distinct arguments, the i-th argument having P_i ($i = 1, 2, ..., n$) as its conclusion and $P_1, P_2, ..., P_{i-1}, P_{i+1}, ..., P_n$ as its premises. On the other hand, to demonstrate that the sentences $P_1, P_2, ..., P_n$ are not logically independent it is sufficient to prove the validity of at least one of the above n arguments.

2.3 Predicate Logic

2.3.1 Universe of discourse, universal and existential sentences

Assertions and sentences arise in the context of particular discussions. We will refer to any single distinct entity which is assumed to exist in the context of a particular discussion as an **individual**. This means that individuals can be tangible, for example, households and units of commodities in a discussion of consumer behaviour, or intangible, for example, the numbers which are considered in the discussion. The collection of all individuals for an ongoing discussion is known as **the universe of discourse** for that discussion.

A **universal sentence** is a sentence which asserts that something is true about every member of a set of individuals without directly referring to any member of the set.[3] For example, the sentence 'All consumers are rational' asserts that for every individual which is a consumer (that is, for every member of the set consisting of all individuals which are consumers) it is true that the individual is rational. In other words, the sentence asserts that, for every individual it is true that if the individual is a consumer then the individual is rational. If there are n individuals (n being a positive integer) which we can denote as $a_1, a_2, a_3, ..., a_n$ then the above sentence asserts that, for every value of i belonging to the set $\{1, 2, 3, ..., n\}$, the sentence 'If a_i is a consumer then a_i is rational' is true. Suppose, for every value of i belonging to $\{1, 2, 3, ..., n\}$, the sentence 'a_i is a consumer' is denoted as C_i and the sentence 'a_i is rational' is denoted as R_i. Then, the given universal sentence can be denoted as $\wedge_{i=1}^{n}(C_i \rightarrow R_i)$.

An **existential sentence** is a sentence which asserts that something is true about at least one member of a set of individuals without directly referring to any member of the set. For example, the sentence 'Some consumers are rational' asserts that for

---

[3] Note, however, that universal sentences do not imply that the set of individuals considered is non-empty.

at least one individual who is a consumer it is true that the individual is rational. In other words, the sentence asserts that, for at least one individual, two claims are simultaneously true: one, the individual is a consumer and two, the individual is rational. An existential sentence, therefore, implies that the subset of individuals considered by it is non-empty. Using the notation in the previous paragraph the given existential sentence can be denoted as $\vee_{i=1}^{n}(C_i \wedge R_i)$.

If n is very large then the notation for universal and existential sentences in sentential logic is not economical but universal and existential sentences can still be considered in sentential logic. However, if the universe of discourse is not finite (we might, for example, be considering a macroeconomic model in which the marginal propensity to consume can take any positive value between 0 and 1 so that all real numbers between 0 and 1 must at least be considered as belonging to the universe of discourse), then no positive integer n exists which would allow us to denote universal and existential sentences in the above manner. Universal and existential sentences can then only be represented by single sentential letters. However, if universal and existential sentences are also represented by sentential letters then there is no way in which arguments such as the following can be proved to be valid in sentential logic.

All consumers are rational. Household 1 is a consumer. Therefore, Household 1 is rational.

Household 1 is rational. Household 1 is a consumer. Therefore, some consumers are rational.

In order for universal and existential sentences to be considered generally it is, therefore, necessary to extend the system of notation and the set of truth rules beyond those in sentential logic.

2.3.2 Individual constants, variables, quantifiers, and predicates

We will define **proper names** as symbols, letters, words, or phrases which *directly refer* to an individual. Proper names uniquely specify individuals but we do not need to consider the meanings that these symbols or words or phrases may have in order to understand which individual is being specified. 'One', 'Two', and 'Three' are the names of three numbers. In a discussion of consumer behaviour, households may be given names such as '1', '2', '3', etc. and commodity bundles may be given names such as A, B, C, etc. In contrast, the expression 'The smallest natural number' describes a unique individual but it is necessary to consider the meaning of the expression to understand which individual is being referred to. Proper names are denoted by lower case letters from the beginning of the English alphabet with or without numerical subscripts, for example, a, b, c, d, a_1, or b_{23}. These are referred to as **individual constants** or **singular terms**.

Variables are lower case letters like v, w, x, y, or z from the end of the English alphabet, with or without numerical subscripts, which are inserted within sequences of words to indicate that each such letter can be replaced in this sequence only by some proper name. Variables, therefore, serve as placeholders for individuals within word sequences. Because, in principle, a given variable can serve as a placeholder for any individual, each individual is referred to as a **value** of the variable.

While we cannot think of 'x is a consumer' or 'x is rational' as being a sentence, consider the expression 'For all values of x it is true that if that value of x is a consumer, then x is rational'. This can be considered a sentence because this is equivalent to stating that 'For all individuals it is true that if that individual is a consumer then that individual is rational' or that 'All consumers are rational'. This universal sentence is, therefore, written as 'For all x it is true that if x is a consumer then x is rational'. Phrases such as 'for all' or 'for every' are known as **universal quantifiers**. The symbol used for a universal quantifier is $\forall$ and the sentence is symbolized as $(\forall x)(x$ is a consumer $\rightarrow x$ is rational).

Similarly, the existential sentence 'Some consumers are rational' is equivalent in meaning to the sentence 'There is at least one value of x for which it is true that the value of x is a consumer and the value of x is rational'. The phrase 'There is at least one' can be replaced with 'There exists an' or 'There is some' without a change in meaning. Phrases such as these are known as **existential quantifiers** and are denoted by the symbol $\exists$. The above sentence is then symbolized as $(\exists x)(x$ is a consumer $\wedge x$ is rational) and is read as 'There exists x for which it is true that x is a consumer and x is rational'. Note that $(\exists x)(x$ is a consumer $\rightarrow x$ is rational) in contrast asserts that there exists an individual for which it is true that if that individual is a consumer then that individual is rational. It does not assert that there exists an individual which is a consumer.

Many of us remember English language exercises in school in which we had to fill in the blanks in a sequence of words and blanks to form a sentence. A sequence of words and blanks which results in an atomic sentence when the blanks in the sequence are filled by proper names is known as a **predicate** in predicate logic. For example, '... is rational' is a sequence of words and blanks and results in atomic sentences when the blank is filled by proper names such as 'Household 1' and 'Alfred Tarski'. '... is at least as good as ...' and '... lies between ... and ...' are respectively examples of a two-place predicate and a three-place predicate. An atomic sentence can be considered to be a zero-place predicate.

In predicate logic, predicates are denoted by upper case letters from the English alphabet with or without numerical subscripts. Thus, we may write $C =$ '... is a consumer' and $R =$ '... is rational'. If we denote the name 'Alfred Tarski' by the individual constant 'a', the sentence 'Alfred Tarski is rational' will be denoted as Ra. The universal sentence 'All consumers are rational' and the existential sentence 'Some consumers are rational' will be denoted as $(\forall x)(Cx \rightarrow Rx)$ and $(\exists x)(Cx \wedge Rx)$, respectively. An n-place predicate with all n places filled with proper names or variables will be denoted by an upper case letter denoting the predicate followed by a sequence of n individual constants and variables (not necessarily distinct), the i-th member of the sequence being the individual constant or variable used to fill the i-th place of the predicate. If Q denotes the predicate '... at least as good as ...' and the names 'A' and 'B' of commodity bundles are denoted respectively by a and b then 'A is at least as good as B' is denoted as Qab. In economic theory, in the case of a two-place predicate, the letter denoting the predicate is often written between the constant or variable in the first place and the constant or variable in the second place. Thus, Qxy may be written as xQy.

2.3.3 Well-formed formulas, scope of a quantifier, bound and free variables

For any positive integer n, an upper case letter denoting a n-place predicate followed by n individual constants or variables is known as an **atomic formula**. For example, if $P = $ '... is preferred to ... by ...' then $Pabc$, $Pxyz$, $Paxb$, $Pxya$, $Paya$, and $Pxxa$ are all examples of atomic formulas.

A number of symbols written down one after the other without any spaces in between, the symbols being either predicate letters, individual constants, variables, sentential connectives, parentheses, or the symbols for universal and existential quantifiers will be called a **PL string**, for example, $a\forall \wedge xPaP \rightarrow (\exists z, (\forall x)((Hx \wedge Px) \rightarrow (\exists y)(Cy \wedge Bxy))$. We will denote a PL string by lowercase letters from the Greek alphabet like ψ, γ, ϕ, and φ, with or without numerical subscripts. A PL string will be called a **well-formed formula in predicate logic (PLwff)** iff its being so can be justified solely by one or more of the following rules:

1. Every atomic formula is a PLwff.
2. For any PLwff ψ, $\sim (\psi)$ is a PLwff.
3. For any two PLwffs ψ and ϕ, $(\psi \wedge \phi)$, $(\psi \vee \phi)$, $(\psi \rightarrow \phi)$, and $(\psi \leftrightarrow \phi)$ are all PLwffs.
4. For any PLwff ψ and any variable ϕ, $(\forall \phi)\psi$ and $(\exists \phi)\psi$ are both PLwffs.

Among the two examples of PL strings considered in the previous paragraph only the second is a PLwff. The reader will, of course, have realized that we have already been using PLwffs to symbolize sentences in preceding sections. Our usage of PLwffs however indicates that, as in the case of SLwffs, there are some conventions which allow us to simplify the writing of PLwffs. Suppose, for any given PLwff with at least one sentential connective or quantifier, we define **the main logical symbol of the PLwff** to be the single quantifier symbol or sentential connective which has to be used with a single PLwff or with a couple of PLwffs to obtain the given PLwff. The common conventions used in writing PLwffs can then be stated as follows:

1. If the main logical symbol of a PLwff is not a '$\sim$', '$\forall$', or '$\exists$' then the outermost parentheses of the PLwff may be omitted.
2. If the main logical symbol of a PLwff is either a '$\rightarrow$' or a '$\leftrightarrow$' then the outermost parentheses of the largest PLwffs to the left and right of the main symbol may be omitted provided the main logical symbol of the concerned PLwff is '$\wedge$' or '$\vee$'.
3. If ψ is an atomic formula, $\sim (\psi)$ may be written as $\sim \psi$ in any PLwff.
4. Convention 2 also applies to any PLwff which is written along with its outermost parentheses as part of any other PLwff.

The **scope of a quantifier** in a PLwff is the smallest PLwff within it which contains the quantifier. For example, the PLwff '$(\exists x)(Fx \wedge (\forall y)\,(Fy \wedge(\forall z)(Fz \wedge Ixz \rightarrow Fyz) \rightarrow Ixy))$' has one existential quantifier and two universal quantifiers. The scope of the existential quantifier in the PLwff is the PLwff itself, the scope of the universal quantifier using the variable y is the PLwff '$(\forall y)(Fy \wedge(\forall z)(Fz \wedge Ixz \rightarrow Fyz) \rightarrow Ixy)$' and the scope of the universal quantifier using the variable z is '$(\forall z)(Fz \wedge Ixz \rightarrow Fyz)$'.

A single instance of writing of a variable within a PLwff, except when it is written immediately after a quantifier symbol, we will refer to as an occurrence of the variable within the PLwff. Consider the PLwff '$(Ix \wedge(\forall y)(Iy \wedge Pyy \rightarrow Pxy)) \wedge (\exists x)(Ix \wedge (Iy \wedge \sim Pyy \rightarrow \sim Pxy))$'. There are four occurrences of x and eight occurrences of y in the PLwff. Note that exactly four occurrences of y lie within the scope of a quantifier using the variable y (those in the PLwff '$(\forall y)(Iy \wedge Pyy \rightarrow Pxy)$') and exactly two occurrences of x lie within the scope of a quantifier using the variable x (those in the PLwff '$(\exists x)(Ix \wedge (Iy \wedge \sim Pyy \rightarrow \sim Pxy))$'.

An **occurrence of a variable in a PLwff is bound** iff it is within the scope of a quantifier using the variable in that PLwff. The **occurrence of a variable in a PLwff is free** iff it is not bound. Therefore, in the PLwff considered in the previous paragraph, four occurrences of y and two occurrences of x are bound and four occurrences of y and two occurrences of x are free.

A **variable is bound in a PLwff** iff every occurrence of that variable in that PLwff is bound. A **variable is free in a PLwff** iff at least one occurrence of that variable in that PLwff is free. The variable x is bound in the PLwff '$(\exists x)(Ix \wedge(\forall y)(Iy \wedge Pyy \rightarrow Pxy)) \wedge (\exists x)(Ix \wedge (Iy \wedge \sim Pyy \rightarrow \sim Pxy))$' but the variable y is free.

In predicate logic, sentences are denoted by PLwffs without any free variables. Moreover, if the individual constants in a given PLwff without free variables are used to denote specific individuals and the predicate letters in atomic formulas within the PLwff are used to denote specific predicates with the requisite number of places indicated in the atomic formulas, then the given PLwff will always be taken to denote a sentence in predicate logic. For example, suppose that $a =$'Household 1', $C = $'$\ldots$ is a commodity bundle' and $R = $'$\ldots$ is rational', then Ca and $(\forall x)(Cx \rightarrow Rx)$ will not be considered as symbolizing meaningless expressions but as symbolizing false sentences or assertions.

2.3.4 Truth rules of predicate logic

The truth rules of logic in predicate logic consist of the truth rules of sentential logic, modified by replacing the expression 'SLwff', wherever it occurs in these rules, with the expression 'PLwff with no free variables' and two other rules: The Truth Rule for Universal Sentences and The Truth Rule for Existential Sentences. In stating these rules we will use a couple of phrases, the meanings of which are clarified by the following definitions.

An individual is defined to **satisfy a PLwff with no free variables** iff the sentence denoted by the PLwff is true. For example, both the numbers One and Two satisfy the PLwff 'Oa' where O = '... is an odd number' and a = 'One'.

An individual is defined to **satisfy a PLwff with one free variable** iff a true sentence is denoted by the PLwff obtained by replacing the free variable in each free occurrence in the PLwff with an individual constant referring to the individual. For example, the number One satisfies the PLwff '$Ox \land Lxb$', where O = '... is an odd number', L = '... is less than ...' and b = 'Two', but the number Three does not satisfy the PLwff.

Truth rule for universal sentences

For any variable v and any PLwff ϕ which has no free variables or has v as the only free variable, the sentence denoted by $(\forall v)\phi$ is true iff every individual satisfies ϕ.

Truth rule for existential sentences

For any variable v and any PLwff ϕ which has no free variables or has v as the only free variable, the sentence denoted by $(\exists v)\phi$ is true iff at least one individual satisfies ϕ.

In sentential logic, we can use the truth rules of sentential logic to determine the truth value of a sentence denoted by an SLwff if we know the truth values of the sentences denoted by the sentential letters occurring in the SLwff. Similarly, in predicate logic, we can use the truth rules of predicate logic to determine the truth of a sentence denoted by a PLwff without free variables if, for every predicate denoted by a predicate letter in the PLwff, we know, for every assignment of individuals (that is, proper names) to the places of the predicate, whether the resulting sentence is true or false and if we know which individuals are referred to by the individual constants occurring in the PLwff. For example, we can determine the truth value of the sentence denoted as $(\forall x)(Fx \rightarrow (\exists y)(\exists z)(Fy \land Mz \land Rxzy))$ if we know, for each individual, whether that individual satisfies the formulae Fx and Mx, and if we know for every triple of individuals, if they are denoted by say 'a', 'b', and 'c', the truth value of sentences denoted by $Rabc$, $Racb$, $Rbac$, $Rbca$, $Rcab$, and $Rcba$. However, to determine the truth value of the sentence denoted as $(\forall x)(Fx \land Ma \rightarrow (\exists y)(\exists z)(Fy \land Mz \land Rxzy))$ we will also need to know which is the individual referred to by the individual constant a. Moreover, given the nature of the truth rules of predicate logic, it can be shown that no information, other than the kind above, is relevant for determining the truth value of sentences denoted by PLwffs without free variables according to the truth rules of predicate logic.

2.3.5 Using multiple quantifiers

We conclude by considering examples of how universal and existential sentences may be written using quantifiers, especially in cases where the use of multiple quantifiers is required to represent a sentence.

Suppose P = '... is a firm', Q = '... is a technology' and A = '... has access to ...'. The sentence 'Every firm has access to every technology' can be symbolized in the first stage as $(\forall x)(Px \rightarrow x$ has access to every technology$)$. The expression 'x has access to every technology' can then be symbolized as $(\forall y)(Qy \rightarrow Axy)$. Therefore, the entire sentence can be finally symbolized as $(\forall x)(Px \rightarrow (\forall y)(Qy \rightarrow Axy))$. Similarly, 'Every firm has access to some technology' may initially be symbolized as $(\forall x)(Px \rightarrow x$ has access to some technology$)$ and then finally as $(\forall x)(Px \rightarrow (\exists y)(Qy \wedge Axy))$. In this way, the sentences 'Some firms have access to every technology' and 'Some firms have access to technologies' can be symbolized respectively as $(\exists x)(Px \wedge (\forall y)(Qy \rightarrow Axy))$ and $(\exists x)(Px \wedge (\exists y)(Qy \wedge Axy))$.

The sentence 'Every firm has access to every technology' has the same meaning as the sentence 'For all values of x and for all values of y, it is true that if the value of x is a firm and the value of y is a technology then the value of x has access to the value of y'. That means one can symbolize the original sentence also as $(\forall x)(\forall y)(Px \wedge Qy \rightarrow Axy)$ or as $(\forall y)(\forall x)(Px \wedge Qy \rightarrow Axy)$. The reader should convince herself that the sentence 'Every firm has access to some technology' can be symbolized as $(\forall x)(\exists y)(Px \rightarrow Qy \wedge Axy)$ but cannot be symbolized as $(\exists y)(\forall x)(Px \rightarrow Qy \wedge Axy)$. This instead represents the assertion that there is at least one technology which can be accessed by every firm.

The sentence 'Some firms have access to technologies' asserts that there exists at least one individual which is a firm and at least one individual which is a technology which can be accessed by the firm. It can, therefore, also be symbolized as $(\exists x)(\exists y)(Px \wedge (Qy \wedge Axy))$ or as $(\exists y)(\exists x)(Px \wedge (Qy \wedge Axy))$. The reader should convince herself that the sentence 'Some firms have access to every technology' can be symbolized as $(\exists x)(\forall y)(Px \wedge (Qy \rightarrow Axy))$ but cannot be symbolized as $(\exists x)(\forall y)(Px \wedge Qy \rightarrow Axy)$ or as $(\forall y)(\exists x)(Px \wedge (Qy \rightarrow Axy))$.

Note that in symbolizing an expression such as 'x has access to every technology' we cannot use x as the variable which is used with the universal quantifier. $(\forall x)(Qx \rightarrow Axx)$ may be read as 'For all values of x it is true that if the value of x is a technology then the value of x has access to itself' which has the same meaning as 'Every technology has access to itself'. However, if we have to symbolize the sentence 'No number is greater than itself' then if we assume $N =$ '...is a number' and $G =$ '...is greater than ...' the sentence may be denoted as $\sim (\exists x)(Nx \wedge Gxx)$.

Universal and existential sentences are further linked by the fact that the negation of an existential sentence can be expressed as a universal sentence and the negation of a universal sentence can be expressed as an existential sentence. This is not surprising. Remember that a universal sentence states that some assertion is true about every member of some set of individuals. This is equivalent to stating that there is not even one member in that set of individuals for which the negation of that assertion is true. This is the negation of an existential sentence which states that there is at least one member in that set of individuals for which the negation of that assertion is true.

The sentence 'The only numbers which are exactly divisible by two are even numbers' can be symbolized as $(\forall x)(Nx \wedge Dxa \rightarrow Ex)$ where $D =$ '...is exactly divisible by ...', $N =$ '...is a number', $E =$ '...is an even number', and $a =$ 'Two'. The sentence can alternatively be expressed as 'There does not exist a value of x for which it is not true that if that value of x is a number exactly divisible by two then that value of x is an even number'. This can be symbolized as $\sim (\exists x) \sim (Nx \wedge Dxa \to Ex)$. This illustrates a general principle that a universal sentence symbolized with '$(\forall x)$' at the beginning is equivalent in meaning to the sentence which has the same symbolic form as the universal sentence except that '$(\forall x)$' is replaced with '$\sim (\exists x) \sim$'.

Similarly, stating that there is at least one member of a set of individuals for which an assertion is true is the same as stating that it is not true that for every member of that set of individuals the negation of that assertion is true. Consider the existential sentence 'There is a positive integer which is less than every even positive integer'. If $P = $ '... is a positive integer', $E = $ '... is even' and $L = $ '... is less than ...' then the sentence may be symbolized as $(\exists x)(Px \wedge (\forall y)(Py \wedge Ey \to Lxy))$. It is, therefore, the negation of the sentence 'For every value of x it is not true that value of x is a positive integer less than every even positive integer'. Therefore the given sentence can be symbolized as $\sim (\forall x) \sim (Px \wedge (\forall y)(Py \wedge Ey \to Lxy))$. This illustrates a general principle for existential sentences whereby an existential sentence symbolized with '$(\exists x)$' at the beginning is equivalent in meaning to the sentence which has the same symbolic form except that '$(\exists x)$' is replaced with '$\sim (\forall x) \sim$'.

EXERCISES

1. Consider the sentence: 'If today is Tuesday then tomorrow is Wednesday'.

   (a) Using sentential letters to denote only atomic sentences denote this sentence by an SLwff.
   (b) Find the days of the week on which the sentence, if pronounced, will be true. Is this sentence a tautology?

2. Asked which day of the week today is, Ravi replied that 'Today is not Wednesday or today is not Friday' while Som replied that 'Today is Tuesday or Wednesday'.

   (a) Denote the statements of both Ravi and Som by SLwffs using sentential letters to denote only atomic sentences.
   (b) According to the truth rules of logic, for which days of the week will Ravi's reply be true but Som's reply be false?

3. Denote the following sentences by SLwffs using sentential letters to denote only atomic sentences:

   (a) Although the rate of unemployment is falling, there is no rise in the rate of inflation.
   (b) Neither is the rate of unemployment falling nor is there a fall in the rate of inflation.
   (c) Either the firm will cut prices or it will undertake lay-offs and close some retail outlets.
   (d) At least one of Outcome A, Outcome B and Outcome C is a Nash equilibrium.
   (e) At most one of Outcome A, Outcome B and Outcome C is a Nash equilibrium.
   (f) Exactly one of Outcome A, Outcome B and Outcome C is a Nash equilibrium.

(g) The fiscal stimulus will not work if the problems in both the financial system and the real estate market are not addressed.

(h) If export demand stagnates economic growth will slow down unless there is a normal monsoon.

(i) A higher rate of economic growth is neither necessary nor sufficient for a higher rate of reduction in poverty.

(j) A higher rate of economic growth is necessary but not sufficient for a higher rate of reduction in poverty.

4. Use truth tables to prove that the sentences denoted by the following SLwffs are tautologies.

(a) $P \wedge Q \rightarrow P$

(b) $P \rightarrow P \vee Q$

(c) $P \wedge (Q \vee R) \leftrightarrow (P \wedge Q) \vee (P \wedge R)$

(d) $P \vee (Q \wedge R) \leftrightarrow (P \vee Q) \wedge (P \vee R)$

(e) $\sim (P \wedge Q) \leftrightarrow \sim P \vee \sim Q$

(f) $\sim (P \vee Q) \leftrightarrow \sim P \wedge \sim Q$

(g) $\sim Q \wedge (P \vee Q) \rightarrow P$

(h) $(P \rightarrow Q) \leftrightarrow \sim P \vee Q$

(i) $\sim (P \rightarrow Q) \leftrightarrow P \wedge \sim Q$

(j) $(P \rightarrow Q) \leftrightarrow (\sim Q \rightarrow \sim P)$

(k) $P \wedge (P \rightarrow Q) \rightarrow Q$

(l) $\sim Q \wedge (P \rightarrow Q) \rightarrow \sim P$

(m) $(P \rightarrow Q \wedge \sim Q) \rightarrow \sim P$

(n) $(P \rightarrow Q) \wedge (Q \rightarrow R) \rightarrow (P \rightarrow R)$

(o) $(P \leftrightarrow Q) \leftrightarrow (P \rightarrow Q) \wedge (Q \rightarrow P)$

5. Assuming that all sentential letters denote atomic sentences, use truth tables to decide whether the sentences denoted by the following SLwffs are tautologies, contradictions, or contingent sentences:

(a) $P \vee \sim P \rightarrow P \wedge \sim P$

(b) $\sim (P \vee (P \wedge Q)) \vee P$

(c) $\sim P \wedge ((P \rightarrow Q) \rightarrow P)$

(d) $\sim (P \leftrightarrow Q) \wedge (P \wedge \sim Q)$

(e) $P \wedge \sim (Q \rightarrow P \wedge Q)$

(f) $(P \vee (\sim P \wedge Q)) \vee (\sim P \wedge \sim Q)$

(g) $\sim ((P \rightarrow Q) \vee (Q \rightarrow P))$

(h) $\sim (P \vee Q) \rightarrow P \wedge Q$

(i) $P \rightarrow (Q \rightarrow (Q \rightarrow P))$

(j) $((P \rightarrow Q) \leftrightarrow Q) \rightarrow P$

6. Establish the validity or invalidity of each of the following arguments without the use of truth tables.

(a) If the organic composition of capital falls then the general rate of profit will increase. If the general rate of profit increases, both the degree of competition and the rate of accumulation will rise. But, the rate of accumulation will not rise. Therefore, the organic composition of capital will not fall.

(b) There will be an increase in investment expenditure if the central bank cuts the interest rate or the government undertakes new infrastructure projects. However, the central bank will not cut the interest rate and the government will not undertake new investment projects. Therefore, there will be no increase in investment expenditure.

(c) If there is a labour shortage then the real wage rate will rise. It is not true that there will be a relaxation in government controls on immigration as well as a rise in the real wage rate. Therefore, if the government relaxes controls on immigration there will be no labour shortage.

7. Establish (without using truth tables) the logical consistency or inconsistency of each of the following sets of sentences denoted by SLwffs.

(a) $\{\sim (P \to Q); \sim (Q \to P)\}$

(b) $\{(P \to (Q \to R)) \to ((P \to Q) \to (P \to R)); P \wedge \sim R; Q \vee \sim R\}$

(c) $\{(P \to Q \wedge R), Q, (\sim R \to S), \sim (S \vee P)\}$

(d) $\{(P \leftrightarrow Q), (Q \to R), (\sim R \vee S), (\sim P \to S), \sim S\}$

8. Which of the following statements are true for arbitrarily given sentences denoted by P and Q?

   (a) P and $\sim P$ are not logically independent.

   (b) If P and Q are logically independent then $\sim P$ and $\sim Q$ are logically independent.

   (c) If P and Q are not logically independent then $\sim P$ and $\sim Q$ are not logically independent.

   (d) If Q is a contradiction then P and Q are not logically independent.

   (e) If Q is a tautology then P and Q are not logically independent.

9. Symbolize the following sentences using the system of notation in predicate logic:

   (a) Every household purchases every commodity.

   (b) Households purchase commodities.

   (c) Some commodities are purchased by every household.

   (d) Every commodity is purchased by some household.

   (e) No household purchases every commodity.

   (f) Only poor households purchase cheap commodities.

   (g) Some wealthy households do not purchase any cheap commodity.

   (h) Every household earns less than some other household.

KEY TERMS

Atomic Sentence	Principle of Bivalence
Bound Variable	Proof
Compound Sentence	Quantifier
Contingent Sentence	SLwff
Contradiction	Sentences
Existential Sentence	Sentential Logic
Free Variable	Tautology
Independence	Truth Rules
Individual Constants	Truth Rules of Logic
Logical Consistency	Truth Tables
Multiple Quantifier	Universal Sentence
PLwff	Variables
Predicate Logic	

USEFUL WEB LINKS

http://courses.umass.edu/phil110-gmh/text/c02_3-99.pdf (accessed on 4 October 2010): truth functional connectives

http://courses.umass.edu/phil110-gmh/text/c03_3-99[09-18-09].pdf (accessed on 2 August 2010): validity in sentential logic, tautologies, contradictions, contingent formulas

http://courses.umass.edu/phil110-gmh/text/c06_3-99.pdf (accessed on 2 August 2010): translations in monadic predicate logic

http://courses.umass.edu/phil110-gmh/text/c07_3-99.pdf (accessed on 2 August 2010): translations in polyadic predicate logic

http://ocw.mit.edu/courses/linguistics-and-philosophy/24-241-logic-i-fall-2005/readings/chp03.pdf (accessed on 2 August 2010); http://ocw.mit.edu/courses/linguistics-and-philosophy/24-241-logic-i-fall-2005/readings/chp04.pdf (accessed on 2 August 2010): sentential calculus

http://ocw.mit.edu/courses/linguistics-and-philosophy/24-241-logic-i-fall-2005/readings/chp13.pdf (accessed on 2 August 2010): translations

http://ocw.mit.edu/courses/linguistics-and-philosophy/24-241-logic-i-fall-2005/readings/chp14.pdf (accessed on 2 August 2010); http://ocw.mit.edu/courses/linguistics-and-philosophy/24-241-logic-i-fall-2005/readings/chp18.pdf (accessed on 2 August 2010): predicate logic

http://oscarhome.soc-sci.arizona.edu/ftp/Logic2.pdf (accessed on 4 October 2010): propositional calculus, sentential connectives

http://oscarhome.soc-sci.arizona.edu/ftp/Logic3.pdf (accessed on 2 August 2010): truth rules, truth tables, tautology

http://oscarhome.soc-sci.arizona.edu/ftp/Logic5.pdf (accessed on 2 August 2010): predicate calculus—predicates, relations, individual constants

http://oscarhome.soc-sci.arizona.edu/ftp/Logic6.pdf (accessed on 2 August 2010): semantics of predicate calculus

http://www.fecundity.com/codex/forallx.pdf (accessed on 2 August 2010): sentential logic, truth tables, quantified logic—sentences and predicates, semantics, independence, logical validity

http://www.math.umn.edu/~jodeit/course/ACaRA01.pdf (accessed on 4 October 2010): mathematical logic

http://www.people.umass.edu/partee/409/Deduction_I.pdf (accessed on 2 August 2010): tautologies, contradictions, contingencies

CHAPTER THREE

Set Theory

This chapter will deal with the basic building blocks for any formal argument using mathematics. In the last chapter we have seen how to formulate an argument analytically; we shall begin by using the things we have learnt.

A collection of objects with a well-defined membership rule is called a set. For example $N = \{1, 2, ...\}$ are the so-called counting numbers or the set of **positive integers**; given any object we may check whether it is a member of this set quite easily; similarly, $-N$ is defined to be the set of negative integers; and so on.

3.1 Operations with Sets

Given two sets A and B, their union $A \bigcup B$ is the collection of objects which are either in A or in B; thus we may define $Z = N \bigcup -N \bigcup \{0\}$ to indicate that we are considering positive integers, negative integers, and zero. Thus Z is the set of all integers.

The intersection of two sets is the set of elements common to both and is denoted by $\bigcap$; thus $A \bigcap B$ is the intersection or the elements common to both the sets.

We shall use the notation $\in$ to mean 'is an element of'; thus for example, $A \bigcup B = \{x : x \in A \text{ or } x \in B\}$; $A \bigcap B = \{x : x \in A \ \& \ x \in B\}$; we shall say that the set A is a subset of the set B provided every element of the former is also a member of the latter and we write $A \subseteq B$; alternatively, $A \subseteq B$ if $x \in A \Rightarrow x \in B$; the symbol $\Rightarrow$ is read as 'implies'. $A \subseteq B \ \& \ B \subseteq A \Longleftrightarrow A = B$; thus we have equality of two sets only when they have identical elements. $\Longleftrightarrow$ is read as 'if and only if' or 'iff'.

Remark 3.1 **The phrase *if and only if* or the abbreviated form iff:** In section 2.2.1, while discussing Principle of Bivalence, a sentential connective 'if and only if' or 'iff' was examined. Recall that, A *iff* B consists of two statements: A implies B and B implies A. Alternatively, one may say that a necessary and sufficient condition for A is B.

As defined in the last chapter, we shall use the notation $\forall$ to denote the quantifiers 'for all'; and we shall use $\exists$ to denote 'there exists'. For instance, $\forall x \in A$ is to be understood as 'for all elements of the set A'; whereas $\exists x \in A$ is to be understood as 'there is some element in the set A'.

We note some of the following rules of operations for any sets A, B, and C:

- $A \cup (B \cup C) = (A \cup B) \cup C$; $A \cap (B \cap C) = (A \cap B) \cap C$ (Associative Laws).

To establish the first, for example, note that $x \in A \cup (B \cup C) \Rightarrow x \in A$ or $x \in (B \cup C)$, that is, $x \in A$ or $x \in B$ or $x \in C$, that is, $x \in A \cup B$ or $x \in C$, that is, $x \in (A \cup B) \cup C$ so $A \cup (B \cup C) \subseteq (A \cup B) \cup C$. A similar argument establishes $(A \cup B) \cup C \subseteq A \cup (B \cup C)$.

- $A \cap (B \cup C) = (A \cap B) \cup (A \cap C)$; $A \cup (B \cap C) = (A \cup B) \cap (A \cup C)$ (Distributive Laws).

To see the former, for example, consider $x \in A \cap (B \cup C) \Rightarrow x \in A \ \& \ x \in (B \cup C) \Rightarrow x \in A \ \& \ x \in B$ or $x \in A \ \& \ x \in C \Rightarrow A \cap (B \cup C) \subseteq (A \cap B) \cup (A \cap C)$; the claim follows by showing the reverse inclusion.

Let $A \subseteq B$; by the set A^c, the complement of A (in B), we shall mean all those elements of B which are not in A; that is, $A^c = \{x \in B : x \notin A\}$. Sometimes we may also write $B - A$ to denote A^c; more generally for any two sets A and B, $B - A$ denotes all those elements of B which are not elements of A. One may check the following: for sets $A, B \subseteq R$

$$(i) \ (A \cup B)^c = A^c \cap B^c \ \text{and}$$

$$(ii) \ (A \cap B)^c = A^c \cup B^c$$

To see the first, consider $x \in (A \cup B)^c$; that is, $x \in R$, $x \notin A \cup B$, that is, $x \notin A$ and $x \notin B$ which means that $x \in A^c \cap B^c$; hence $(A \cup B)^c \subseteq A^c \cap B^c$. The reverse relationship may be shown exactly as above. Property (ii) may be shown similarly.

EXERCISES

Prove the following:

1. $(A \cup B) \cap (A \cup C) = A \cup (B \cap C)$.
2. $A \cap (B - C) = (A \cap B) - (A \cap C)$.
3. $(A - C) \cap (B - C) = (A \cap B) - C$.
4. $(A - B) \cup B = A$ iff $B \subset A$.

3.2 Binary Relations

Next, we introduce the notion of Cartesian product of two sets A and B:

$$A \times B = \{(a, b) : a \in A, b \in B\}$$

the set of ordered pairs of elements; in each pair the first is from the set A while the second is from the set B; in general, thus, $A \times B \neq B \times A$. Any subset $\mathcal{R}$ of ordered pairs $A \times B$ defines a **relation**; if $(a, b) \in \mathcal{R}$ we shall say that '*a is in the relation $\mathcal{R}$ to b*'; we may also use the notation

$$a \mathcal{R} b$$

in such cases. In many instances, we shall require $A = B$; then $A \times B$ is written as A^2 and any relation $\mathcal{R}$ is said to be a **binary relation** defined on the set A. $\mathcal{R}$ is said to be **reflexive** on A if for every $a \in A$, $(a, a) \in \mathcal{R}$; the relation is said to be **complete** on A if for any $a, b \in A$ either $(a, b) \in \mathcal{R}$ or $(b, a) \in \mathcal{R}$. The relation is said to be **transitive** on A if for any triple $a, b, c \in A$, $(a, b) \in \mathcal{R}$, $(b, c) \in \mathcal{R} \Rightarrow (a, c) \in \mathcal{R}$. A reflexive, complete, and transitive binary relation on a set A is said to be an **ordering** over A.

Given the binary relation $\mathcal{R}$, two other binary relations may be defined in the following manner. We shall define $\forall x, y \in A$, $x\mathcal{P}y$ if $x\mathcal{R}y$ and $\sim y\mathcal{R}x$ (we use the symbol $\sim$ to denote the negation of the statement following the symbol; so in this case, this means that it is not the case that $y\mathcal{R}x$). $\mathcal{P}$ is called the **asymmetric** part of the relation $\mathcal{R}$; similarly the **symmetric** part of the relation $\mathcal{R}$ is denoted by $\mathcal{I}$ and is defined by the following: $\forall x, y \in A$, $x\mathcal{I}y$ if $x\mathcal{R}y$ and $y\mathcal{R}x$ holds.

3.3 Even and Odd Integers

Two subsets of the set N will occupy us next: the even and odd integers. We define $x \in N$ as even if $x = 2r$ for some $r \in N$; $x \in N$ is said to be odd if it is not even; or more directly, if $x = 2r + 1$ for some $r \in N$. We note the following:

1. If $a, b \in N$ are even then so is $a + b$.
2. If $a \in N$ is even, then $a \times b$ is even for any $b \in N$.
3. If $a, b \in N$ are such that $c = a \times b$ is odd, then so are a, b. Hence if a^2 is odd so is a.
4. If a^2 is even so is a.

Proofs of the above should be easy to establish (use the definitions). A word about the description of odd integers as those which may be represented as $2r + 1$ for some integer r: the idea is that an odd integer is one which upon division by 2 leaves a non-zero remainder; thus the only such remainder is unity. By analogy if a is a multiple of another integer m, $a = mr$ for some integer r; on the other hand if a is not a multiple of m then by dividing by m there is a non-zero remainder d which could be any integer satisfying $1 \leq d \leq m - 1$.

3.4 Real Numbers

We cannot restrict ourselves to the consideration of integers only; as soon as we need to divide, we run out of integers and we need to admit a wider class of numbers. If $a, b \in N$, $b \neq 0$ then a/b is said to be a rational number; some numbers are rational numbers since they can be put in this form; others cannot be, for example, $\sqrt{3}$. To show that this is not a rational number (an irrational): suppose to the contrary it is; then $\sqrt{3} = a/b$ for some $a, b \in N$, $b \neq 0$; **assume too that a and b have no common factor between them.** Hence: $3 = \dfrac{a^2}{b^2}$ or $3b^2 = a^2 \Rightarrow a$ is a multiple of 3 for otherwise $a = 3r + d, 1 \leq d \leq 2$ for some integer d; so that $a^2 = 9r^2 + 6rd + d^2$ $\Rightarrow b^2 = 3r^2 + 2rd + d^2/3$ which cannot be an integer since the last term is either $1/3$ or $4/3$: contradiction. Hence as claimed, a is a multiple of 3. But then, $a = 3r \Rightarrow b^2 = 3r^2 \Rightarrow b$ is a multiple of 3, which is a contradiction of our choice of a and b.

Rational and irrational numbers together constitute the set of real numbers.

Alternatively, an axiomatic definition of real numbers could be provided as follows: any set R of elements with the following properties will be called the set of real numbers. Two operations $+$ (SUM) and $\cdot$ (PRODUCT) and a binary relation $\geq$ are defined on R such that $\forall x, y \in R, x + y, x \cdot y \in R$ and further:

1. $\forall x, y \in R, x + y = y + x; \ x \cdot y = y \cdot x$ (Commutative Laws)
2. $\forall x, y, z \in R, x + (y + z) = (x + y) + z; \ x \cdot (y \cdot z) = (x \cdot y) \cdot z$ (Associative Laws)
3. $\forall x, y, z \in R, x \cdot (y + z) = x \cdot y + x \cdot z$ (Distributive Law)
4. $\forall x, y \in R, \exists z \in R$ such that $x + z = y$ (z is denoted by $y - x$); $\forall x \in R, x - x = 0 \in R$; 0 is independent of x; $0 - x$ will be denoted by $-x$.
5. There is at least one $x \in R, x \neq 0$. If $x, y \in R, x \neq 0$, then there is $z \in R$ such that $z \cdot x = y$ and we shall denote z by y/x; $\forall x \neq 0, x/x = 1 \in R$; 1 is independent of x.
6. $\forall x, y \in R$ either $x \geq y$ or $y \geq x$ ($\geq$ is complete on R); $x \geq x \forall x \in R$ ($\geq$ is reflexive on R).
7. $\forall x, y, z \in R, x \geq y, y \geq z \Rightarrow x \geq z$ ($\geq$ is transitive on R).
8. $\forall x, y \in R, x > y \ (x \geq y, \sim y \geq x) \Rightarrow x + z > y + z$ for any $z \in R$.
9. $\forall x, y \in R, x > 0, y > 0 \Rightarrow x \cdot y > 0$

    Before we introduce the next property we need to define the term **bounded**. A subset $S \subseteq R$ is said to be bounded above if there is $z \in R$ such that $\forall x \in S, z \geq x$, z is called an upper bound for the set. The subset S is said to be bounded below if there is $w \in R$ such that $\forall x \in S, x \geq w$, w is called a lower bound for the set. If the subset if bounded above and below then the subset is said to be bounded.
10. Any non-empty subset S of R which is bounded above has a least upper bound (lub) (that is, z is ℓub of S, if it is an upper bound and for any upper bound $z', z' \geq z$)

The above completes the list of axioms required for the real number system. Note that the last also implies that if there is a non-empty subset of R which is bounded below then there will be a greatest lower bound (glb) (that is, w is the $g\ell b$ of S, if it is a lower bound and for any other lower bound $w', w \geq w'$).

It should be noted that the axioms 1–10 serve to introduce the various properties of real numbers that we are familiar with; for example, axiom 4 introduces subtraction; axiom 10 introduces irrational numbers. Consider, for instance, the set $S = \{x \in R : x^2 \leq 2\}$: clearly the set is bounded above, non-empty, and hence must possess (by axiom 10) a ℓub and it is $\sqrt{2}$.

We shall refer to the set R of real numbers as the real line; subsets of this set will be denoted by intervals of the type $\{x \in R : a \leq x \leq b, a, b \in R\}$ or $[a, b]$: called a closed interval; we also have the open interval (a, b) or the subset $\{x \in R : a < x < b, a, b \in R\}$ or intervals of the type $\{x \in R : a < x \leq b, a, b \in R\}$ denoted by $(a, b]$ or intervals of the type $\{x \in R : a \leq x < b, a, b \in R\}$ denoted by $[a, b)$. All subsets of the real line are made up of the union of such intervals. R^2, R^n are referred to as the Euclidean plane or the Euclidean space; their subsets are the Cartesian products of the intervals described above.

3.5 Infimum and Supremum $\{g\ell b, \ell ub\}$

There is another aspect of a ℓub (also called the **supremum**) which should be noted. Consider a non-empty set $S \subset R$ which is bounded above and let s^* denote the ℓub; then **for any number** $\varepsilon^k > 0, \exists\ x^k \in S, x^k \neq s^*$ such that $x^k > s^* - \varepsilon^k$. By choosing smaller and still smaller ε^ks, we can get elements of the set S in the shrinking interval $[s^* - \varepsilon^k, s^*]$. Thus, in a sense to be made precise, there are elements of the set S arbitrarily 'close' to the number s^*. We shall return to this basic fact later. A similar property holds for $g\ell b$ (also called the **infimum**).

EXERCISES

1. Show that $\sqrt{2} + \sqrt{3}$ is irrational.
2. If a, b, c, d are rational and x is irrational, show that $\dfrac{ax + b}{cx + d}$ is usually irrational. Construct an example where this is not so.
3. If $a/b < c/d$ and if $b, d > 0$ show that $a/b < (a + c)/(b + d) < c/d$.
4. Show that $g\ell b$ and ℓub of a set are unique whenever they exist.

5. Compute the supremum and infimum for the following sets:
   (a) $S = \{x | 3x^2 - 10x + 3 < 0\}$ and
   (b) $S = \{x | (x - a)(x - b)(x - c) < 0\}$ where $0 < a < b < c$.
6. Let A, B be two sets of real numbers bounded above. Let $\alpha = \ell ub(A)$ and $\beta = \ell ub(B)$. Define $C = \{z | z = x.y, x \in A, y \in B\}$. Investigate whether $\alpha.\beta = \ell ub(C)$.

3.6 Functions: Preliminaries

Let F be a relation; that is, $F \subset S \times T$; if $(a, b), (a, c) \in F$ implies $b = c$ for every $a \in S$, $b, c \in T$ then F is a function with **domain** of definition $A = \{a \in S : \exists b \in T, (a, b) \in F\}$; $B = \{b \in T : (a, b) \in F$ for some $a \in S\}$ is the **range** of the function F and we shall write $F : S \to T$; we shall also say that b is the value of the function F at a; sometimes we shall write $b = F(a)$; we shall also write $A = D(F)$ (domain of F) and $B = R(F)$ (the range of F). If $B = T$, we shall say that F is **onto** T; F is **one to one on** $D(F)$ iff $F(a) = F(b) \Rightarrow a = b$, that is, distinct points have distinct values.

1. Given any relation $R \subset S \times T$, define $R^{con} = \{(b, a) \in T \times S : (a, b) \in R\}$; note that if R is a function, it is not necessary that R^{con} is a function; if

however R^{con} is a function, then R^{con} is called the inverse function and is written as R^{-1}.

2. If F is one to one on its domain then F^{con} is also a function. It is called the inverse function of F.

3. If F, G are two functions such that $R(F) \subset D(G)$, then we may define the composite function GF by $GF(x) = G(F(x))$.

Let N denote the set of positive integers and let N_n denote the set of the first n positive integers where n is finite. If F is a function such that $D(F) = N_n$ for some finite n then we shall call F a finite sequence and its range will be written as $\{F_1, F_2, ...F_n\}$; F_is are called the terms of the sequence. By an infinite sequence we shall mean a function F whose domain of definition is N; F_i is called the i-th term of the sequence. For example, $F_i = \dfrac{(-1)^i}{i}, i = 1, 2, 3, ...$ defines an infinite sequence.

3.7 Countable Sets

Two sets A, B are said to be equivalent iff there is a one to one function F with domain A and range B. In this situation, we write $A \leftrightarrow B$. Thus, a set A is finite if $N_n \leftrightarrow A$; the set A is said to be countable iff $N \leftrightarrow A$; otherwise, the set A is uncountable.

Thus, using the map which establishes the one to one property with the set of positive integers, all the elements of an infinite countable set may be written down as elements of an infinite sequence. For example, if we have $\phi : N \leftrightarrow S$ then $\phi(1) \in S, \phi(2) \in S, \cdots$; writing $\phi(n) = s_n$, the elements of the set S may be written as a sequence $\{s_n\}$. Note that it follows trivially that if a set A is finite, then any subset of A is also finite. We have in addition, the following:

Claim 3.7.1 Any subset of a countable set is countable.

Proof Let S be countable and let $A \subseteq S$; we shall show that A is countable. Since S is countable, we can write the elements of S as $s_1, s_2, s_3, \cdots$. If $A \subseteq S$ is finite, then there is nothing to prove; suppose then that A is infinite; consequently, S must be infinite as well. Define the function $k : N \to N$ recursively as follows:

$k(1) =$ smallest integer m such that $s_m \in A$; $k(n) =$ smallest integer $r > k(n-1)$ such that $s_r \in A$ $n = 2, 3, \cdots$. Thus $s_{k(n)} \in A$ for all n. Note that by construction, $k(m) < k(n)$ iff $m < n$; next define the map $\psi : N \to A$ by $\psi(n) = s_{k(n)}$; one may check that ψ is one to one and $N \leftrightarrow A$ and hence, A is countable as claimed.

Thus it follows that any infinite subset of the set N is also countable. We shall use this result later on. However, not all sets are countable. For example, while the set of rationals is countable, the set of reals is uncountable. We shall provide proofs for these assertions next. First we shall prove:

Claim 3.7.2 The set of reals in the interval $[0, 1]$ is uncountable.

Suppose to the contrary that the set of reals in the interval $[0, 1]$ is countable: then the set of reals in $[0, 1]$ may be written as a sequence $s_1, s_2, ..., s_n, ...$ and moreover each $s_i = 0.s_{i1}s_{i2}.....s_{in}....$where s_{ij} is a number between 0 and 9. Next, consider the number $t = 0.t_1t_2....t_n....$ where each t_i is an integer between 0 and 9 and further $t_i \neq s_{ii}$ for each i. Hence, first of all, $t \in [0, 1]$ and is a real number; second, $t \neq s_i$ for all i. Thus the sequence s_i cannot be a complete listing of all reals in $[0, 1]$. This completes the demonstration.

To prove that the set of rationals is countable, we take the following route. We first establish the following:

Claim 3.7.3 Let $F = \{A_1, A_2, ...A_n, ...\}$ be a countable collection of countable sets with $A_i \cap A_j = \emptyset, i \neq j$. Then $A = \cup_{i=1}^{\infty}A_i$ is countable.

Let us write the elements of each set $A_i = a_{i1}, a_{i2}, ..., a_{in}, ...$ Now by definition, $x \in A \Rightarrow x \in A_i$ for some unique i which in turn implies that $x = a_{ij}$ for some unique pair i, j. Thus $F(x) = (i, j)$ defines the one to one function $F : A \to N \times N$. Next, see that the function $G(i, j) = 2^i 3^j$ defines a one to one function $G : N \times N \to N$; thus the composite function $GF : A \to N$ is one to one and hence by definition A is countable.

Claim 3.7.4 Let $F = \{A_1, A_2, .., A_n, ..\}$ be a countable collection of sets; let $G = \{B_1, B_2, .., B_n, ..\}$ where $B_1 = A_1$ and $B_n = A_n - \cup_{i=1}^{n-1}A_i$. Then $B_i \cap B_j = \emptyset$ and $\cup_{i=1}^{\infty}A_i = \cup_{i=1}^{\infty}B_i$.

Note that $B_i \cap B_j = \emptyset$ by construction. Let $A = \cup_{i=1}^{\infty}A_i$, $B = \cup_{i=1}^{\infty}B_i$; and let $x \in A$; thus $x \in A_i$ for some i; consider the smallest such i that is $x \in A_i$ but $x \notin \cup_{k=1}^{i-1}A_k$ thus, $x \in B_i$ or $x \in B$. Hence $A \subset B$. It is immediate that $B \subset A$ and hence, $A = B$ as claimed.

Combining the last two we may state the following: a countable collection of countable sets is countable. Now identify A_k with all positive rationals with denominator k; then each A_k is countable and the set of all positive rationals is seen to be countable. Similarly the set of all negative rationals may be seen to be countable and hence the set of all rational numbers is countable.

EXERCISES

1. A claim goes as follows: The set of all intervals $\mathcal{I}$ of positive length is countable. This may be shown as follows: let the set of all rational numbers be denoted by $\{x_1, x_2, \cdots, x_n \cdots\}$; let $I \in \mathcal{I}$ then I contains infinitely many rationals x_n but there is only one with the smallest index n; define $F(I) = n$ if x_n is the rational with the smallest index n in I. This function establishes a 1-1 correspondence between $\mathcal{I}$ and a subset of N, the set of positive integers and the claim follows. Is the 'proof' correct?

2. Let S be the set of all sequences whose terms are either 0 or 1. Show that S is uncountable.

3. Let S be a set of finite number of elements n; consider the set T of all possible subsets of S. Prove that T has finitely many elements and find out the number of elements of T.

3.8 Open and Closed Sets

Let $S \subseteq R, x, y \in S$; we define the distance between x, y by $d(x, y) = |x - y|$, interpreted as 'the absolute value of the difference between x and y'; if we are considering sets in higher dimensions, that is $x, y \in S \subseteq R^n, n > 1$ then $x = (x_1, x_2, ..., x_n), y = (y_1, y_2, ..., y_n)$ and we define the distance between x, y by the following: $d(x, y) = \sqrt{\{\sum_{i=1}^{n}(x_i - y_i)^2\}}$. We shall also have to use the notion of the length or norm or absolute value of $x \in R^n$ (notation $|x|$) by $|x| = d(x, 0)$, where 0 is the origin. (The term absolute value is used in the context of R, that is, when $n = 1$.) The following properties relating to $|x|, x \in R^n$ should be noted:

(i.) $|x| \geq 0$ for all $x \in R^n$ and $|x| = 0$ iff $x = 0$.
(ii.) $|x - y| = |y - x|$ for all $x, y \in R^n$.
(iii.) $|x + y| \leq |x| + |y|$ for all $x, y \in R^n$.
(iv.) $|x - y| \geq |x| - |y|$ for all $x, y \in R^n$.

Note that assertions (i) and (ii) follow from the definitions. Property (iii) depends upon the Cauchy-Schwartz inequality:

$$\left(\sum_{k=1}^{n} a_k b_k\right)^2 \leq \sum_{k=1}^{n} a_k^2 \sum_{k=1}^{n} b_k^2$$

We first provide a quick proof of this inequality: first, note that any square can never be negative; hence for any real number z, $\sum_{k=1}^{n}(a_k z + b_k)^2 \geq 0$ which means that $Az^2 + 2Bz + C \geq 0$ where $A = \sum_{k=1}^{n} a_k^2, B = \sum_{k=1}^{n} a_k b_k, C = \sum_{k=1}^{n} b_k^2$. If $A = 0$, the Cauchy-Schwartz inequality is trivially true. If $A \neq 0$, let $z = -B/A$ then we have $B^2 - AC \leq 0$ which is the desired inequality. Hence, now note that writing $x = (x_k), y = (y_k)$

$$|x + y|^2 = \sum_{k=1}^{n}(x_k + y_k)^2 = |x|^2 + |y|^2$$

$$+2\sum_{k=1}^{n} x_k y_k \leq |x|^2 + |y|^2 + 2|x||y| = (|x| + |y|)^2.$$

The last assertion follows exactly as above.

We are now ready to define the notion of points which are close to one another. Consider $x \in S \subseteq R^n$; by the term neighbourhood of the point x in S we shall mean the set $N_\varepsilon(x)$ (or $N(x, \varepsilon)) = \{y \in R^n : d(y, x) < \varepsilon\}$ for some number $\varepsilon > 0$. The neighbourhood of the point x is thus a circle with centre x and radius ε. The smaller the number ε chosen, the closer are the points to the point x.

$x \in S \subseteq R^n$ is an interior point of S, if there is number $\varepsilon > 0$ such that $N_\varepsilon(x) \subseteq S$; if no such number exists then the point x is said to be a boundary point of S.

$S \subseteq R^n$ is an **open subset in** R^n if **every point** of S is an interior point. Note that every neighbourhood defined above is an open set. (*As a consequence of this* *definition, we are able to claim the following: the set R^n is an open subset of itself; the empty set $\emptyset$ is open subset of R^n. For the moment, accept the latter as an axiom. In fact, in point set topology, where open sets are the basic building blocks, these two facts together with the claims 3.8.1–3.8.4 demonstrated below are taken to be the definition for sets to be called open.)*

Claim 3.8.1 Let $\{O_\alpha\}, \alpha \in \Omega$ be some family of open subsets in R^n; then $O = \cup_{\alpha \in \Omega} O_\alpha$ is an open subset of R^n. (This assertion is often expressed by saying that property of being open is preserved under arbitrary union.)

For a proof, observe that $x \in O \Rightarrow x \in O_\alpha$ for some $\alpha \in \Omega$; consequently, there is some number $\varepsilon > 0$ such that $N_\varepsilon(x) \subseteq O_\alpha$; but then, $N_\varepsilon(x) \subseteq O$ so that x is an interior point of the set O: thus, O is open as claimed.

Claim 3.8.2 Let $\{O_{\alpha_i}\}, \alpha_i \in P_n$ for some integer n be a finite collection of open subsets in R^n; then $O = \cap_{\alpha_i \in P_n} O_{\alpha_i}$ is an open subset of R^n.

(This assertion is referred to as the property of being open and is preserved under finite intersection.)

For a proof, observe that $x \in O \Rightarrow x \in O_{\alpha_i}$ for all $\alpha_i \in P_n$; consequently, there exist $\varepsilon_i > 0$ corresponding to each $\alpha_i \in P_n$ such that $N_{\varepsilon_i}(x) \subseteq O_{\alpha_i}$. Let $\varepsilon = Min_i \, \varepsilon_i > 0$ and consider $N_\varepsilon(x) \subseteq N_{\varepsilon_i}(x)$ for each ε_i. Thus $N_\varepsilon(x) \subseteq O_{\alpha_i}$ for each α_i and hence $N_\varepsilon(x) \subseteq O$ which proves that every point is an interior point and hence O is open as claimed.

$S \subseteq R^n$ is a **closed subset in R^n** iff $S^c = R^n - S$ is an open subset of R^n. Now note the following: the empty set $\emptyset$, the whole set R^n are closed subsets. In fact these sets are both closed as well as open. Further the counterpart of properties 3.8.1 and 3.8.2 are the following

Claim 3.8.3 Let $\{O_\alpha\}, \alpha \in \Omega$ be some family of closed subsets in R^n; then $O = \cap_{\alpha \in \Omega} O_\alpha$ is a closed subset of R^n.

Claim 3.8.4 Let $\{O_{\alpha_i}\}, \alpha_i \in P_n$ for some integer n be a finite collection of closed subsets in R^n; then $O = \cup_{\alpha_i \in P_n} O_{\alpha_i}$ is a closed subset of R^n.

Thus, the property of closed sets is preserved under arbitrary intersection and finite union. The proofs of these assertions follow from the definition of closed sets and the property of complements and the fact that $O_{\alpha_i} = R^n - S_{\alpha_i} \Rightarrow \cap O_{\alpha_i} = R^n - \cup S_{\alpha_i}$ and $\cup O_{\alpha_i} = R^n - \cap S_{\alpha_i}$.

Another route to defining closed sets (independent of the definition of open sets) depends on the notion of a **limiting point or limit point or the point of accumulation** for a set. Given a set $S \subseteq R^n, \ell \in R^n$ is a limiting point for the set S, iff any neighbourhood $N_\varepsilon(\ell) \cap S$ contains a point of S different from ℓ. Thus in any neighbourhood of a limit point of a set, there is an infinite number of elements of that set.

By way of example, recall the definition of ℓub and $g\ell b$ for any set: these are points of accumulation or limit points for the sets. Note from the definition of limiting point, it is not necessary that such points be members of the set; however, in the special situation when all limit points of a set are contained in the set, the set is defined to be closed. That these two definitions are equivalent may be seen thus:

Claim 3.8.5 Let $S \subseteq R^n$. $R^n - S$ is an open subset of $R^n \Leftrightarrow$ all limit points of S are members of S.

For a proof, assume first that $R^n - S$ is an open subset of R^n; in addition, let ℓ be a limit point of S; and if possible let $\ell \notin S$. Then $\ell \in R^n - S$ and hence there is some neighbourhood $N_\varepsilon(\ell) \subseteq R^n - S$ or $N_\varepsilon(\ell) \cap S = \emptyset \Rightarrow \ell$ cannot be a limit point for the set S: a contradiction. Hence $\ell \in S$.

Assume next that all limit points of S are members of S. If possible, let $R^n - S$ not be open, that is, there is a point $x \in R^n - S$ and x is not an interior point of $R^n - S$, that is, for any neighbourhood $N_\varepsilon(x)$, $N_\varepsilon(x) \cap S \neq \emptyset$ or x is a limit point of S: a contradiction. Hence no such point can exist and $R^n - S$ must be open. ●

We are now ready to state and prove a very important result, the first major mathematical result we have encountered. This result provides a condition under which limit points exist for infinite sets. We need an additional definition first. Let $S \subset R^n$; we shall say that S is **bounded** if there exist intervals $[\alpha_i, \beta_i]$ such that $\alpha_i, \beta_i \in R$, $i = 1, 2, \cdots, n$ such that $S \subset \times_{i=1}^n [\alpha_i, \beta_i]$. The same notion may be captured by requiring that there be $\alpha, \beta \in R^n$ such that $x \in S \Rightarrow \alpha \leq x \leq \beta$, that is, for each component k, we must have $\alpha_k \leq x_k \leq \beta_k$. Yet another way of representing this notion is to require the existence of some real number α such that $S \subset N(0, \alpha)$.

Proposition 3.1 Bolzano-Weirstrass Theorem: Any non-empty bounded infinite subset of R^n has a limit point.

(For $n = 1$: ℓub or $g\ell b$ will surely exist for a bounded non-empty infinite subset of the real line and these are limit points.) So consider R^2; any set non-empty, bounded $S \subset R^2$ must be such that there is some real number $\alpha > 0$, $I = [-\alpha, \alpha]$ and $S \subset I^2$. Consider $I_{11} = [-\alpha, 0]$, $I_{12} = [0, \alpha]$; and consider all Cartesian products of the form $A \times B$ where A, B could each be either I_{11} or I_{12}: note that there are four such products and the union of all of them make up I^2 and hence contain S; since S is non-empty and infinite, at least one of these Cartesian products would contain an infinite subset of S. Call that particular set I_1^*; note that $I_1^* \subset I^2$ and is made up of the Cartesian product of two intervals of length α. Divide these intervals into half and consider Cartesian products of all possible combinations of these halves. Again, as before, one of these Cartesian products must contain an infinite subset of S: call it I_2^*: it is made up by considering the Cartesian products of two intervals each of length $\alpha/2$. These intervals are of the type $[-\alpha/2, 0]$ or $[-\alpha, -\alpha/2]$ or $[0, \alpha/2]$ or $[\alpha/2, \alpha]$. Continue this process, in each case subdividing the intervals into equal halves, considering all possible Cartesian products (4) and then choosing one such product which has an infinite subset of the set S; consequently at the k-th stage, we would have chosen intervals with length $\alpha/2^k$ whose Cartesian product make up I_k^*, say. Let I_k^* be written as $[\gamma_{1k}, \beta_{1k}] \times [\gamma_{2k}, \beta_{2k}]$; where $-\alpha \le \gamma_{ik} < \beta_{ik} \le \alpha$ for all k and for each $i = 1, 2$; in addition, $|\beta_{ik} - \alpha_{ik}| = \alpha/2^k$. Let $\gamma_i^* = \sup_k \gamma_{ik}$ and $\beta_i^* = \inf_k \beta_{ik}$; these exist, and are well defined, being supremum and infimum of a non-empty and bounded sets of real numbers. Note also that it must be the case that $\gamma_i^* = \beta_i^*$ for otherwise, $\gamma_i^* < \beta_i^*$ and let $|\beta_i^* - \gamma_i^*| = \delta$ say, and then the length of the interval $I_k^* \ge \delta\ \forall k$: a contradiction. Let this common value $(\gamma_i^* = \beta_i^*)$ be denoted by $x_i^*, i = 1, 2$. Now we claim that $(x_1^*, x_2^*) \in I$ and is a limit point of the set S.

The first follows from the fact that $-\alpha \le \gamma_{ik} < \beta_{ik} \le \alpha$ for all k and for each $i = 1, 2$; next consider any neighbourhood of (x_1^*, x_2^*) say N; then for k large enough, $I_k^* \subset N$ and by definition, I_k^* contains an infinite subset of S: thus the claim follows. •

The extension to R^n follows by performing exactly the same steps as mentioned earlier, except that we need to consider the Cartesian product of n intervals at each stage.

3.9 Compactness

Let $S \subset R^n$ and let $\mathcal{F} = \{F_\alpha\}$ be a collection of subsets of R^n such that $S \subseteq \cup_\alpha F_\alpha$. In this situation, we call $\mathcal{F}$ to be a **covering** for the set S; in case all the members of the collection $\mathcal{F}$ are open subsets, we call the covering to be an **open covering**. If the collection $\mathcal{F}$ forms a covering of the set S and contains a countable collection, then we have a countable covering. We shall deduce a fundamental fact about open coverings:

Claim 3.9.1 Any open covering of a set S in R^n contains a countable sub-collection which is also a covering (called a sub-covering).

To see this, first, we need to show the following:

Claim 3.9.2 Let $\mathcal{G} = \{A_1, A_2, ...\}$ denote a countable collection of neighbourhoods in R^n with rational radii and centres at points with rational coordinates. Let $x \in R^n$ and let $S \subset R^n$ be an open set containing x. Then at least one of the neighbourhoods in $\mathcal{G}$ contains x and is contained in S.

First, note that the collection $\mathcal{G}$ is countable; this is so since each neighbourhood in the collection is identified uniquely by n+1 rational numbers $(x_1, ..., x_n; r)$ where the first n refer to the rational coordinates of the centre and the last is the rational radius; we also know that such n+1-tuples of rational numbers will be countable since rational numbers are countable as we have shown above. If $x \in R^n$, then there is some neighbourhood $N(x, r) \subset S$, since the set S is open; consequently, there must be $y \in N(x, r)$ such that y has rational coordinates; in particular, find a rational number y_k such that $|y_k - x_k| < r/4n$ for each $k = 1, 2, ..., n$ and let $y = (y_1, .., y_n)$; then $|y - x| \le |y_1 - x_1| + ... |y_n - x_n| < r/4$. Next, find a rational number q such that $r/4 < q < r/2$. Now note that $x \in N(y, q) \subset N(x, r) \subset S$ and $N(y, q) \in \mathcal{G}$ and hence the claim is proved. •

Claim 3.9.3 Let $S \subset R^n$ and let $\mathcal{F}$ denote an open covering of the set S; then there is a countable sub-collection of $\mathcal{F}$ which also covers S.

We recall the collection $\mathcal{G} = \{A_1, A_2, ...\}$: a countable collection of neighbourhoods in R^n with rational radii and centres at points with rational coordinates. Let $x \in S$; then since $\mathcal{F}$ is an open covering of S, there is a member of the collection, F_α, an open set such that $x \in F_\alpha$; hence by the previous result, there is a member of the collection $\mathcal{G}$, say A_K such that $x \in A_K \subset F_\alpha$; since $\cup_{A_K \in \mathcal{G}} A_K \supset S$, the claim is proved. •

*Claim 3.9.4 Let $\{Q_1, ..Q_n, ..\}$ denote a countable collection of **non-empty and closed** subsets of R^n such that $Q_{k+1} \subset Q_k$ and Q_1 is bounded. Then the intersection $S = \cap_{k=1}^{\infty} Q_k$ is closed and non-empty.*

First, note that S, being the intersection of closed sets, is closed. Next, we assume that each Q_k contains infinitely many points, since otherwise the proof is trivial. (Examine why this is so.) Now define the collection $A = \{x_1, x_2, ..., x_k, ...\}$ where each $x_k \in Q_k$; since $A \subset Q_1$ which is bounded so is A. Hence, by the Bolzano-Weirstrass Theorem, A has a limit point $\bar{x}$. Then by definition, every neighbourhood of $\bar{x}$ contains infinitely many points of A; further, all but at most a finite number of points of A belong to the set Q_k; hence $\bar{x}$ is also a limit point for each Q_k and hence must belong to Q_k, since it is closed. This true for every Q_k and hence, $\bar{x} \in S$, as claimed. •

*Proposition 3.2 Heine Borel Theorem: Let $\mathcal{F}$ denote an open covering of a **closed and bounded** set $S \subset R^n$; then there is a finite sub-collection of $\mathcal{F}$ which also covers the set S.*

(This finite sub-collection which covers the set S is called a finite sub-cover for S.)

First, we note that there is a **countable** sub-collection of $\mathcal{F}$, say $F_1, F_2, ..., F_n, ...$ which also covers the set S. Define $B_n = \cup_{i=1}^{n} F_i$; each B_n being a union of a finite number of open sets is open. Define $Q_1 = S$; and for $n > 1$, $Q_n = (R^n - B_n) \cap S$; then Q_n is closed; also $Q_1 = S$ is closed and bounded. Note too that we must have $Q_{n+1} \subset Q_n \subset S$; thus if no Q_n is empty, then we have just seen that $\cap_{i=1}^{\infty} Q_i$ is non-empty; that is, there is $x \in \cap_{i=1}^{\infty} Q_i$ or $x \in Q_i$ for all i, that is, $x \in S$ and $x \notin B_n$ for all n: a contradiction since $x \in S \Rightarrow x \in F_i$ for some $i \Rightarrow x \in B_i$ for some i: a contradiction. Thus, there must be some m such that $Q_m = \emptyset$ that is, $\{R^m - B_m\} \cap S = \emptyset \Rightarrow B_m \supset S$ and we have exhibited the existence of a finite sub-collection which is a covering for the set S, as claimed. •

We define a set as being **compact**, if every open covering of it, contains a finite sub-cover. The Heine Borel Theorem shows that in R^n, closed and bounded sets are compact. In fact, the link is strong and goes the other way as well:

Claim 3.9.5 Let S be a subset of R^n. Then the following statements are equivalent:

   (i) *S is compact.*
   (ii) *S is closed and bounded.*
   (iii) *Every infinite subset of S has a limit point in S.*

By the claim that two statements are equivalent, we mean that the two statements are each necessary and sufficient for the other. The Heine Borel Theorem shows that (ii) $\Rightarrow$ (i). We shall prove that (i) $\Rightarrow$ (ii) $\Rightarrow$ (iii) $\Rightarrow$ (ii). This would establish equivalence (check how).

So let us assume that (i) holds, that is, $S \subset R^n$ is compact; then for each $x \in S$, consider the neighbourhood $N(x, r)$ where $r > 0$; note then that $\{N(x, r); x \in S\}$ constitutes an open covering of S since $S \subset \cup_{x \in S} N(x, r)$. Hence, by compactness of S, there is a finite sub-collection which also covers S, that is, there is $\{N(x_i, r), i = 1, ..m, x_i \in S \forall i\}$ which also covers S; let $d(0, x_k) = \max_{i=1,...,m} d(0, x_i) = d$, say; then consider δ appropriately chosen in relation to d, r such that $N(0, \delta) \supset \cup_{i=1}^m N(x_i, r) \supset S$ and this proves that S is bounded.

Suppose next, that there is a limit point of S: y which is not in S; let $x \in S$ and define $r_x = |x - y|/2$; note that for each x $r_x > 0$ so that the neighbourhood $N(x, r_x)$ is well defined for every $x \in S$; further, $\cup_{x \in S} N(x, r_x) \supset S$ and hence there is a finite sub-collection $\{N(x_i, r_{x_i}), i = 1, ..., m\}$ such that $\cup_{i=1}^m N(x_i, r_{x_i}) \supset S$; let $\bar{r} = \min_{i=1,...m} r_{x_i}$ Then see that $N(y, \bar{r}/2) \cap N(x_i, r_{x_i}) = \emptyset$; this follows since

$$x \in N(y, \bar{r}/2) \Rightarrow |x - y| < \bar{r}/2$$

and $|x - x_i| = |x - y + y - x_i| \geq |y - x_i| - |x - y| > 2r_{x_i} - \bar{r}/2 \geq r_{x_i}$

Thus it follows that $N(y, \bar{r}/2) \cap S = \emptyset$. Hence, y cannot be a limit point for the set S. Thus, S must be closed. Thus, (i) $\Rightarrow$ (ii).

Next, let (ii) hold. (iii) is immediate (check definitions.) Finally, let (iii) hold; if possible, let S be unbounded, that is, for any integer n, there is $x_n \in S$ such that $|x_n| > n$. Consider the infinite subset $T = \{x_1, x_2, ..., x_n, ...\}$ of the set S; then T has a point of accumulation $y \in S$; for any $n > 1 + |y|$, we have

$$|x_n - y| \geq |x_n| - |y| > n - |y| > 1$$

This shows that y cannot be a point of accumulation for the set T. Hence, S must be bounded. To complete our claim we must show that S is closed. Suppose then that S has a limit point y. Consider neighbourhoods $N(y, 1/K)$, $K = 1, 2, 3, ...$; each such neighbourhood contains a point of S and we can generate distinct points of S, denoted by $T = \{x_1, x_2, ...x_n, ...\}$ such that $x_j \in N(y, 1/j)$; now T being an infinite subset of S has a limit point x in S by virtue of our hypothesis (iii). If possible let $x \neq y$; then

$$|y - x| \leq |y - x_k| + |x_k - x| < |y - x_k| + 1/k \text{ since } x_k \in T$$

Consider N large enough so that $1/N < |y-x|/2$; then for all $k \geq N$, $|y-x_k| > |y-x|/2$ so that y cannot be a limit point for the set T; hence $x = y$ and the claim is established. $\bullet$

We also note the following rather useful property, named the **finite intersection property**, of a family of non-empty closed subsets of a compact set:

Claim 3.9.6 Let $S \subset \mathfrak{R}^n$ be a non-empty compact set and let $\{C_\alpha\}$ be a family of non-empty closed subsets of S such that for any finite number $\alpha_1, \alpha_2, ...\alpha_k$, $\bigcap_{i=1}^k C_{\alpha_i} \neq \emptyset$; then $\bigcap_\alpha C_\alpha \neq \emptyset$.

Suppose then to the contrary that

$$\bigcap_\alpha C_\alpha = \emptyset$$

This means that

$$\bigcup_\alpha C_\alpha^c \supseteq S;$$

where C_α^c denotes the complement of C_α and is thus, by definition, an open set. We also know that this means we have an open covering of a compact set and hence there must be a finite sub-cover, that is, there must exist finite number $\alpha_i, \alpha_2, ...\alpha_n$ such that

$$\bigcup_{i=1}^n C_{\alpha_i}^c \supseteq S;$$

this in turn implies that

$$\bigcap_{i=1}^n C_{\alpha_i} = \emptyset$$

for otherwise, $x \in \bigcap_{i=1}^n C_{\alpha_i} \Rightarrow x \in S,$ and $x \notin \bigcap_{i=1}^n C_{\alpha_i}^c$: a contradiction to the finite sub-cover claim. But then we have arrived at a contradiction to the hypothesis about the sets C_α. This establishes the claim. •

EXERCISES

1. Prove that in $\Re$, the real line, the only subsets which are both open and closed are the empty set and $\Re$ itself. Is the same statement true for $\Re^2$?

2. Show that every open set S in $\Re$ is the union of a countable collection of disjoint open intervals. (Hint: To approach this problem, first consider the notion of a 'component interval' of an open set S: this is an open interval $(a, b) \subset S$ such that the end points a, b are not elements of S; next note that if S is bounded, then every point $x \in S$ belongs to a **uniquely determined component interval of S.** To see this note that $x \in S \Rightarrow \exists$ open interval $I \subset S, x \in I$; using the bounded nature of S consider $a = g\ell b$ of the left end points of all such intervals I containing x and let $b = \ell ub$ of the right end points of all such intervals I containing x. Next note that $a, b \notin S$ and hence (a, b) is a component interval of S and it contains x; also note that it is the only component interval containing x; call it I_x; note that $\cup_{x \in S} I_x = S$; the collection is countable may be finally seen as follows: let $x_1, x_2, \cdots, x_n \cdots$denote the sequence of rational numbers. For each I_x consider the rational with the lowest index n such that $x_n \in I_x$; define $f(I_x) = n$ and note that $f(I_x) = f(I_y) = n$ implies $I_x = I_y$ thus f is one to one between the collection of intervals $I_x, x \in S$ and a subset of the positive integers. For any open set S define $S_n = S \cap (-n, n)$ where n is an

integer; note S_n is bounded and $S = \cup_{n=1}^{\infty} S_n$ and S_n is the union of a countable collection disjoint open intervals: hence the claim.)

3. Show that every closed set in $\Re$ is the intersection of a countable collection of closed sets.

4. Consider $S \subset \Re$ and $x \in S$ such that x is **not** a limit point of S; such a point x is called an isolated point of S; consider the collection of isolated points of S: is it countable?

KEY TERMS

Binary Relations	Heine Borel Theorem
Bolzano-Weirstrass Theorem	Infimum
Bounded Set	Limit Point
Closed Set	Odd Integers
Compact Sets	Open Set
Countable Sets	Range
Domain	Real Numbers
Even Integers	Rules of Operation with Sets
Finite Intersection Property	Set
Functions	Supremum

CHAPTER FOUR
Functions of a Single Variable

4.1 Limits

Recall the notion of a *lub* for a subset S; the feature we wish to recall is that if x is the *lub* of the set S, then in any neighbourhood of the point x there is an element of the set S; thus if we keep on shrinking the neighbourhoods, we shall get points of the set S as close to the point x as we wish. This idea is often captured by saying that there is a sequence of elements $\{x_n \in S\}, n = 1, 2, ...$, such that $\lim_{n\to\infty} x_n = x$ and we say that the **limit of the sequence** $\{x_n\}$ is x. That is, for any $\epsilon > 0$ there is N such that for $n > N$, $|x_n - x| < \epsilon$.

We have already seen the notion of a limit point; note that a limit of a sequence is a limit point for the sequence: in fact, it must be the only limit point. In general, we have the following result:

Claim 4.1.1 Let $S \subset R^n$ and $x^o \in S$ where x^o is a limit point of the set S. Then, there is a sequence $\{x_n\}, x_n \in S \ \forall n$, such that $\lim_{n\to\infty} x_n = x^o$.

A necessary and sufficient condition for the existence of limits is contained in the following 'Cauchy criterion':

Claim 4.1.2 Let $\{x_n\}$ be a sequence of elements from $S \subset R^n$; then there exists $y \in R^n$ such that $\lim_{n\to\infty} x_n = y$ if and only if for every $\epsilon > 0$ there is an integer N such that $n > N, m > N \Rightarrow |x_n - x_m| < \epsilon$.

Thus from a finite stage onwards, terms of the sequence are quite close to one another.

One of the most well-known conditions for the convergence of sequences occurs in the case of monotonic real sequences, that is, sequences $\{x_n\}, x_n \in \Re$ such that $(x_{n+1} - x_n)(x_n - x_{n-1}) \geq 0 \forall n$; consider, for example, monotonic increasing sequence, that is, $x_{n+1} \geq x_n \forall n$. A necessary and sufficient condition for the convergence of such a sequence is:

Claim 4.1.3 A monotonic increasing sequence of real numbers is convergent iff it is bounded above.

Consider $\{x_n\}$ monotonic increasing and bounded above; let $S = \{x_1, x_2, \cdots\}$ and let M be such that $x_n \in S \Rightarrow x_n \leq M$; let $\overline{M}$ denote the *lub* S: existence follows from the fact that S is a non-empty set bounded above. For any $\epsilon > 0$ there is $x_n > \overline{M} - \epsilon$ for $n \geq N$, say. Since $\{x_n\}$ is monotonic increasing it follows that $x_n \in [\overline{M} - \epsilon, \overline{M}] \forall n \geq N$, that is, $|x_n - \overline{M}| < \epsilon \forall n \geq N$ and the sequence converges to $\overline{M}$. Conversely, suppose that the sequence converges to $\overline{M}$, that is, for any $\epsilon > 0$, there is $N > 0$ such that $|x_n - \overline{M}| < \epsilon \forall n \geq N$; that is, $\overline{M} - \epsilon < x_n < \overline{M} + \epsilon \forall n \geq N$; thus the sequence is bounded above. •

It is also easy to check that the Cauchy criterion is equivalent to demanding that the sequence is bounded above if the sequence is monotonic increasing. (It may be instructive for the readers to prove this.)

Recall also the notion of a real valued function $f : S \to R$; consider $x^o \in S$; in fact, let x^o be a limit point for the set S; we shall say that $\lim_{x \to x^o} f(x) = A$ if for every $\epsilon > 0$, there is $\delta > 0$ such that $|f(x) - A| < \epsilon$ whenever $|x - x^o| < \delta$. Note that here $S \subset R^n$ is also admissible since the distance between x, x^o in R^n has been defined and we then define $|x - x^o|$ appropriately. In this situation one also uses the following phrases 'the limit of $f(x)$, as x approaches x^o, is A'. If there is such a number A, we often say that the '$\lim f(x)$ as x approaches x^o exists'. We also have the following restatement of the Cauchy criterion for functions:

Claim 4.1.4 Let $f : S \subset R^n \to R^k$ and let a be a limit point of the set S; then there exists a point $b \in R^k$ such that $\lim_{x \to a} f(x) = b$ iff, for every $\epsilon > 0$ there is neighbourhood $N(a)$ such that $x, y \in N'(a) \cap S \Rightarrow |f(x) - f(y)| < \epsilon$.

(Here $N'(a) = \{x \neq a : |x - a| < \delta\}$ for some $\delta > 0$: deleted neighbourhood.)

Also, from definition of limits, it follows that we have the following algebra of limits:

Claim 4.1.5 Let $f, g : S \subset R^n \to R$ and let a be a limit point of the set S and let $A = \lim_{x \to a} f(x)$, $B = \lim_{x \to a} g(x)$. We have:

(i) $\lim_{x \to a}(f(x) \pm g(x)) = A \pm B$
(ii) $\lim_{x \to a} f(x).g(x) = A.B$
(iii) $\lim_{x \to a} f(x)/g(x) = A/B$ *provided $B \neq 0$.*

4.2 Continuity

The definitions provided above, allow us to define the notion of continuity of a function at a point. Let the function $f : S \subset \Re \to \Re$; and let a be a limit point for the set S. We shall define the function $f(x)$ to be continuous at $x = a$ if

(i) $f(a)$ is well defined and
(ii) $\lim_{x \to a} f(x) = f(a)$.

In case the point a is not a limit point of the set S, condition (i) is sufficient for the continuity of the function at $x = a$; if the function is continuous at every point of

S, we shall say the function is continuous on the set S. It should be noted that the property of continuity of the function at the point $x = a$ allows us to write:

$$\lim_{x \to a} f(x) = f(\lim_{x \to a} x)$$

Also the algebra of limits that we have noted above allows us to conclude that the sum or difference, product, and quotient of continuous functions is continuous (care must be taken to ensure in the case of quotient of two functions that the denominator function does not vanish anywhere in the domain that we shall consider). Also note that on R, for example, polynomial functions, exponential functions, and logarithmic functions and some trigonometric functions such as sin and cos are continuous functions.

Examine what happens with the trigonometric function $\tan x$ when $-\pi \leq x \leq \pi$.

More importantly, for continuous functions of a single variable a good check is to draw a graph of the function and if the graph can be drawn without lifting the pencil from the paper on which it is being drawn, the test for continuity is passed. (Check that this is indeed so.)

Some examples of continuous functions, on the real line R are: polynomial functions, logarithmic functions (what is their domain of definition?), exponential functions, and so on. Continuous functions have various interesting properties: the crucial one among them is the preservation of the property of compactness, that is, if the domain is compact then so is the image if the function is continuous. Formally, we have:

Claim 4.2.1 Let $S \subset \Re$ and $f : S \to \Re$ that is, $f(S) \subset \Re$. If S is a compact set then so is $f(S) = \{y \in \Re : y = f(x),$ for some $x \in S\}$.

Recall the definition of compact sets; and let $\mathcal{F} = \{O_\alpha\}$ denote an open covering of the set $f(S)$. Consider $x^o \in S$; then $f(x^o) \in O_{\alpha(x^o)}$ for some α, say $\alpha(x^o)$; by the definition of an open set, there is a neighbourhood $N(f(x^o)) \subset f(S)$; let the radius for this neighbourhood be $\epsilon > 0$; that is, $y \in N(f(x^o)) \Rightarrow |y - f(x^o)| < \epsilon$ from the definition of continuity at x^o, we know that there is some $\delta > 0$ such that $x \in N(x^o) = \{x : |x - x^o| < \delta\} \Rightarrow |f(x) - f(x^o)| < \epsilon \Rightarrow f(x) \in N(f(x^o))$; Consider the collection of all such neighbourhoods $N(x^o)$ as x^o varies over all of S: they form an open covering for the set S and hence they must contain a finite sub-cover $N(x^i)|i = 1, 2, ...N$, say; that is, $S \subset \cup_{i=1}^n N(x^i)$. By construction, it follows that $f(N(x^i)) \subset N(f(x^i)) \subset O_{\alpha(x^i)}$. It thus follows that $f(S) \subset \cup_{i=1}^n O_{\alpha(x^i)}$ and demonstrates the existence of a finite sub-cover for any arbitrary open cover for the set $f(S)$ and this proves the claim.

Claim 4.2.2 If $f : [a, b] \subset \Re \to \Re$ is continuous on $[a, b]$ with $f(a).f(b) < 0$ then there is $c \in (a, b)$ such that $f(c) = 0$.

Proof For the sake of definiteness, let $f(a) < 0$ and $f(b) > 0$; define $S = \{x \in [a, b] : f(x) \geq 0\}$. Note that $x \in S \Rightarrow x \geq a$ and $b \in S$: thus S is non-empty and bounded below. Hence

$$c = \inf_{x \in S} x$$

exists. Also note that by definition $c \in (a, b)$. We claim that $f(c) = 0$; for if not then either $f(c) > 0$ in which case $c \neq \inf_{x \in S} x$ since $f(c - \epsilon) > 0$ and hence $c - \epsilon \in S$ for some $\epsilon > 0$; or $f(c) < 0$ in which case too, there is $\epsilon > 0$ such that $\inf_{x \in S} \geq c + \epsilon > c$ and hence the claim follows.

Claim 4.2.3 Intermediate Value Theorem: If $f : [a, b] \to \Re$ is continuous on $[a, b]$ and if $m = \inf_{x \in [a,b]} f(x)$ and $M = \sup_{x \in [a,b]} f(x)$ and $m < M$ then for any $m_o \in (m, M)$ there is $c \in [a, b]$ such that $f(c) = m_o$.

Proof We have already noted that there is $x_1, x_2 \in [a, b]$ such $f(x_1) = m$ and $f(x_2) = M$; define $g(x) = f(x) - m_o$ and note that $g(x_1).g(x_2) < 0$ and hence there is c in the interval $[x_1, x_2]$ or $[x_2, x_1]$ such that $f(c) = 0$. Since $c \in [a, b]$ the claim is established.

4.2.1 Uniform continuity

Consider the function $f(x) = 1/x$ defined over the set $S = (0, 1]$; the function is continuous over the set S; thus if we choose any $\epsilon > 0$ and any $x \in S$, there is $\delta > 0$ such that $|f(y) - f(x)| < \epsilon$ whenever $|x - y| < \delta$. For instance, choose $\epsilon = 10, x = 1/2$ then $|1/y - 2| < 10$, whenever $1/12 \leq y \leq 1/8$; let us say, therefore, that there is a $\delta, 1 > \delta > 0$ such that $0 < |y - 1/2| < \delta \Rightarrow |f(y) - f(1/2)| < \epsilon$. Consider, however, $x = \delta, y = \delta/11$; note that $x, y \in S$ and $|x - y| = 10\delta/11 < \delta$ but $|f(x) - f(y)| = 10/\delta > 10 > \epsilon$: thus the δ which 'worked for' $x = 1/2$ no longer 'works' when $x = \delta$. We express this by saying that while $f(x) = 1/x$ is continuous on the set S, it is not uniformly continuous there.

Accordingly, we define: Let $f : S \subset \Re^n \to \Re^m$; then f is said to be **uniformly continuous** on S if, for any $\epsilon > 0$, there is $\delta > 0$ such that if $x, y \in S$ then $|x - y| < \delta \Rightarrow |f(x) - f(y)| < \epsilon$.

We also have the following result:

Proposition 4.1 Let S be a compact subset of $\Re^n$; iff is a continuous function on S, then f is uniformly continuous on S.

Proof Consider $\epsilon > 0$. Let $a \in S$; associate with each a, a neighbourhood $N(a; r) = \{y \| a - y| < r\}$ such that $x \in N(a; r) \cap S \Rightarrow |f(x) - f(a)| < \epsilon/2$, where r in general depends on a; consider then, $\{N(a; r(a)/2\}, a \in S$; this collection forms an open covering of S a compact set. Hence there is finite sub-covering $\{N(a_i; r_i/2\}, r_i = r(a_i), i = 1, 2, \cdots, n$; let 2δ be the smallest of the numbers r_i; consider next any $x, y \in S$ and $|x - y| < \delta$: then by definition $x \in N(a_k, r_k/2)$ for some k: then $|f(x) - f(a_k)| < \epsilon/2$. Also note that

$$|y - a_k| = |(y - x) + (x - a_k)| < \delta + r_k/2 \leq r_k/2 + r_k/2 = r_k \Rightarrow y \in N(a_k, r_k) \cap S$$

Hence, we also have $|f(y) - f(a_k)| < \epsilon/2$; thus $|f(x) - f(y)| \leq |f(x) - f(a_k)| + |f(a_k) - f(y)| < \epsilon$. This completes the demonstration. ●

We consider another example to indicate the nature of uniform continuity. Consider $f(x) = x^2$ defined on $S = \{x : 0 < x \le 1\}$. It will be instructive to note that $f(x)$ is uniformly continuous on the set S; to see this note that:

$$|x^2 - y^2| = |x - y||x + y| \le 2|x - y| \text{ on } S$$

Consequently, for any $\epsilon > 0$, choosing $\delta = \epsilon/2$, we have that $|x - y| < \delta \Rightarrow |f(x) - f(y)| < \epsilon$. Thus the requirements of the definition are satisfied. However if instead of S, we consider $\Re_+ = \{x : 0 \le x < +\infty\}$ then $f(x)$ is not uniformly continuous on $\Re_+$. This follows since we have that for $y = x + \epsilon/4$, $|x - y| < \epsilon/2 = \delta$, $y^2 - x^2 = x.\epsilon/2 + \epsilon^2/16 = \epsilon(x/2 + \epsilon/16) > \epsilon$ for $x > 2$.

As we shall see later, the property of uniform continuity will be useful in many places.

4.2.2 Existence of extrema

Given a set $S \subset R^m$, and $f: S \to R$, the function f is said to have an absolute (or global) maximum on the set S if there is a point $a \in S$ such that $f(x) \le f(a) \forall x \in S$. If on the other hand, there is a neighbourhood $N(a)$ such that $y \in N(a) \cap S \Rightarrow f(y) \le f(a)$ then a is said to be a relative (local) maximum. Global and local minimum for the function $f(x)$ correspond to the global and local maximum for the function $-f(x)$. A very important point worth noting in this connection is the following:

Claim 4.2.2 Let $S \subset R^m$ be a compact set, and $f : S \to R$ be a continuous function. Then f has an absolute maximum and minimum on the set S.

We already know that $f(S)$ is compact and in particular, we have $glbf(S) \le f(x) \le lubf(S) \forall x \in S$; we have shown that any non-empty, bounded, closed subset of real numbers contains its *lub* and *glb* and so the claim is demonstrated.

EXERCISES

1. Let $f(x)$ be continuous on the interval $[a, b] \subset \Re$ and further let $f(x) = 0 \forall x$ rational in $[a, b]$. Show that $f(x) = 0 \forall x \in [a, b]$.

2. Construct an example to show that

$$\lim_{x \to a} f(x) + g(x) = A + B$$

need not imply that $\lim_{x \to a} f(x) = A(\text{ or } B)$ and $\lim_{x \to a} g(x) = B(\text{ or } A)$, respectively.

3. Let $f(x)$ be continuous on $[a, b] \subset \Re$; define $g(a) = f(a)$ and for $a <$ $x \le b$ define

$$g(x) = \max_{y \in [a,x]} f(y);$$

Investigate the continuity of the function $g(x)$ on $[a, b]$.

4. Let $f(x) = [x], x \in \Re$ denote the greatest integer not greater than x. Examine the continuity of $f(x)$ at $x = 2$. What, if any, are the points of continuity of $f(x)$?

5 Consider a sequence of intervals $S_n = (a_n, b_n)$, $n = 1, 2, \cdots$, $S_{n+1} \subseteq S_n$ for all n with $\lim_{n\to\infty}(b_n - a_n) = 0$. Show that there is a unique point common to all the intervals.

6 Show that the function

$$f(x) = \begin{cases} x & \text{when } 0 \le x < 1/2 \\ 1 & \text{when } x = 1/2 \\ 1 - x & \text{when } 1/2 < x < 1 \end{cases}$$

is discontinuous at $x = 1/2$.

7 Analyse the continuity of the following function at $x = 0$:

$$f(x) = \begin{cases} \dfrac{1}{1 - e^{\frac{1}{x}}} & \text{when } x \ne 0 \\ 0 & \text{when } x = 0 \end{cases}$$

8 Consider the function $f(x) = x^2$ on $\Re$ and determine whether the function is uniformly continuous on $\Re$.

9 If a function $f(x)$ is uniformly continuous on a bounded set $S \subset \Re$, show that there exists a real M such that $|f(x)| < M$ on S.

10 Let $f: [a, b] \to [c, d]$ be one to one and continuous on $[a, b]$. Show that the function $f(x)$ must be strictly monotonic on $[a, b]$.

4.3 Differentiability

First, we consider functions $f(x), x \in R$: functions of a single variable. Let F be defined over the (a, b), then for any two **distinct points** $x, x^o \in (a, b)$ we can define the quotient:

$$\frac{f(x) - f(x^o)}{x - x^o}$$

Keeping the point x^o fixed, we vary x so as to assume values arbitrarily close to x^o: geometrically, we are doing the following: on the graph of the function $y = f(x)$ we plot the points $((x, y = f(x))$ and $(x^o, y^o = f(x^o))$; the quotient measures the slope of the chord joining these two points. As the point x approaches x^o along the graph, y approaches y^o and the chord tends to become a tangent to the curve $y = f(x)$ at (x^o, y^o); if this tangent is well defined, then the quotient approaches the slope of this tangent and we say that

$$\lim_{x \to x^o} \frac{f(x) - f(x^o)}{x - x^o}$$

exists and is $f'(x^o)$: the derivative of the function $f(x)$ at x^o. The other notations for the derivative at $x = x^o$ are given by the following:

$$\frac{df(x^o)}{dx} \text{ or } \frac{dy}{dx}\Big|_{x=x^o}$$

It is easy to check from the definition of the derivative that if the derivative exists at a point x^o, the function is continuous at the point: this follows since unless the $f(x) \to f(x^o)$ as $x \to x^o$, the quotient will not approach any definite value. Further, whenever the derivative exists at the point $x = x^o$, the function is said to be **differentiable** at that point. The sign $(+, -)$ of the derivative at a point signifies the sign of the quotient as well, for x close to x^o. Thus the sign signifies whether the value of the function is increasing or decreasing as x increases. A $+$ (respectively $-$) derivative at a point x^o thus signifies that in some neighbourhood $N(x^o)$, the function increases (respectively decreases) with x. If the derivative does not change sign in an interval (a, b), the function is said to be monotonic in that interval.

Claim 4.3.1 If a function $f(x)$ is constant in an interval (a, b) then the function has a zero derivative at all points of (a, b).

In fact, if we recall the earlier definition of a local maximum and minimum, the following claim is immediate:

Claim 4.3.2 If $f(x)$ is defined over (a, b) and has a derivative in that interval and if further $x^o \in (a, b)$ is point of local maximum (or minimum), then $f'(x^o) = 0$.

This result is sometimes re-stated thus: a **necessary** condition for an extremum to exist at an interior point is that the derivative should vanish, provided derivatives exist. Among results used in economics, this perhaps is the most well-worked result. To collect our results together, we note the following:

Let $f : [a, b] \to R$ be a continuous function; then $f[a, b]$ is compact and the function attains a maximum and minimum over $[a, b]$. The points of maxima or minima may be tracked down thus, if derivatives of the function $f(x)$ exist in the interval (a, b). Solve the equation $f'(x) = 0$ in (a, b); say you obtain $x_i, i = 1, 2, ..m$; then compute the values of the function $f(.)$ at $a, b, x_i, i = 1, 2, ...m$; this will identify where the maxima and minima are located.

Treating the derivative of the function $f(x)$ at a generic point x, $f'(x)$, as a function of x, we may look for its derivative too: the second derivative if it exists is defined exactly as above, replacing $f(x)$ by $f'(x)$ everywhere. We denote this second derivative by $f''(x)$. In the same fashion, we can define its derivative, the third derivative and so on. Note that the sign of the second derivative indicates what happens to the first derivative as the value of x increases, say. In fact, using the second derivative, we may define a test for identifying whether a point of local extrema is actually a minima or maxima. The geometry of a local point of a maxima (the top of a hill) should convince readers that the following must be true:

*Claim 4.3.3 A set of **sufficient** conditions for function $f(x)$ to attain a local maxima (minimum) at an interior point $x^o \in (a, b)$ is that $f'(x^o) = 0$, $f''(x^o) < 0$ (respectively, > 0), provided these derivatives exist.*

Remark 4.1 Note that we are describing conditions for a local extremum. Sometimes we shall be interested in locating a global maximum of a function $f(x)$ over the set S, that is, locate $\overline{x} \in S$ such that $f(x) \leq f(\overline{x}) \forall x \in S$. We shall see later that for some class of functions, obtaining a local maximum will also simultaneously help locate global maximum.

Remark 4.2   One should note that vanishing derivative at a local maximum (or minimum) is only necessary when the local maximum or minimum is at an interior point. Consider for example, $f(x) = 4 - x$ in the interval $[0, 2]$: the local maximum (also the global maximum in $[0, 2]$) occurs at $x = 0$ where the derivative is -1. Considering the function $f(x) = x + 4$ over the same interval, note now that the local maximum and also the global maximum is at $x = 2$ where the derivative is $+1$. So non-zero derivative is consistent with an extremum at boundary points.

The condition relating to the requirement of a zero first order derivative at an interior point of local extrema is known as a **necessary first order condition**; the sign of the second derivative which, together with the former, ensures the nature of the extremum is called a **sufficient second order condition**.

The derivative of a function, if it exists at a point, may be used to approximate the value of the function in a neighbourhood of the point. For example, visualize the derivative as the slope of the tangent to the function $y = f(x)$ at the point x^o; the approximation is obtained by moving along the tangent with a constant slope of $f'(x^o)$. We shall make these ideas explicit as we shall proceed. First some quick properties of derivatives of sum, product, and quotient of functions and of composite functions.

Claim 4.3.4

(i) $(f \pm g)' = f' \pm g'$
(ii) $(f.g)' = f'.g + f.g'$
(iii) $(f/g)' = \dfrac{g.f' - f.g'}{g^2}$ *(valid only at points where the function g does not become zero)*
(iv) $f(g(x^o))' = f'(g(x^o)).g'(x^o)$.

4.3.1 Approximations

Consider next, a function $f(x)$ defined over an interval $[a, b]$ with derivatives in the interval, crossing any horizontal line parallel to the x-axis at the end points: geometry should convince us that the function must have turned around at some intermediate point; note that turning around would imply that the derivative is zero at the turning point. These notions are made explicit by:

Proposition 4.2   Rolle's Theorem: Let $f(x)$ be continuous over $[a, b]$ and possess derivatives in (a, b) and suppose further that $f(a) = f(b)$, then there is $c \in (a, b)$ such that $f'(c) = 0$.

Suppose, first, that for all $c \in (a, b)$, $f(c) = f(a) = f(b)$ then $f(x)$ is constant over $[a, b]$ and hence $f'(c) = 0 \forall c \in (a, b)$; if this is not so, then there exists either $c \in (a, b)$ such that $f(c) < f(a) = f(b)$ or $f(c) > f(a) = f(b)$. In the first case, $f(x)$ attains an interior minimum and hence there is $c \in (a, b)$ such that $f'(c) = 0$; in the second case, $f(x)$ attains an interior maximum and hence, the claim follows. ●

Remark 4.3 We may weaken the requirements of Rolle's Theorem by dropping the requirement that f be continuous at the end points. For example, we may claim that: Let f be defined over (a,b) with a derivative at each interior point of (a,b); assume also that $f(a+) = \lim_{x \to a+} f(x)$ and $f(b-) = \lim_{x \to b-} f(x)$ exist and are finite, then if $f(a+) = f(b-)$, there is an interior point $x_o \in (a,b)$ such that $f'(x_o) = 0$

We have seen above that the derivative of a function at a point provides us with information about the instantaneous rates of change at that point; this, in turn, may be used to analyse 'local' changes in the values of the function. At times, we need to study global changes and one may ask whether the derivative will provide us with some information about such changes. One such result is provided by the following result where a global change $f(b) - f(a)$ is related to a derivative at an intermediate point:

Proposition 4.3 Mean Value Theorem: Let $f(x)$ be as continuous over $[a,b]$ and possess derivatives over (a,b); then there is $c \in (a,b)$ such that:

$$f'(c) = \frac{f(b) - f(a)}{b - a}$$

The claim follows by noting that the function

$$g(x) = f(b) - f(x) + \frac{f(b) - f(a)}{b - a}(x - b)$$

satisfies all the conditions of Rolle's Theorem since $g(a) = g(b) = 0$ and consequently there must be some $c \in (a,b)$ such that $g'(c) = 0$; this yields the claim. •

Remark 4.4 The requirements in the Mean Value Theorem may be weakened to accommodate the weaker requirements of Rolle's Theorem stated in Remark 4.3. Let f be defined over the interval $[a,b]$ and possess a derivative in (a,b); assume further that the limits $f(a+)$ and $f(b-)$ both exist and satisfy $f(a) - f(a+) = f(b) - f(b-)$ then there is $x_o \in (a,b)$ such that $f(b) - f(a) = f'(x_o)(b-a)$.

We may restate the claim of the Mean Value Theorem thus:

$$f(b) = f(a) + (b - a)f'(c)$$

and this leads to our first approximation result:

Claim 4.3.4 In case b is close to a, an approximation for $f(b)$ is given by $f(a) + (b - a)f'(a)$.

We provide approximations which are better by using polynomials of higher order through the following result:

Proposition 4.4 Taylor's Theorem: Let $f^{(n)}(x)$ denote the n-th derivative of the function $f(x)$ at x; let this n-th derivative exist everywhere in (a,b) and that $f^{(n-1)}(x)$ is continuous everywhere on $[a,b]$. Consider $x_o \in [a,b]$; then for any $x \in [a,b]$, there is $x_1 = \lambda x + (1 - \lambda)x_o, 0 < \lambda < 1$ *such that*

$$f(x) = f(x_o) + \sum_{k=1}^{n-1} \frac{f^{(k)}(x_o)}{k!}(x - x_o)^k + \frac{f^{(n)}(x_1)}{n!}(x - x_o)^n$$

The approximation to $f(x)$ by an *n-th* degree polynomial is provided by replacing x_1 by x_o in the above expression. The proof of the claim follows from defining a function

$$F(t) = f(t) + \sum_{k=1}^{n-1} \frac{f^{(k)}(t)}{k!}(x - t)^k$$

and taking $x, x_o \in [a, b], x_o < x$, say; observe that $F(t)$ is continuous over the interval $[x_o, x]$ and has a derivative over the interval (x_o, x) and hence by the Mean Value Theorem, there is $x_1 \in (x_o, x)$ such that

$$F'(x_1) = \frac{F(x) - F(x_o)}{x - x_o}$$

The claim for Taylor's Theorem follows noting that $F(x) = f(x)$ and that $F'(t) = \frac{(x - t)^{n-1}}{(n - 1)!} f^{(n)}(t)$. •

EXERCISES

1. A function $f(x)$ satisfies $|f(x)| < A$ and $|f''(x)| < B$ for some constants A, B whenever $x > a$; show that in this range of x-values, $|f'(x)| < 2\sqrt{(AB)}$.
2. Discuss the applicability of Rolle's Theorem to the function $f(x) = |x|$ in the interval $(-1, +1)$.
3. Show that if $x \in [0, 1]$ then

$$|\log(1 + x) - x + 1/2.x^2| \le 1/3.x^3$$

4. Show that $a^x > x^a$ if $x > a > e$. (Hint: consider the function $f(x) = x \log a - a \log x$.)

4.4 Integration: Fundamental Theorem of Calculus

4.4.1 Introduction

We have seen above how we may use a limiting process (the derivative) to arrive at the slope of a function; another limiting process (the integral) enables us to arrive at the area under a curve. It may be useful to provide a somewhat more general treatment than what one may have been accustomed to in elementary calculus. In this section, we shall be concerned with **finite** intervals $[a, b]$ and **bounded** functions f, g, α, β defined on the finite interval.

We consider $[a, b] \subset \Re$ and a set of points $P = \{x_0, x_1, \cdots, x_n\}$ such that $a = x_0 < x_1 < \cdots, x_{i-1} < x_i < \cdots, x_n = b$; then P is a **partition** of $[a, b]$. If P, P' are two partitions of $[a, b]$ then P is said to be **finer** than P' if $P' \subset P$. Let P denote a partition of $[a, b]$; then the **norm** of P is the length of the largest of the numbers $\Delta x_k = x_k - x_{k-1}$ and is written as $|P|$. Let f be any real valued function defined on

$[a, b]$, that is, $f : [a, b] \to \Re$, and P be some partition of it; then $\Delta f_k = f(x_k) - f(x_{k-1})$ so that $\sum_{k=1}^{n} \Delta f_k = f(b) - f(a)$. Given two real valued functions f, α on $[a, b]$ and a partition $P = \{x_0, x_1, \cdots, x_n\}$ of $[a, b]$, with $t_k \in [x_{k-1}, x_k]$, we define:

$$S(P, f, \alpha) = \sum_{k=1}^{n} f(t_k) \Delta \alpha_k$$

to be the **Riemann-Stieltjes** sum of f with respect to α on $[a, b]$.

We say that f is **Riemann-integrable** with respect to α on $[a, b]$ ($f \in R(\alpha)$ on $[a, b]$) if there is a number A such that for every $\epsilon > 0$, there exists a partition P_ϵ of $[a, b]$ such that for every partition P finer than P_ϵ and for every choice of points $t_k \in [x_{k-1}, x_k]$, we have $|S(P, f, \alpha) - A| < \epsilon$. In case such a number exists, it is uniquely determined and we write:

$$A = \int_a^b f \, d\alpha \text{ or as } \int_a^b f(x) d\alpha(x)$$

and call it the Riemann-Stieltjes integral of f with respect to α on $[a, b]$; in such cases we say $f \in R(\alpha)$. We shall refer to the functions f and α as the **integrand** and the **integrator**, respectively.

In the special case when $\alpha(x) = x$, we write $S(P, f)$ for $S(p, f, \alpha)$ and the number A defined above, whenever it exists as

$$A = \int_a^b f(x) d(x)$$

and call A the Riemann integral of f over $[a, b]$ and write $f \in R$. This is, of course, the integral encountered in elementary calculus. The generality obtained will not be of great use for our purpose but the choice of discontinuous α enables any finite sum (or infinite) to be expressed as a Stieltjes integral and has applications to probability theory where it plays an important rule.

4.4.2 Functions of bounded variation

We shall require the properties of **functions of bounded variation** on some interval $[a, b]$. Let P be any partition of an interval $[a, b]$; recall $\Delta f_k = f(x_k) - f(x_{k-1})$ where $\{x_0, x_1, x_2, \cdots, x_n\}$ denote the partition P; if there is some number $M > 0$ such that $\sum_{k=1}^{n} |\Delta f_k| \leq M$ for all partitions of $[a, b]$ then the function f is said to be of bounded variation on $[a, b]$. Thus, for example, if the function $f(x)$ is monotonic over $[a, b]$ then f must be of bounded variation over $[a, b]$: this follows since monotonic f implies that Δf_k are of the same sign, note that the sum is either $f(b) - f(a)$ or $f(a) - f(b)$ and the claim follows. In case the function f is continuous over $[a, b]$ and $f'(.)$ exists and is bounded in (a, b), then one may show that $f(x)$ is of bounded variation over $[a, b]$. However, the most interesting feature of functions of bounded variation is contained in the following:

Claim 4.4.1 f is a function of bounded variation over $[a, b]$ iff f can be expressed as the difference of two increasing functions.

We consider next, the concept of total variation for a function of bounded variation f over $[a, b]$. For any partition $P = \{x_0, x_1, \cdots, x_n\}$ of $[a, b]$, let $V_f(a, b)$ denote $\sup\{\sum(P) | P \in \mathcal{B}[a, b]\}$, where $\sum(P)$ denotes the sum $\sum_{k=1}^{n} |\Delta f_k|$ corresponding to the partition P and $\mathcal{B}[a, b]$ denote the set of all possible partitions of the interval $[a, b]$. The $V_f(a, b)$ is called the **total variation** of f over $[a, b]$.

For any function of bounded variation f over $[a, b]$, define the function V over $[a, b]$ as follows:

$$V(x) = V_f(a, x) \text{ when } a < x \leq b; \ V(a) = 0$$

Note that:

(i) V is increasing on $[a, b]$;
(ii) $V - f$ is increasing on $[a, b]$.

To see these properties we proceed as follows. If $a < x < y \leq b$, then $V_f(a, x) + V_f(x, y) = V_f(a, y)$ and consequently, $V_f(a, y) - V_f(a, x) = V_f(x, y) \geq 0$, that is, we have $V_f(a, y) \geq V_f(a, x)$ or $V(y) \geq V(x)$ which is (i). For (ii) define $D(x) = V(x) - f(x), x \in [a, b]$; consider $a \leq x < y \leq b$; then $D(y) - D(x) = V(y) - V(x) - (f(y) - f(x)) = V_f(x, y) - (f(y) - f(x)) \geq 0 \Rightarrow D(y) \geq D(x)$ which is (ii). Note that the inequality follows from the definition of $V_f(x, y)$. This proves claim 4.4.1. •

4.4.3 Basic properties of the integral

The basic linearity of the integral needs to be noted:

Claim 4.4.2 If $f \in R(\alpha)$ and $g \in R(\alpha)$ on $[a, b]$ then $c_1 f + c_2 g \in R(\alpha)$ for any constant c_1, c_2. Moreover, if $f \in R(\alpha)$ and $f \in R(\beta)$ on $[a, b]$ then $f \in R(c_1\alpha + c_2\beta)$ on $[a, b]$.

Proof Let $h = c_1 f + c_2 g$. Given some partition P of $[a, b]$ we have:

$$S(P, h, \alpha) = \sum_{k=1}^{n} h(t_k)\Delta\alpha_k = c_1 S(P, f, \alpha) + c_2 S(P, g, \alpha)$$

Given any $\epsilon > 0$, choose a partition P_ϵ' such that for any P, satisfying $P_\epsilon' \subset P$ implies that $|S(P, f, \alpha) - \int_a^b f d\alpha| < \epsilon$; such a P_ϵ' exists because the integral exists. Again choose P_ϵ'' such that $P_\epsilon'' \subset P$ implies that $|S(P, g, \alpha) - \int_a^b g d\alpha| < \epsilon$; now just consider $P_\epsilon = P_\epsilon' \cup P_\epsilon''$, then since P finer than P_ϵ is also finer than P_ϵ' and P_ϵ'', we have

$$|S(P, h, \alpha) - c_1 \int_a^b f d\alpha - c_2 \int_a^b g d\alpha| < |c_1|\epsilon + |c_2|\epsilon$$

which completes the proof.

The second part of the claim follows exactly as above. •

Next, we note that as a consequence of the above, the integral is 'additive' over the interval $[a, b]$ in the following sense:

Claim 4.4.3 Let $c \in (a, b)$ then if any two of the following integrals exist, then so does the third, and they are related as stated:

$$\int_a^b f\, d\alpha = \int_a^c f\, d\alpha + \int_c^b f\, d\alpha$$

Proof Let P be any partition of $[a, b]$; let $P' = P \cap [a, c]$ and $P'' = P \cap [c, b]$ be the corresponding partitions of $[a, c]$ and $[c, b]$, respectively. Then, it follows that

$$S(P, f, \alpha) = S(P', f, \alpha) + S(P'', f, \alpha) \qquad (4.1)$$

which in fact provides the basis of the claim. For example, suppose that the two integrals on the right exist; then given any $\epsilon > 0$, there exists a partition P'_ϵ of $[a, c]$ and a partition P''_ϵ of $[c, b]$ such that

$$|S(P', f, \alpha) - \int_a^c f\, d\alpha| < \epsilon/2$$

whenever P' is finer than P'_ϵ and

$$|S(P'', f, \alpha) - \int_c^b f\, d\alpha| < \epsilon/2$$

whenever P'' is finer than P''_ϵ. Note that $P_\epsilon = P''_\epsilon \cup P'_\epsilon$ is a partition of $[a, b]$ and if any partition P is finer than P_ϵ, then P' is finer than P'_ϵ and P'' is finer than P''_ϵ and hence we can combine the above inequalities to get

$$|S(P, f, \alpha) - \int_a^c f\, d\alpha - \int_c^b f\, d\alpha| < \epsilon$$

so that $\int_a^b f\, d\alpha$ exists and is equal to $\int_a^c f\, d\alpha + \int_c^b f\, d\alpha$. In a similar fashion the basic relationship (4.1) may be exploited in the remaining cases. ●

Definition 4.1 If $a < b$, $\int_b^a f\, d\alpha$ is defined to be $-\int_a^b f\, d\alpha$ whenever $\int_a^b f\, d\alpha$ exists. Further we define $\int_a^a f\, d\alpha = 0$.

4.4.4 Integration by parts

Here we investigate the relationship between $\int_a^b f\, d\alpha$ and $\int_a^b \alpha\, df$. We shall show that whenever one exists, so does the other and that there is a relationship between their values.

Claim 4.4.4 If $f \in R(\alpha)$ on $[a, b]$ then $\alpha \in R(f)$ on $[a, b]$ and

$$\int_a^b f\, d\alpha + \int_a^b \alpha\, df = f(b)\alpha(b) - f(a)\alpha(a)$$

Proof Let $\epsilon > 0$ and let the integral $\int_a^b f\, d\alpha$ exist; then there is a partition P_ϵ of $[a, b]$ such that for any partition P finer than P_ϵ we have

$$|S(P, f, \alpha) - \int_a^b f\, d\alpha| < \epsilon. \qquad (4.2)$$

Consider, next, some partition P' of $[a, b]$ and an arbitrary Riemann-Stieltjes sum

$$S(P', \alpha, f) = \sum_{k=1}^{n} \alpha(t_k)\Delta f_k = \sum_{k=1}^{n} \alpha(t_k)f(x_k) - \sum_{k=1}^{n} \alpha(t_k)f(x_{k-1})$$

Further writing $A = f(b)\alpha(b) - f(a)\alpha(a)$, we have $A = \sum_{k=1}^{n} f(x_k)\alpha(x_k) - \sum_{i=1}^{n} f(x_{k-1})$ $\alpha(x_{k-1})$; hence

$$A - S(P', \alpha, f) = \sum_{k=1}^{n} f(x_k)(\alpha(x_k) - \alpha(t_k)) + \sum_{k=1}^{n} f(x_{k-1})(\alpha(t_k) - \alpha(x_{k-1}))$$

Let P be the partition of $[a, b]$ obtained by taking all the points x_k, t_k together: thus $x_0 = y_0 \le y_1 = t_1 \le y_2 = x_1 \le \cdots \le t_n = y_{2n-1} \le x_n$ and

$$S(P, f, \alpha) = \sum_{r=0}^{2n} f(\xi_r)(\alpha(y_r) - \alpha(y_{r-1})) = \sum_{k=1}^{n} f(\xi_k)(\alpha(t_k) - \alpha(x_{k-1}))$$

$$+ \sum_{k=1}^{n} f(\xi_k)(\alpha(x_k) - \alpha(t_k))$$

for any choice of points $\xi_r \in [y_{r-1}, y_r]$; choosing $\xi_r = y_{r-1}$ for r even and $\xi_r = y_r$ for r odd, we have

$$S(P, f, \alpha) = \sum_{k=1}^{n} f(x_{k-1})(\alpha(t_k) - \alpha(x_{k-1})) + \sum_{k=1}^{n} f(x_k)(\alpha(x_k) - \alpha(t_k)) = A - S(P', \alpha, f)$$

Since P is finer than (P' and hence) P_ϵ, and hence (4.2) is valid so that we have

$$\left| A - S(P', \alpha, f) - \int_a^b f\, d\alpha \right| < \epsilon$$

which completes the proof of the claim that $\int_a^b \alpha\, df$ exists and is equal to $A - \int_a^b f\, d\alpha$. •

4.4.5 The Riemann-Stieltjes integral as a Riemann integral

In case we have $\alpha(x)$ continuously differentiable, we then have:

$$\int_a^b f\, d\alpha = \int_a^b f\alpha'(x)dx$$

whenever the integral on the left exists. To see the validity of the above, write $g(x) = f(x)\alpha'(x)$ and consider (the Riemann sum):

$$S(P, g) = \sum_{k=1}^{n} g(t_k)\Delta x_k$$

Note that this is a sum of areas of rectangles of height $g(t_k)$ built on the partition P. The finer the partition, the closer the sum is to the area under the curve of the function $g(x)$. Now for the same partition P, and the same set of points $\{t_k\}$, consider the Riemann-Stieltjes sums:

$$S(P, f, \alpha) = \sum_{k=1}^{n} f(t_k)\Delta\alpha_k$$

Note that we may apply the Mean Value Theorem to $\Delta\alpha_k$ to obtain $\Delta\alpha_k = \alpha'(v_k)\Delta x_k$, where $v_k \in (x_{k-1}, x_k)$.

Thus, we have:

$$S(P, f, \alpha) - S(P, g) = \sum_{k=1}^{n} f(t_k)[\alpha'(t_k) - \alpha'(v_k)]\Delta x_k$$

Recall that we are considering only bounded functions; thus, there is $M > 0$ such that $|f(x)| \leq M$ on $[a, b]$; next we are given that $\alpha'(.)$ is continuous on the interval $[a, b]$: hence it is **uniformly continuous** on that interval, that is, for any $\epsilon > 0$ there is $\delta > 0$ such that $\forall x, y \in [a, b]$, $0 < |x - y| < \delta$ implies that $|\alpha'(x) - \alpha'(y)| < \epsilon/2M(b - a)$. Consider then any partition P'_ϵ with norm $< \delta$; for any finer partition P, it will be the case that $|\alpha'(x) - \alpha'(y)| < \epsilon/2M(b - a)$ and hence for all such P, we have that $|S(P, f, \alpha) - S(P, g)| < \epsilon/2$; now we also know that since the integral $\int_a^b f d\alpha$ exists, there is some partition P''_ϵ such that P finer than P''_ϵ implies that $|S(P, f, \alpha) - \int_a^b f d\alpha| < \epsilon/2$. Hence, when P is finer than $P_\epsilon = P''_\epsilon \cup P'_\epsilon$, we have $|S(p, g) - \int_a^b f d\alpha| < \epsilon$ and this completes the demonstration. ●

4.4.6 The Riemann-Stieltjes integral as a finite sum

It is easy to see that if the function α is constant on the interval $[a, b]$, then since the sum $S(P, f, \alpha)$ for any partition P is zero, the integral $\int_a^b f d\alpha$ exists and is 0. But if the function α is constant except for a step discontinuity at the point $c \in [a, b]$ so that α is constant on $[a, c)$ and on $(c, b]$ but the two values are not the same, the integral $\int_a^b f d\alpha$ need not even exist. It all depends upon the behaviour of the function f at the point c: *if f is defined so that at the point c, at least one of f or α is continuous from the left and at least one of f or α is continuous on the right* then $\int_a^b f d\alpha$ exists and is given by $f(c)[\alpha(c+) - \alpha(c-)]$. (It may be recalled that $\alpha(c+) = \lim_{h\to 0}\alpha(c+h)$ and $\alpha(c-) = \lim_{h\to 0}\alpha(c - h)$.) In fact, it may be shown that the integral $\int_a^b f d\alpha$ does not exist if both the functions f, α are discontinuous from the left or from the right at c. Thus, the value of a Rieman-Stieltjes integral can be altered by changing the value of the function f at a single point; one may even affect the existence of the integral.

The function α defined above is a *step function*. Formally, let α be defined on $[a, b]$ such that α is discontinuous at a finite number of points c_k where $a \leq c_1 < \cdots < c_n \leq b$. If α is constant on each open interval (c_{k-1}, c_k), then α is a step function and the difference $\alpha(c_k+) - \alpha(c_{k-1}-)$ is said to be the jump at c_k. Step functions provide the following link between Riemann-Stieltjes integrals and finite sums.

Claim 4.4.5 Let α be a step function on $[a, b]$ with jump α_k at x_k where $a \leq x_1 < \cdots < x_n \leq b$. Let f be defined on $[a, b]$ such that not both f and α are discontinuous from the left or from the right at each x_k. Then $\int_a^b f\,d\alpha$ exists and we have

$$\int_a^b f\,d\alpha = \sum_{k=1}^n f(x_k)\alpha_k$$

The proof follows by noting that the integral can be written as the sum of the integrals over (x_{k-1}, x_k) which, in turn, exist by virtue of the result stated above and hence the integral reduces to the sum claimed. More importantly, we have a kind of converse to the above too.

Claim 4.4.6 Every finite sum can be written as a Riemann-Stieltes integral.

Consider a finite sum $\sum_{k=1}^n \alpha_k$; define a function f on $[0, n]$ as follows:

$$f(x) = \alpha_k \text{ if } k - 1 < x \leq k, \ k = 1, 2, \cdots, n, \ f(0) = 0$$

Then, by virtue of what we have indicated above, it follows that

$$\sum_{k=1}^n \alpha_k = \sum_{k=1}^n f(k) = \int_a^b f\,d[x]$$

where $[x]$ is the greatest integer $\leq x$.

For the remaining section, we shall confine our attention to integrators (that is, functions α which are increasing and we shall use the notation $\alpha \uparrow [a, b]$ to signify that the integrator α is increasing on the interval $[a, b]$. The significant effect of assuming the integrators to be increasing is that the differences $\Delta\alpha_k$ which appear in the Riemann-Stieltjes sum become non-negative. We define, for any partition P of the interval $[a, b]$ and for functions f defined on $[a, b]$, $\alpha \uparrow$ on $[a, b]$, the functions $M_k(f) = \sup\{f(x)|x \in [x_{k-1}, x_k]\}$ and $m_k(f) = \inf\{f(x)|x \in [x_{k-1}, x_k]\}$. Then the numbers $U(P, f, \alpha) = \sum_{k=1}^n M_k(f)\Delta\alpha_k$ and $L(P, f, \alpha) = \sum_{k=1}^n m_k(f)\Delta\alpha_k$ are called the upper and lower Stieltjes sum, respectively, of the function f with respect to α for the partition P. Note then that since $m_k(f) \leq M_k(f)$ and since $\Delta\alpha_k \geq 0$, it follows that $L(P, f, \alpha) \leq U(P, f, \alpha)$ and in particular, we have $L(P, f, \alpha) \leq S(P, f, \alpha) \leq U(P, f, \alpha)$.

Further, if the partitions become finer, the lower sums increase and the upper sums decrease; this follows from the property of infimum and supremum being taken over larger sets of points when the partition becomes finer; also, for any two partitions P_1, P_2, we must have $L(P_1, f, \alpha) \leq U(P_2, f, \alpha)$. (Consider the partition $P = P_1 \cup P_2$: P is finer than both P_1, P_2. Now apply the previous inequalities.) The upper and lower Stieltjes sums allow us to define the **upper** and **lower** integrals, respectively, thus: $\overline{I}(f, \alpha) = \inf\{U(P, f, \alpha)|P \in \mathcal{P}[a, b]\}$, $I(f, \alpha) = \sup\{L(P, f, \alpha|P \in \mathcal{P}[a, b]\}$. From the definition and from the properties of the upper and lower sums, it follows that $\overline{I}(f, \alpha) \geq I(f, \alpha)$. We are now ready to state a condition which is strong enough to ensure that $f \in R(\alpha)$.

The function f is said to satisfy *Riemann's condition* with respect to the integrator α on $[a, b]$ if for every $\epsilon > 0$ there is a partition P_ϵ such that for all partitions finer that P_ϵ, we have $0 \le U(P, f, \alpha) - L(P, f, \alpha) < \epsilon$; then we have:

Proposition 4.5 Let α be $\uparrow$ over $[a, b]$. Then the following conditions are equivalent:

1. $f \in R(\alpha)$ on $[a, b]$.
2. f satisfies Riemann's condition with respect to α on $[a, b]$.
3. $\underline{I}(f, \alpha) = \overline{I}(f, \alpha)$.

Proof We prove that $1 \Rightarrow 2$ first. In case $\alpha(a) = \alpha(b)$, 2 holds trivially so assume that $\alpha(a) < \alpha(b)$; consider $\epsilon > 0$; then since *1* holds, there is a partition P_ϵ such that for all finer partitions P, and for all choices of t_k and t'_k in $[x_{k-1}, x_k]$ we have

$$|\sum_{k=1}^{n} f(t_k)\Delta\alpha_k - A| < \epsilon/3 \text{ and } |\sum_{k=1}^{n} f(t'_k)\Delta\alpha_k - A| < \epsilon/3$$

where $A = \int_a^b f d\alpha$. Thus combining these two inequalities we have

$$|\sum_{k=1}^{n}[f(t_k) - f(t_{k-1})]\Delta\alpha_k| < 2/3\epsilon$$

Recall too that from definition, $M_k(f) - m_k(f) = \sup|f(x) - f(x'|x, x' \in [x_{k-1}, x_k]$, it follows that for every $h.0$ we can choose t_k, t'_k such that $f(t_k) - f(t'_k) > M_k(f) - m_k(f) - h$. Consider $h = \epsilon/3(\alpha(b) - \alpha(a))$, we have

$$U(P, f, \alpha) - L(P, f, \alpha) < \sum_{k=1}^{n}[f(t_k) - f(t_{k-1})]\Delta\alpha_k + h\sum_{k=1}^{n}\Delta\alpha_k < \epsilon$$

which establishes the claim.

Now suppose that 2 holds. We already know that $\underline{I}(f, \alpha) \le \overline{I}(f, \alpha)$. Consider $\epsilon > 0$; then there exists a partition finer than P_ϵ such that for the finer partition P, we have $U(P, f, \alpha) < L(P, f\alpha) + \epsilon$. Hence for such partitions P, we have: $\overline{I}(f, \alpha) \le U(P, f, \alpha) < L(p, f, \alpha) + \epsilon \le \underline{I}(f, \alpha)$ where the ϵ is arbitrary. Hence $2 \Rightarrow 3$.

Finally suppose 3 holds. and let the common value be A. For any $\epsilon > 0$, consider partitions P_ϵ, P'_ϵ such that for partitions P finer that P_ϵ, $U(P, f, \alpha) < \overline{I}(f, \alpha) + \epsilon$ and for partitions P finer that P'_ϵ, we have $L(P, f, \alpha) > \underline{I}(f, \alpha) - \epsilon$. Thus consider $P^\epsilon = P_\epsilon \cup P'_\epsilon$; then for partitions P finer that P^ϵ, we must have

$$\underline{I}(f, \alpha) - \epsilon < L(P, f, \alpha) \le S(P, f, \alpha) \le U(P, f, \alpha) < \overline{I}(f, \alpha) + \epsilon$$

which implies that $|S(P, f, \alpha) - A| < \epsilon$ whenever P is finer than P^ϵ which is *1*. This proves the claim. ●

The above may be used to provide some sufficient conditions for the existence of the integral.

Claim 4.4.7 If f is continuous (bounded variation) over $[a, b]$ and α is of bounded variation (continuous, respectively) on $[a, b]$ then $f \in R(\alpha)$ on $[a, b]$.

Proof We shall prove it for $\alpha \uparrow$ over $[a, b]$. Since α is of bounded variation on $[a, b]$, we know there is $M > 0$ such that $\sup_{x, y \in [a,b]} |\alpha(x) - \alpha(y)| \leq M$; also since f is continuous over $[a, b]$, a compact set, f is uniformly continuous on $[a, b]$. Hence for any $\epsilon > 0$ there is $\delta > 0$ such that $|x - y| < \delta \Rightarrow |f(x) - f(y)| < \epsilon/M$; let P_ϵ be a partition of $[a, b]$ with norm $< \delta$; then for all partitions finer than P_ϵ, we must have $M_k(f) - m_k(f) \leq \epsilon/M$; this follows, since $M_k(f) - m_k(f) = \sup_{x, y \in [x_{k-1}, x_k]} |f(x) - f(y)|$. Now using the fact that $\Delta \alpha_k \geq 0$ we have:

$$0 \leq U(P, f, \alpha) - L(P, f, \alpha) = \sum (M_k(f) - m_k(f)).\Delta \alpha_k \leq \epsilon/M. \sum \Delta \alpha_k < \epsilon$$

for all partitions P finer than P_ϵ. But that implies that $f \in R(\alpha)$. If we know that α is of bounded variation, then we know that $\alpha = \beta - \gamma$ where β, γ are both increasing functions: consequently by virtue of what we have just shown, $f \in R(\beta)$ and $f \in R(\gamma)$ and hence $f \in R(\alpha)$. •

When f is of bounded variation and α is continuous, the proof of the claim follows, interchanging the role of f and α in the above. This result has an immediate application for the special case when $\alpha(x) = x$. Note that the Riemann integral $\int_a^b f(x) dx$ exists if **either** f is continuous over the interval $[a, b]$ or if f is of bounded variation over $[a, b]$.

4.4.7 The integral as a function

We first provide an estimate of the integral through the following result:

Claim 4.4.8 Let $\alpha \uparrow$ over $[a, b]$ and $f \in R(\alpha)$ on $[a, b]$; also let M, m be respectively the supremum and infimum of the set $\{f(x) : a \leq x \leq b\}$. Then there is c such that $m \leq c \leq M$ and $\int_a^b f d\alpha(x) = c \int_a^b d\alpha(x)$. In case the function f is continuous, $c = f(x_o), x_o \in [a, b]$.

Proof In case $\alpha(a) = \alpha(b)$ then the function α must be a constant (recall that it is increasing) and hence the claim is trivially true with both integrals being zero. So consider the case when $\alpha(a) < \alpha(b)$; now note that the upper and lower sums must lie in between $m(\alpha(b) - \alpha(a))$ and $M(\alpha(b) - \alpha(a))$, that is,

$$m(\alpha(b) - \alpha(a)) \leq L(P, f, \alpha) \leq U(P, f, \alpha) \leq M(\alpha(b) - \alpha(a))$$

Consequently,

$$c = \frac{\int_a^b f d\alpha}{\int_a^b d\alpha}$$

lies between m and M and hence when f is continuous, since any intermediate value must be attained, we have the existence of $x_o \in [a, b]$ such that $f(x_o) = c$. •

*Proposition 4.6   Let α be of bounded variation in $[a, b]$ and let $f \in R(\alpha)$ on $[a, b]$.
Define F by*

$$F(x) = \int_a^x f \, d\alpha \text{ for } x \in [a, b]$$

Then we have:

1. *F is of bounded variation in $[a, b]$.*
2. *Every point of continuity of α is also a point of continuity of F.*
3. *If $\alpha \uparrow$ on $[a, b]$ the derivative $F'(x)$ exists at every point $x \in (a, b)$ such that $\alpha'(x)$ exists and $f(x)$ is continuous and moreover, $F'(x) = f(x).\alpha'(x)$.*

Proof Consider $\alpha \uparrow$ on $[a, b]$, without any loss of generality, for reasons similar to the ones considered above. Now consider $x \neq y$ and the fact that $F(y) - F(x) = \int_x^y f \, d\alpha = c[\alpha(y) - \alpha(x)]$, where $m \leq c \leq M$, using the notation of the last claim; this allows us to conclude *1* and *2*; dividing by $y - x$ and allowing $x \to y$, $c \to f(x)$ which is *3*; recall that since $f(.)$ is continuous, $c = f(x_o)$, $x_0 \in [x, y]$; hence when $y \to x$, $c \to f(x)$, once more appealing to the continuity of f at x. •

When the integrator $\alpha(x) = x$, the integral reduces to the Riemann integral and when this integral exists, we say that $f \in R$ on $[a, b]$; for such functions we have the **Fundamental Theorem of Integral Calculus**:

Proposition 4.7   Let $f \in R$ on $[a, b]$ and g be a function defined on $[a, b]$ such that $g'(x) = f(x)$ for every $x \in (a, b)$; at the endpoints assume that $g(a+)$ and $g(b-)$ exist and satisfy $g(a) - g(a+) = g(b) - g(b-)$. Then we have:

$$\int_a^b f \, dx = g(b) - g(a)$$

Proof For any partition of $[a, b]$, we have:

$$g(b) - g(a) = \sum_{k=1}^n [g(x_k) - g(x_{k-1})] = \sum_{k=1}^n g'(t_k)(x_k - x_{k-1}) = \sum_{k=1}^n f(t_k)\Delta x_k$$

where $t_k \in (x_{k-1}, x_k)$ determined by the Mean Value Theorem of Differential Calculus; but given any $\epsilon > 0$, the partition can be taken so fine such that:

$$|g(b) - g(a) - \int_a^b f \, dx| = |\sum_{k=1}^n f(t_k)\Delta x_k - \int_a^b f \, dx| < \epsilon$$

which proves the claim. •

Sometimes the function g is called the anti-derivative of f and a search for anti-derivatives thus enables one to compute values for Riemann integrals. As we shall see, the computation of Riemann integrals will play an important role in many situations and consequently the study of anti-derivatives too will figure prominently. Before passing on to other points of interest, we should indicate that we shall denote the anti-derivative of the function f by the following notation:

$$\int f dx$$

Thus, the above notation will mean a function $g(x)$ such that $g'(x) = f(x)$.

Next, we consider a result which allows us to take derivatives under the integral sign. We confine attention to Riemann integrals.

Proposition 4.8   Consider $A = \{(x, y)|a \leq x \leq b, c \leq y \leq d\}$ and assume that for each fixed $y \in [c, d]$ the integral

$$F(y) = \int_a^b f(x, y)dx$$

exists. If the partial derivative f_y is continuous on A, the derivative $F'(y)$, $y \in [c, d]$ exists and is given by

$$F'(y) = \int_a^b f_y(x, y)dx.$$

Using the above notation, we also have:

Proposition 4.9   Moreover if $p, q : [c, d] \rightarrow [a, b]$ with finite derivatives $p'(y), q'(y)$ and given the continuity of $f(x, y)$ and f_y over the region A, define

$$\phi(y) = \int_{q(y)}^{p(y)} f(x, y)dx$$

for every $y \in [c, d]$, then $\phi'(y)$ exists and is given by

$$\phi'(y) = \int_{q(y)}^{p(y)} f_y(x, y)dx + f(p(y), y).p'(y) - f(q(y), y).q'(y).$$

For a demonstration of the above propositions, consider the following. Let $y_o \in [c, d]$; for $y \neq y_o$, we have:

$$\frac{F(y) - F(y_o)}{y - y_o} = \int_a^b \frac{f(x, y) - f(x, y_o)}{y - y_o}dx = \int_a^b f_y(x, \overline{y})dx$$

where $\overline{y} = \lambda y + (1 - \lambda)y_o$ for some $\lambda \in [0, 1]$; now using the continuity of f_y on A, we conclude that the first proposition is valid. As far as the second part is concerned, define the function $G(x_1, x_2, x_3) = \int_{x_1}^{x_2} f(t, x_3)dt$, whenever $x_1, x_2 \in [a, b]$, $x_3 \in [c, d]$. Then, $\phi(y) = G(q(y), p(y), y)$ and consequently we have:

$$\phi'(y) = G_1(q(y), p(y), y).q'(y) + G_2(q(y), p(y), y).p'(y) + G_3(q(y), p(y), y)$$

From Proposition 4.6, we have $G_1(x_1, x_2, x_3) = -f(x_1, x_3)$, $G_2(x_1, x_2, x_3) = f(x_2, x_3)$ while from the previous proposition,

$$G_3(x_1, x_2, x_3) = \int_{x_1}^{x_2} f_{x_3}(t, x_3)dt$$

Making these substitutions, we obtain the desired expression. ●

4.4.8 Improper integrals

In all the earlier discussions the limits of the integrals a, b were considered to be finite. Thus when we said that $f \in R(\alpha)$ over $[a, b]$, the basic presumption was that f, α were defined and bounded on a finite interval $[a, b]$. For such functions if the integral $\int_a^b f \, d\alpha$ exists, we shall write $f \in R(\alpha; a, b)$; in case $\alpha(x) = x$, we write $f \in R(a, b)$; and $\alpha \uparrow$ on $[a, +\infty)$ would mean that the function α is monotonic increasing on the interval $[a, +\infty)$

We shall consider the behaviour of the integral $\int_a^b f \, d\alpha$ as $b \to \infty$. Consider $f \in R(\alpha; a, b)$ for every $b \geq a$. Keeping a, α, f fixed, define a function on $[a, +\infty)$:

$$I(b) = \int_a^b f \, d\alpha, b \geq a;$$

then $I(b)$ is called an indefinite integral or an improper integral of the first kind and is denoted by $\int_a^\infty f \, d\alpha$; if the $\lim_{b \to +\infty} I(b)$ exists and is equal to, say, A, then the improper integral $\int_a^\infty f \, d\alpha$ is said to converge and the value of the improper integral $\int_a^\infty f \, d\alpha$ is defined to be A; if the limit of $I(b)$ does not exist, then the improper integral is said to diverge.

Example 4.1 *Consider $f = x^{-p}, \alpha(x) = x, a = 1$; then $\int_1^b x^{-p} dx = \dfrac{1 - b^{1-p}}{p - 1}$ provided $p \neq 1$. When $p = 1$, the integral has the value of $\ln b$. It is now easy to see that the integral diverges whenever $p \leq 1$ and converges for $p > 1$.*

Most of the results stated earlier for ordinary integrals thus have counterparts for improper integrals. We may define:

$$\int_{-\infty}^a f \, d\alpha = \lim_{b \to \infty} \int_{-b}^a f \, d\alpha$$

and hence

$$\int_{-\infty}^{+\infty} f \, d\alpha = \lim_{b \to +\infty} \int_{-b}^b f \, d\alpha$$

Essentially, therefore, when we encounter improper integrals, we need to be sure that they converge. We provide some easy tests for convergence. Consider the case when $\alpha \uparrow$ on $[a, +\infty)$, $f \in R(\alpha; a, b), b \geq a$ and further $f(x) \geq 0, x \geq a$; in this situation we have:

Claim 4.4.9 The integral $\int_a^{+\infty} f \, d\alpha$ converges iff there is a number M such that $\int_a^b f \, d\alpha \leq M, b \geq a$.

Proof Recall the definition of $I(b)$ and note that under the conditions specified, $I(b)$ is monotonic increasing and hence the condition follows. ●

We have in addition a Comparison Test. Suppose, we have for a function such as f (described above), another function $g(x)$ such that for every $b \geq a$, $0 \leq f(x) \leq g(x) \forall x \geq a$; then if $\int_a^{+\infty} g(x) d\alpha$ converges then so does $\int_a^{+\infty} f(x) d\alpha$. Moreover, we have:

$$\int_a^{+\infty} f(x)d\alpha \leq \int_a^{+\infty} g(x)d\alpha$$

The claim follows by observing that given the conditions we must have:

$$\int_a^b f(x)d\alpha \leq \int_a^b g(x)d\alpha \leq \int_a^{+\infty} g(x)d\alpha$$

Finally, we have a Limit Comparison Test, as well. For a function f as described above, suppose there is a function $g \in R(\alpha; a, b)$ for every $b \geq a$ where $g(x) \geq 0$ if $x \geq a$. If

$$\lim_{x \to +\infty} \frac{f(x)}{g(x)} = 1$$

then the integrals $\int_a^{+\infty} f(x)d\alpha$ and $\int_a^{+\infty} g(x)d\alpha$ both converge or diverge together. Note that under the condition specified, we have $1/2 \leq f(x)/g(x) \leq 5/4$ whenever $x \geq N$, say. This immediately suggests that the conclusion follows by way of the Comparison Test.

EXERCISES

Unless specified otherwise, integrals are to be interpreted to be Riemann integrals.

1. If $f \in R(\alpha)$ on $[a, b]$ and if $\int_a^b f d\alpha = 0$ for every f which is monotonic on $[a, b]$, show that α must be a constant.

2. Consider the function $f(x) = 1/2^n$ when $1/2^{n+1} < x \leq 1/2^n$, $n = 0, 1,$ 2, $\cdots$ and $f(0) = 0$. Show that $f \in R$ on $[0, 1]$.

3. If $f \in R$ on $[a, b]$ then for $t \in [a, b]$, let $\phi(t) = \int_t^b f(x)dx$; then show that $\phi'(t)$ exists and is equal to $-f(t)$.

4. Let $[x]$ denote the greatest integer not greater than x; investigate whether $\int_0^2 x[x]dx$ exists.

KEY TERMS

Bounded Variation
Cauchy Criterion
Continuity
Differentiability
Fundamental Theorem of Integral
    Calculus
Global Maximum and Minimum
Improper Integrals
Integrand
Integration
Integration by Parts
Integrator
Limits

Local Maximum and Minimum
Mean Value Theorem
Monotonic Function
Necessary First Order Condition
Properties of Integrals
Riemann-Stieltjes Integral
Riemann's Condition
Rolle's Theorem
Step function
Sufficient Second Order Condition
Taylor's Theorem
Total Variation
Uniform Continuity

USEFUL WEB LINKS

http://math.furman.edu/~dcs/book/c3pdf/sec37.pdf (accessed on 2 August 2010); http://www.maths.abdn.ac.uk/~igc/tch/ma1002/diff/node39.html (accessed on 2 August 2010): Rolle's and Mean Value Theorems

http://math.furman.edu/~dcs/book/c4pdf/sec46.pdf (accessed on 2 August 2010); http://science.kennesaw.edu/~plaval/math2202/intimp.pdf (accessed on 2 August 2010): improper integrals

http://math.stanford.edu/~jlee/math51/extremity.pdf (accessed on 2 August 2010): Hessian, Taylor's Theorem, Lagrange, quadratic forms

http://people.maths.ox.ac.uk/flynn/genus2/AnI0506/analysisI-wk6.pdf (accessed on 2 August 2010); http://www.math.ubc.ca/~cass/courses/m220-00/cauchy.pdf (accessed on 2 August 2010);

http://www.maths.qmul.ac.uk/~ig/MAS111/Cauchy%20Criterion.pdf (accessed on 2 August 2010): Cauchy criterion for convergence

http://www.math.hmc.edu/calculus/tutorials/int_by_parts/ (shows integration by parts with relevant examples) (accessed on 2 August 2010); http://www.sosmath.com/calculus/integration/byparts/byparts.html (accessed on 2 August 2010): integration by parts

http://www.math.hmc.edu/calculus/tutorials/taylors_thm/ (accessed on 2 August 2010); http://www.maths.abdn.ac.uk/~igc/tch/ma2001/notes/node46.html (accessed on 2 August 2010): Taylor's Theorem

http://www.math.ucdavis.edu/~emsilvia/math127/chapter7.pdf (accessed on 2 August 2010): Riemann-Stieltjes integration

CHAPTER FIVE

Economic Applications I
Choice, Utility, and Aggregation

5.1 Introduction

Any decision-maker when confronted with a problem of choice among different alternatives, has to be able to compare the various alternatives, discard some, and arrive at a chosen alternative. In other words, the decision-maker must have a criterion with which the choice is to be made. This criterion is best thought of as a binary relation $\mathcal{R}$ defined over the **non-empty** set of alternatives $\mathcal{X} \subset \Re^n$ (one should say that $\mathcal{R}$ is defined over the Cartesian product $\mathcal{X} \times \mathcal{X}$). As we have seen before, if the relation $\mathcal{R}$ satisfies reflexivity, completeness, and transitivity, we say that $\mathcal{R}$ constitutes an ordering over the set $\mathcal{X}$.

It will be convenient to consider the following: let $\mathcal{P}$ be defined as $x\mathcal{P}y \Leftrightarrow x\mathcal{R}y$ & $\sim y\mathcal{R}x$: the asymmetric part of the $\mathcal{R}$ whereas $\mathcal{I}$ as $x\mathcal{I}y \Leftrightarrow x\mathcal{R}y$ & $y\mathcal{R}x$, the symmetric part of $\mathcal{R}$.

5.2 Possibility of Choosing the 'Best'

Does the fact that $\mathcal{R}$ is an ordering imply that the decision-maker is able to choose from the set $\mathcal{X}$, using this relation or criterion $\mathcal{R}$? In other words, is the decision-maker able to identify the 'best' alternative where the 'best' is defined by means of the relation $\mathcal{R}$? This is the question which we shall seek to explore in this section.

Note that we are actually asking whether $\mathcal{C}_{\mathcal{X}} = \{x \in \mathcal{X} : \forall y \in \mathcal{X}, \ x\mathcal{R}y\} \neq \emptyset$.

The set $\mathcal{C}_{\mathcal{X}}$ is such that if $x \in \mathcal{C}_{\mathcal{X}}$ then there cannot be any $y \in \mathcal{X}$ such that $y\mathcal{P}x$: thus there cannot be a 'better alternative' to $x \in \mathcal{C}_{\mathcal{X}}$ and it is in this sense that we may consider the set $\mathcal{C}_{\mathcal{X}}$ as being the set of **chosen** alternatives.

This set may consist of a **unique** alternative, in which case we can call this the chosen alternative, or the 'best' (according to the criterion $\mathcal{R}$) alternative; on the other hand, the set may consist of many alternatives. In the latter case, note that $x, y \in \mathcal{C}_{\mathcal{X}} \Rightarrow x\mathcal{I}y$.

A general argument showing the non-emptiness of the set $\mathcal{C}_{\mathcal{X}}$ may be ensured provided the set $\mathcal{X}$ is compact and the ordering $\mathcal{R}$ is in addition, upper semi-continuous.

By upper semi-continuity of $\mathcal{R}$, we shall mean that $\mathcal{R}_x = \{y \in \mathcal{X} : y\mathcal{R}x\}$ is a **closed** subset of $\mathcal{X}$ for every $x \in \mathcal{X}$.

First, we note that:

$$C_{\mathcal{X}} = \bigcap_{x \in \mathcal{X}} \mathcal{R}_x$$

Thus, the set of chosen alternatives is the *intersection of a family of closed subsets of a compact set $\mathcal{X}$.*

Next, consider any finite number of alternatives $x_1, x_2, ..., x_k$ in $\mathcal{X}$; by renaming, if required, and hence without any loss of generality, we may take $x_1\mathcal{R}x_2\mathcal{R}x_3.....\mathcal{R}x_k \Rightarrow x_1\mathcal{R}x_j \forall j$: thus

$$x_1 \in \bigcap_{j=1}^{k} \mathcal{R}_{x_j} \neq \emptyset;$$

consequently the set of chosen alternatives is the *intersection of a family of closed non-empty subsets of a compact set $\mathcal{X}$,* any **finite number** of which have a **non-empty intersection**. Thus, by the finite intersection property of a family of compact sets, the set of chosen alternatives is non-empty. However, this set need not be a singleton that is, there may be more than one member of this set.

To insist on a unique choice, we need to ensure additional conditions. One such is to ensure that $\mathcal{R}$ satisfies **strict quasi-concavity** on a **convex set** $\mathcal{X}$: that is, $x, y \in \mathcal{X}, x \neq y, x\mathcal{R}y \Rightarrow \lambda x + (1 - \lambda)y\mathcal{P}y, 0 < \lambda < 1$. Note that not only do we need to strengthen the binary relation but also the domain needs to be convex. It is straightforward to check that now, given that $\mathcal{R}$ is an ordering and is strictly quasi-concave over a non-empty convex compact set of alternatives, $\mathcal{X}$ implies that the set of chosen alternatives is a singleton.

We then have a rather general condition for the existence of a non-empty set of chosen alternatives. In fact, we may weaken these conditions a bit if we are merely interested in having a non-empty set of chosen alternatives. Consider the binary relation $\mathcal{R}$ and suppose that it satisfies reflexivity and upper semi-continuity (in other words, we have given up completeness and transitivity). We then have the following:

Proposition 5.1 Given a non-empty, compact set of alternatives $\mathcal{X} \subset \Re^n$ and a binary relation $\mathcal{R}$ defined on $\mathcal{X}$ satisfying upper semi-continuity and reflexivity, a necessary and sufficient condition for $C_{\mathcal{X}} \neq \emptyset$ is that for every finite set of alternatives $x_1, x_2, ..., x_k \in \mathcal{X}$, there exists $z \in \mathcal{X}$ such that $z\mathcal{R}x_j$ for all $j = 1, 2, ..., k$ (the domination of every finite set of alternatives).

Note that z could be one of these alternatives x_j but need not necessarily be so; when $\mathcal{R}$ is complete and transitive, one may choose one of the x_js for z, as we have shown above. The proof is a repetition of the last claim since the domination of every finite set essentially guarantees that the sets $\mathcal{R}_x$ have the finite intersection property. That then is the weakest condition under which one may demonstrate the existence of a chosen alternative.

In some cases it is not enough to be able to choose from the set of alternatives; the decision-maker should be able to choose from **every finite subset** of alternatives. Our claim may then be applied to every pair of alternatives to imply that this means that $\mathcal{R}$ must be complete and **acyclic**. $\mathcal{R}$ is said to be acyclic if for any finite number of alternatives $x_1, x_2, ...x_n$, $x_1 \mathcal{P} x_2 \mathcal{P} ......\mathcal{P} x_n \Rightarrow x_1 \mathcal{R} x_n$. Note that violation of acyclicity, given complete $\mathcal{R}$ means that there exists a finite set of alternatives such that $x_1 \mathcal{P} x_2 \mathcal{P} ......\mathcal{P} x_n \Rightarrow \sim x_1 \mathcal{R} x_n$ or that $x_n \mathcal{P} x_1$ so that the ordering is a **cycle**: for every alternative, there is a 'better' alternative and hence no choice is possible. Insisting on the ability to choose from all finite subsets requires the ordering to be not only complete but also acyclic. Note that acyclicity once again ensures that the sets $\mathcal{R}_x$ have the finite intersection property.

5.3 The Construction of a Continuous Utility Indicator Function

Consider the following situation: a binary relation $\mathcal{R}$ defined over the set of alternatives $\mathcal{X} \subset \mathfrak{R}^n$ (one should say that $\mathcal{R}$ is defined over the Cartesian product $\mathcal{X} \times \mathcal{X}$) which satisfies: reflexivity, completeness, and transitivity. Thus, $\mathcal{R}$ constitutes an ordering over the set $\mathcal{X}$. The problem that we shall try to solve is whether under some conditions we can use a continuous real valued function $U(.) : \mathcal{X} \to \mathfrak{R}$ which reflects this ordering. In other words we wish to investigate whether there is a continuous real-valued function $U(.)$ such that $\forall x, y \in \mathcal{X}, U(x) \geq U(y) \Leftrightarrow x \mathcal{R} y$. We shall call this function $U(.)$, if it exists as a **utility indicator function**, or a utility function, in short, for the decision-maker. The problem was analysed and solved by Debreu (1959) and we shall provide a simple version of this argument.

We show that if in addition the relation $\mathcal{R}$ satisfies two other properties, that is, **monotonicity** and **continuity**, such a utility function may be constructed. By monotonicity, we mean that $x \geq y \Rightarrow x \mathcal{R} y$; and by continuity we mean that the sets $\{y \in \mathcal{X} : y \mathcal{R} x\}$, $\{y \in \mathcal{X} : x \mathcal{R} y\}$ are **closed** subsets of § for every $x \in \mathcal{X}$; note that the first set was only involved when we considered the upper semi-continuity of the relation $\mathcal{R}$ in the previous section.

Some immediate consequences of this definition may be noted. First, let $\mathcal{P}$ be defined as $x \mathcal{P} y \Leftrightarrow x \mathcal{R} y \& \sim y \mathcal{R} x$: the asymmetric part of the $\mathcal{R}$ whereas $\mathcal{I}$ as $x \mathcal{I} y \Leftrightarrow x \mathcal{R} y \& y \mathcal{R} x$, the symmetric part of $\mathcal{R}$.

It follows then that $\mathcal{R}$ is continuous means that if $x \mathcal{P} y$ then for any alternative x' close to x, $x' \mathcal{P} y$ and for every y' close to y, $x \mathcal{P} y'$. The reader may formulate this idea by using appropriately defined open sets.

It should also be immediate that if $x > y$ (that is, each component of x is greater than the corresponding component of y), then we must have $x \mathcal{P} y$.

That continuity may be violated, may easily be seen from the example of lexicographic orderings: consider $\mathfrak{R}^2$ and let $x = (x_1, x_2), y = (y_1, y_2) \in \mathcal{X} \subset \mathfrak{R}^2$; define $\mathcal{R}$ as follows $x \mathcal{R} y \Leftrightarrow$ either $x_1 > y_1$ or $x_1 = y_1 \& x_2 > y_2$. Such an ordering is called a lexicographic ordering on the set $\mathcal{X}$; in fact, the ordering of words in a dictionary

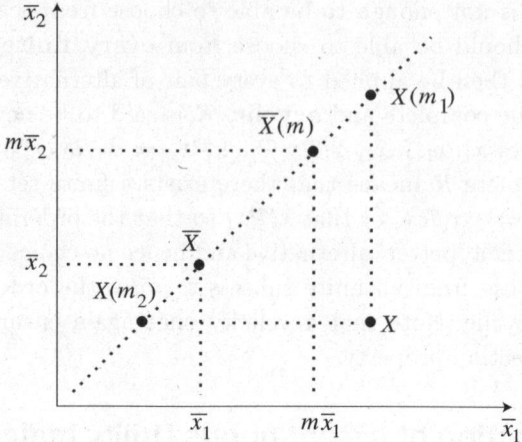

FIGURE 5.1 Construction of Utility Indicator Function

follows the same pattern, given the natural ordering of the alphabets with 'a' being the first, 'b' the second and so on. It would be instructive to construct the set $\{y \in \mathcal{X}, y \mathcal{R} x\}$ for some x and check that such a set cannot be closed.

If the ordering satisfies all the assumptions, an easy construction shows that it is possible to have a continuous real-valued function representing it.

We shall present below the argument for $\mathcal{X} = \mathfrak{R}^2$ although the construction for an arbitrary $\mathcal{X} \subset \mathfrak{R}^n$ is just as straightforward.

Consider any bundle $\overline{X} = (\overline{x}_1, \overline{x}_2)$ and assign $U(\overline{X}) = 1$ (see Figure 5.1); consider the ray through the origin passing through the point $\overline{X}$. Any point on this ray has coordinates: $(m\overline{x}_1, m\overline{x}_2)$ for some number m; denote this point by $\overline{X}(m)$; as the number m increases, $\overline{X}(m)$, moves up the ray; for example, $\overline{X}(1/2)$ is exactly the midpoint of the line segment $O\overline{X}$ and $\overline{X}(1)$ is the point $\overline{X}$, etc. By virtue of monotonicity, it follows that $\overline{X}(m)\mathcal{P}\overline{X}(n)$ where m, n are two numbers if and only if (iff) $m > n$. Now for every point on the ray, we define: $U(\overline{X}(m)) = m$.

Consider any point X which is not a point on the ray as in Figure 5.1; note that there are numbers m_1, m_2 such that $X(m_1)\mathcal{P}X$ and $X\mathcal{P}X(m_2)$. As m shrinks from m_1 and reduces to m_2, the point $X(m)$ moves on the ray towards $X(m_2)$: first, when m is near m_1 the points $X(m)$ are preferred to X; and for m near m_2, X is preferred to $X(m)$. Check the following:

1. There is a value for m, $\tilde{m}$, $m_1 > \tilde{m} > m_2$ such that $X(\tilde{m})\mathcal{I}X$.
2. This value of m is unique.

Now given the above, it is straightforward to define $U(X) = U(X(\tilde{m})) = \tilde{m}$ and the construction of the utility indicator is complete. Note that this assignment of numbers to bundles represents the preferences. Suppose that for some bundles X and Y, $U(X) \geq U(Y)$; the way we have assigned numbers this means that if $U(X) = m \& U(Y) = n$

then there are points $X(m)$, $X(n)$ on the ray such that $X \mathcal{I} X(m)$, $Y \mathcal{I} X(n)$ and since $m \geq n$, $X(m) \mathcal{R} X(n)$ and hence $X \mathcal{R} Y$. Check the converse argument too.

Steps 1 and 2 are crucial. If step 1 does not hold then preferences will have to jump and continuity will be violated. If the claim in step 2 is violated, check what happens.

A formal proof of step 1 may be constructed thus: define $m^* = $ **glb** $\{m : X(m) \mathcal{P} X\} = $ **glb** X^m, say; and $m_* = $ **lub** $\{m : X \mathcal{P} X(m)\} = $ **lub** X_m, say; these bounds are well defined. Note that X^m corresponds to the points on the ray which are preferred to X; whereas X_m corresponds to those on the ray which are such that X is preferred to them. Note that $m^* \notin X^m$. (Hint: use the implications of the continuity of $\mathcal{R}$, mentioned above). Thus $X \mathcal{R} X(m^*)$. Similarly, $X(m_*) \mathcal{R} X$: thus $m_* \geq m^*$. Again, there is m_3 arbitrarily close to (and greater than) m^* such that $X(m_3) \mathcal{P} X$; similarly, there is m_4 arbitrarily close to (and less than) m_* such that $X \mathcal{P} X(m_4)$; hence $m_3 > m_4$ and it follows that $m^* \geq m_*$; thus $m^* = m_* = \tilde{m}$, say and hence $X(\tilde{m}) \mathcal{I} X$.

As an additional exercise, the interested reader should try to see why continuity of the ordering $\mathcal{R}$ is also **necessary** for the existence of a continuous utility indicator; for example, it should be checked that for a lexicographic ordering, there can be no continuous utility indicator.

5.4 Arrow's Theorem

Consider a society made up of many (at least two) individuals; consider also two alternatives A and B; suppose some of the individuals prefer A while others prefer B. What should be society's ranking over A and B? Or consider the problem outlined in the Introductory section: a group of persons, a committee, has to decide on who is the healthiest among a class of 30 students. Each member of the committee has a ranking over the individuals; is it possible to devise a ranking which reflects the ranking by individual members? If so what properties should such an aggregation procedure exhibit? These questions are taken up for consideration next.

An exercise of the type outlined above may be carried out by introducing a societal utility function of the type $\sum_i U^i(.)$, for example, where each $U^i(.)$ is of the type that we constructed in the last section. Here i stands for individuals and each i's real valued $U^i(.)$ for the alternative A, could be added up to obtain an aggregate number for A; a similar exercise could be carried out for B and whichever alternative had the higher associated aggregate score, could be identified as society's preferred alternative. Unfortunately, this approach has been seen to raise more problems than it solved.

A frontal attack on this problem was launched by Kenneth Arrow in 1951 in his justly famous monograph, *Social Choice and Individual Values*. In this section, we discuss Arrow's central result. The purely axiomatic treatment is of fundamental importance both on account of form as well as content.

5.4.1 Notation and definitions

Let S be the set of alternatives, with $\#S \geq 3$; N the finite set of individuals with $\#N \geq 2$; T the set of all orderings (reflexive, transitive, and connected or complete binary relations) over the set S; $R_i \in T$ denotes an ordering for the individual $i \in N$; $\Re = (R_1, \ldots, R_n) \in T^n$. Let $D \subseteq T^n$, $f : D \rightarrow T$, a social welfare function (SWF) and we shall write $f(\Re) = R$. The relations P, P_i are derived respectively from R, R_i in the usual manner. It is the SWF f which we shall study. The SWF is said to satisfy the following conditions:

1. Condition U (Unrestricted Domain): *iff* $D = T^n$.
2. Condition P (Weak Pareto Rule): *iff* $(\forall \Re \in T^n)(\forall x, y \in S)[(\forall i \in N)(x P_i y) \rightarrow x P y]$.
3. Condition I (Independence of Irrelevant Alternatives): *iff* $(\forall \Re, \Re' \in T^n)$ $(\forall x, y \in S)$

$$[(\forall i \in N)[(x R_i y \longleftrightarrow x R_i' y) \wedge (y R_i x \longleftrightarrow y R_i' x)]$$
$$\rightarrow I[(x R y \longleftrightarrow x R' y) \wedge (y R x \longleftrightarrow y R' x)]]$$

   $j \in N$ is a **dictator** *iff* $(\forall \Re \in T^n)(\forall x, y \in S)\left[x P_j y \rightarrow x P y\right]$.

4. Condition ND (No Dictator): The SWF is said to satisfy ND *iff* there is no $j \in N$ who is a dictator.

In the above statements, some symbols have been introduced and although these are standard, it may be helpful to have a glossary:

$\forall$: *for all* ; $\wedge$: *and* ; $\vee$: *or* ; $\rightarrow$: *implies* ; $\longleftrightarrow$: *iff*

Condition U thus signifies that the SWF f must be an ordering for every logically possible configuration of individual orderings; P requires that if for any two alternatives $x, y \in S$, **every** individual prefers x to y then society should prefer x to y too. Condition I requires that society's ranking between the alternatives x and y must depend **only** on individual ordering over these two states. Thus if $\Re, \Re'$ are two configurations which agree on the individual rankings over the two states x and y, then the social ranking obtained from $\Re, \Re'$ must also agree over the ranking of x *vis-à-vis* y. Finally, note that a dictator is a person whose strict preferences over **every** pair of alternatives determine societal strict preferences **regardless** of the ranking of other individuals in society. The condition ND rules out the existence of such a person.

Let $V \subseteq N$; V is said to be **almost decisive** for $(x, y) \in S \times S$ *iff*

$$(\forall \Re \in T^n) \quad [(\forall i \in V)(x P_i y) \wedge (\forall i \in N - V)(y P_i x) \rightarrow x P y]$$

we shall write $D(x, y)$; thus when individuals in the set V prefer x to y and everyone else has the opposite ranking, then society prefers x to y.

V is said to be **decisive** for $(x, y) \in S \times S$ *iff* $(\forall \Re \in T^n)$ $[(\forall i \in V)(x P_i y) \rightarrow x P y]$; we shall then write $\overline{D}(x, y)$; thus in contrast to the definition of almost decisive set, individuals in V are decisive for (x, y) if whenever individuals in V prefer x to y, society too prefers x to y; the preferences of individuals who are not in V are of no concern.

V is said to be **decisive** *iff* V is decisive for all $(a, b) \in S \times S$. $V \subseteq N$ is a **minimally decisive** set *iff* V is decisive and no proper subset of V is decisive.

5.4.2 A lemma

The above definitions have set the ground for concluding that if a group is decisive over some ordered pair of distinct elements, then the group is decisive. A formal statement is contained in the following:

Let $f : T^n \to T$ satisfy P and I. Then if $V \subseteq N$ is almost decisive for some ordered pair of distinct elements then V is a decisive set.

Proof Let V be $D(x, y), x \neq y$; let z be a distinct element from x and y. Consider the following:

$(\forall i \in V) \quad (x P_i y \quad \wedge \ y P_i z)$

$(\forall i \in N - V) \quad (y P_i x \wedge y P_i z)$

Thus $x P y$ (V is $D(x, y)$); $y P z$ (by virtue of condition P); consequently $x P z$. Thus V is $\overline{D}(x, z)$ by virtue of Condition I since V is the only set of persons whose choice over the alternatives x and z has been specified. Thus, we have: $D(x, y) \to \overline{D}(x, z)$.

Similarly, by considering

$(\forall i \in V) \quad (x P_i y \quad \wedge \ z P_i x)$

$(\forall i \in N - V) \quad (y P_i x \wedge z P_i x)$

we may conclude that $D(x, y) \to \overline{D}(z, y)$.

Hence, by using the above results, we have $D(x, y) \to \overline{D}(a, b) \ \forall (a, b) \in \{x, y, z\} \times \{x, y, z\}, a \neq b$.

Next consider $(a, b) \in S \times S, a \neq b$. If $(a = x \vee a = y) \vee (b = x \vee b = y)$, then $\overline{D}(a, b)$ has been shown already. Suppose then $(a \neq x, a \neq y, b \neq x, b \neq y)$. Now apply the argument to $\{x, y, a\}$ first and conclude that $\overline{D}(x, a)$. Next apply the argument to the triple $\{x, a, b\}$ and next conclude that $\overline{D}(a, b)$. This proves the claim.●

Thus, the source of the problem stands revealed; our definitions and axioms have allowed us to conclude that being decisive over some distinct pair implies being decisive over all pairs. This prepares us for the last step in the argument, which we present next.

5.4.3 The theorem

There is no SWF satisfying U,P,I, and ND.

Proof Suppose to the contrary, there is a SWF satisfying U,P, I and ND. Note that N is a decisive set by virtue of condition P. Thus $\exists V \subseteq N$ such that V is minimally decisive: follows by virtue of finiteness. V is non-empty by P; by ND, $\#V > 1$. Consider a partition of V into $V_1, V_2 (V_i \neq \Phi, V_1 \cap V_2 = \Phi, V_1 \cup V_2 = V)$ and consider the following preferences over three alternatives $x, y, z \in S$:

$\forall i \in V_1 \qquad x P_i y \quad \wedge \ y P_i z$

$\forall i \in V_2 \qquad y P_i z \quad \wedge \ z P_i x$

$\forall i \in N - V \quad z P_i x \quad \wedge \ x P_i y$

Note that by virtue of the decisiveness of the set V, yPz must hold.

One has $xRy \vee yPx$. In the first case, note that one must have xPz by virtue of the transitivity of the social R and hence one may conclude that V_1 is $D(x, z)$ and hence, from the Lemma, V_1 is decisive, which is a contradiction to the minimality of V. Check that in the second case we must have that V_2 is $D(y, x)$ and hence as above, must be decisive: again a contradiction to the minimality of V. Thus both xRy and yPx lead to a contradiction and the theorem is established. •

Thus, a reasonable set of conditions U, P, and I imply that the only possible SWF will be dictatorial; consequently 'democratic' decision-making when many individuals are involved may be difficult. This result gave rise to the area of social choice theory and also was one of the items mentioned in the citation to the Nobel Prize received by Kenneth Arrow in 1972.

Further Readings for Section I

There are two sets of concerns.

First, the purely Mathematical ones; the readers may find additional material from the following: for Chapter 2, Suppes (1964), Copi and Cohen (1997); for Chapters 3 and 4, since the material is standard, any good book on Mathematical Analysis would be adequate. We have found the relevant chapters of Apostol's (1974) classic book to be very useful.

We would recommend Werner Hildenbrand's (1983b) Introduction to Gerard Debreu's collected papers, 'Mathematical Economics' to be sampled at some stage, to provide an appreciation by a very skilled practitioner of one of the most skilled in the area. The reason why an axiomatic approach must be used is very well documented in this essay.

The construction of a continuous utility indicator is based on a simplification of Debreu (1954). The possibility of choosing the best is based on Mukherji (1977). Arrow's Theorem is naturally based on Arrow (1951, 1963); also see Sen's (1970) account. In particular, Chapter 1 and 1* in the latter would be relevant as references for Chapters 2 and 5.

For a very elegant application of Integration by parts, see Hildenbrand (1983a).

KEY TERMS

Arrow's Theorem	Continuous Utility Indicator Function
Choosing the 'Best' Alternative	Monotonicity
Continuity	Social Welfare Function

SECTION TWO

- Introduction
- Real Linear Algebra
- Functions of Several Variables
- Static Optimization
- Economic Applications II: Demand and Supply
- Decision-making under Alternative Scenarios

CHAPTER SIX

Introduction

6.1 The Objective of Section II

If we reconsider the problems of choice discussed earlier, it should be clear that we have not in any sense, solved the entire problem. It is as if we have found out the conditions under which the problem, that is, choosing the best, may yield a solution. We have not really established any property that the solution may have. We try to set matters straight in this respect in the present section.

First, we specify matters in this connection. Two generic types of decision-makers are households and firms. They are distinguished by their main activities. The households consider choosing a best from among the various alternatives that may be affordable while the firm may wish to choose among various feasible possibilities, a line of action which is the most profitable, given the nature of the technology. In both these contexts, one may note that typically there are many goods and services which may be involved. Therefore, we need to develop the relevant mathematics which can tackle many variables simultaneously. In other words, we need to learn how to look at functions when there are many independent variables. We do so before reverting to the problems of the household and of the firm.

KEY TERMS

Firm
Household
Independent Variables

CHAPTER SEVEN

Real Linear Algebra

7.1 Preliminaries: Vector Spaces, Sub-spaces, Linear Dependence, Rank of a Sub-space, Matrices

Given the real line $\Re$, consider

$$\Re^n = \underbrace{\Re \times \Re \cdots \times \Re}_{n \text{ times}}, n > 1.$$

Elements of $\Re^n$ will be called n-vectors as opposed to elements of $\Re$, which are *scalars*. An n-vector will be written as

$$x = (x_1, x_2, \cdots, x_n) \text{ or } x = (x_i).$$

In addition, $\Re^n$ will be called the n-space. Using the operations on $\Re$, we define the following operations on $\Re^n$:

1. **Sum:** If $x = (x_i), y = (y_i) \in \Re^n$, then $x+y = (x_i+y_i) = (x_1+y_1, \cdots, x_n+y_n)$ is the sum of the two n-vectors.
2. **Scalar Multiple:** If $x \in \Re^n$, $n > 1, \alpha \in \Re$, then the scalar multiple $\alpha x = (\alpha x_1, \cdots, \alpha x_n)$.

To return to our definitions of sum and scalar multiple, the first operation allows us to add two n-vectors. The number of components in each must be the same, so that one can form the sum by adding component by component; in forming the scalar multiple, each component is multiplied by the same scalar. These operations allow us to generate other vectors by a process of taking scalar multiples and performing vector sums; for example, if $x^1, x^2, \cdots, x^r$ are elements of $\Re^n$, then given any scalars $\beta_i, i = 1, 2, \cdots r,$

$$\sum_{i=1}^{r} \beta_i x^i$$

is a **linear combination** of the x^is. We shall use 0 for the scalar 0 as well as the n-vector 0 (the null vector), each component being the scalar 0: the context should make clear which of the two is being used.

Using the standard properties of addition and multiplication, we note that for the operations mentioned earlier among the elements $x, y, z \in \mathfrak{R}^n$, $\lambda, \mu \in \mathfrak{R}$, the following rules apply immediately.

For addition:

1. Associative Law (A1): $(x + y) + z = x + (y + z)$;
2. Commutative Law (A2): $x + y = y + x$;
3. Subtraction Law (A3): For any x, y there is z such that $x + z = y$.

For multiplication:

4. Vector Distributive Law (M1): $\lambda(x + y) = \lambda x + \lambda y$;
5. Scalar Distributive Law (M2): $(\lambda + \mu)x = \lambda x + \mu x$;
6. Scalar Associative Law (M3): $\lambda(\mu x) = (\lambda \mu)x$;
7. Identity Law (M4): $1x = x$.

In fact we could have begun by postulating objects over which we can define addition and scalar multiples, satisfying A1–A3 and M1–M4 and this would have been adequate for our purpose; that is, the specific use of $\mathfrak{R}$ or $\mathfrak{R}^n$ is not really necessary for us. An abstract system of objects with A1–A3 and M1–M4 as axioms would have served our purpose. Such an algebraic system is called a **vector space**. $\mathfrak{R}^n$ is a vector space.

Let $x^1, x^2, \cdots, x^r \in \mathfrak{R}^n$. We shall say that such a collection $\{x^j\}$ is linearly dependent if there exists a set of scalars β_j, $j = 1, \cdots, r$ **not all zero** such that:

$$\sum_{j=1}^{r} \beta_j x^j = 0.$$

If no such β_j exist, for the collection $\{x^j\}$ then we shall say that the collection $\{x^j\}$ is **linearly independent**.

For example, consider $e^j = (0, 0, \cdots, 1, \cdots, 0)$ where the 1 appears in the j-th component and every other component is zero. Note that $\{e^j\}$ is linearly independent. However, $(1, 2), (2, 3), (4, 5)$ a collection in $\mathfrak{R}^2$ is linearly dependent. Also note that if any collection of n-vectors has 0, or the null vector, as a member, then the collection must be linearly dependent; this follows since one may choose the scalars for all the non-zero n-vectors to be 0 and choose $\beta \neq 0$ for the null vector.

Also see that if a collection is linearly dependent, then adding another member will make the new collection linearly dependent as well; dropping a member from the collection may, however, make the collection linearly independent: for example, one may try to drop any member from the collection of three 2-vectors given above; and try adding any 2-vector to the collection of three 2-vectors. A linearly independent collection may become linearly dependent when the collection is enlarged; while a linearly dependent collection remains linearly dependent if the collection is enlarged.

We have in this connection the following:

Claim 7.1.1 If each member of the collection of vectors $\{x^1, \cdots, x^{r+1}\}$ is a linear combination of the vectors $\{y^1, \cdots, y^r\}$ then the collection $\{x^1, \cdots, x^{r+1}\}$ is linearly dependent.

Proof The proof shall be by induction on the number r; consider $r = 1$; then we have a collection $\{x^1, x^2\}$, each a linear combination of y^1; thus we have scalars β_i such that $x^i = \beta_i \, y^1, i = 1, 2$. If $\beta_i = 0$ for some i then there is nothing to prove, any collection containing the null vector is linearly dependent. Suppose then that $\beta_i \neq 0, i = 1, 2$; then consider:

$$1/\beta_1.x^1 + (-1/\beta_2)x^2 = y^1 - y^1 = 0;$$

this shows that the collection $\{x^1, x^2\}$ is linearly dependent. Hence, the result is true for $r = 1$. We make the **Induction Hypothesis** that the result is true for $r = k$. Consider then the case $r = k + 1$. Now we have:

$$x^j = \sum_{i=1}^{r} \beta_{ij} \, y^i, \quad j = 1, 2, \cdots, k+1$$

Once again, if $\beta_{ij} = 0$ for all i, j then there is nothing to prove; let $\beta_{11} \neq 0$. Next define

$$z^j = -\frac{\beta_{1j}}{\beta_{11}}x^1 + x^j, \quad j = 2, \cdots k+1$$

Thus,

$$z^j = -\frac{\beta_{1j}}{\beta_{11}} \sum_{i=1}^{k} \beta_{i1}y^i + \sum_{i=1}^{k} \beta_{ij}y^i = (-\beta_{1j} + \beta_{1j})y^1 + \sum_{i=2}^{k} \{\frac{-\beta_{1j}.\beta_{i1}}{\beta_{11}} + \beta_{ij}\}y^i$$

and each z^j, $j = 2, \cdots, k+1$ is a linear combination of the $k - 1$ vectors $y^2, \cdots, y^k$ and hence by virtue of the Induction Hypothesis, the vectors z^j, $j = 2, \cdots, k+1$ is linearly dependent. Hence there exist scalars α_j not all zero such that

$$\sum_{j=2}^{k+1} \alpha_j z^j = 0,$$

that is, we have scalars γ_j not all zero such that:

$$\sum_{j=1}^{k+1} \gamma_j x^j = 0$$

where

$$\gamma_1 = -\frac{\alpha_2 \beta_{1j}}{\beta_{11}}, \gamma_j = \alpha_j, j = 2, \cdots, k+1$$

which proves the result is true for $r = k + 1$; hence the result claimed is true. •
  Some immediate consequences of this result are noted next.

1. Any collection of $n + 1$ or more vectors in $\Re^n$ must be linearly dependent. This follows since any element $x + (x_i) \in \Re^n$ may be written as $x = \sum_{i=1}^{n} x_i.e^i$ and hence the claim follows.
2. Any n homogenous linear equations in $n + 1$ unknowns always has a non-zero solution. To see this note that such a system of homogenous equations

may be written thus:

$$\alpha_{i1}x_1 + \alpha_{i2}x_2 + \cdots + \alpha_{in+1}x_{n+1} = 0 \ , i = 1, 2, \cdots, n.$$

In other words, writing

$$a^j = \begin{pmatrix} \alpha_{1j} \\ \alpha_{2j} \\ \vdots \\ \alpha_{nj} \end{pmatrix}$$

we are trying to solve the equation $\sum_{j=1}^{n+1} a^j.x_j = 0$ which will always have a non-zero solution since the collection of n-vectors $\{a^j, j = 1, 2, \cdots, n+1\}$ is a linearly dependent set.

Let V be a vector space; $U \subset V$ is a **sub-space** of V if:

(i) $x, y \in U \Rightarrow x + y \in U$

(ii) $x \in U$ and α any scalar $\Rightarrow \alpha.x \in U$

Thus in $\Re^2$, the only sub-spaces are straight lines through the origin; more importantly, U is a sub-space of $\Re^m$ implies that $0 \in U$. The maximum number of linearly independent vectors in a sub-space U is known as the **rank** of the sub-space U. For instance, by virtue of Claim 7.1.1, it follows that the maximum number of linearly independent vectors in $\Re^n$ is n. This is so since any n-vector $x = (\alpha_1, \alpha_2, \cdots, \alpha_n) = \sum_i \alpha_i e_i$ where $e_i = (0, 0, \cdots, 1, \cdots, 0)$ where the 1 occurs in the i-th position, all other components being zero. Hence, any collection of $n + 1$, n-vectors in $\Re^n$ is linearly dependent and since the e_is form a linearly independent collection, the rank of $\Re^n$ is n.

If for a sub-space U, rank $= r$ and $x_i, i = 1, 2, \cdots, r$ is a collection of linearly independent vectors in U, then the collection is $\{x_i\}$ is said to be a set of **basis vectors** of U; consider any $y \in U$; then the collection $\{x_i\}, y$ must be linearly dependent and we must have $y = \sum_i \alpha_i x_i$ for some collection of scalars α_i. Thus any element of a sub-space may be written as a linear combination of the basis vectors; note that this representation must be unique as well. In fact, one may show:

Claim 7.1.2 If $\{x_1, x_2, \cdots, x_n\}$ is a collection of linearly independent vectors in a sub-space S then the collection forms a basis if and only if every $y \in S$ is a linear combination of vectors $\{x_i\}$.

Thus given a basis we should be able to construct the entire sub-space.

Given two n-vectors $x, y \in U$ where U is some sub-space of $\Re^n$, $x = (\alpha_1, \cdots, \alpha_n)$, $y = (\beta_1, \cdots, \beta_n)$ we define the **scalar product** or the **inner product** of x, y as $x.y = \sum_i \alpha_i \beta_i$.

A rectangular array of numbers $[\alpha_{ij}]$, $i = 1, \cdots, m; \ j = 1, \cdots, n$ constitutes a **matrix** of **order** $m \times n$; m is the number of **rows** and n the number of **columns** of the matrix, denoted by, say $A = [\alpha_{ij}]$. The i-th row is denoted by $a_i = [\alpha_{i1}, \cdots, \alpha_{in}] \in \Re^n$

while the j-th column is denoted by $a^j \in \Re^m$ where

$$a^j = \begin{pmatrix} \alpha_{1j} \\ \alpha_{2j} \\ \vdots \\ \alpha_{nj} \end{pmatrix}.$$

Sometimes we consider matrices $A = (a_{ij})$ which are **square**, that is, the number of rows is equal to the number of columns. In such cases we refer to the elements a_{jj} as being the **diagonal** of the matrix; other elements $a_{ij}, i \neq j$ will be referred to as **off-diagonal** elements. A square matrix, all of whose off-diagonal elements are zero, will be called a **diagonal matrix**. The diagonal matrix with all diagonal elements equal to unity is denoted by I, called the **identity matrix**.

Using the operations defined above, we may define the sum and product of matrices: let $A = [\alpha_{ij}]$, $B = [\beta_{ij}]$ denote matrices. If they are of the same order $m \times n$ then the **sum** $A + B = [\gamma_{ij}]$ where $\gamma_{ij} = \alpha_{ij} + \beta_{ij}$; if the order of a be $m \times r$ and the order of B be $r \times n$ (that is, if the number of columns of $A =$ number of rows of B) then the product $A.B$ is defined to be a matrix of order $m \times n = [\gamma_{ij}]$ where $\gamma_{ij} = a_i.b^j$ and where a_i is the i-th row of A and b^j is the j-th column of B.

If A is $m \times n$ and B is $n \times r$; note that while we can form the matrix product $A.B$ as indicated above, it is not necessary that we are able to form the matrix product $B.A$ if $r \neq m$. If $r = m$ then, of course, both the products $A.B$ as well as $B.A$ are defined but $A.B \neq B.A$ since while $A.B$ is $m \times m$, since $(r = m)$, $B.A$ is $n \times n$. We note further that given any matrix A, we can transform rows into columns and construct another matrix called the **transpose** of A and denoted by A^T. Note also that $(A.B)^T = B^T.A^T$. For square matrices of order n, $A.I = I.A = A$ for any square matrix A.

Two of the more common matrix products used are the following: $x.A$ and $A.y$ where x, y are respectively of order $1 \times m$ and $n \times 1$ and hence, are called **row vectors** and **column vectors**, respectively. Using the matrix product rule $x.A$ is a row vector and is a linear combination of the rows of the matrix A; while $A.y$ is column vector and a linear combination of the columns of A.

Given any matrix A, its rows $\{a_i\}$ may be used to define the following sub-space $X_A = \{x : x = \sum_i \theta_i a_i$ for any scalars $\theta_i\}$; the reader should check that this indeed is a sub-space and it is called the **row space** for the matrix A. One may similarly define the **column space** for the matrix as follows: $X^A = \{x : x = \sum_j \theta_j a^j$ for any scalars $\theta_j\}$ writing the columns of the matrix A as a^j. A fundamental property of these sub-spaces is contained in the following result:

Claim 7.1.3 For any matrix A, the rank of the row space = rank of the column space.

Proof Let the row rank be r and column rank be s and suppose that $r < s$; further, let a_i, $i = 1, 2, \cdots, r$ form a row basis; whereas a^j, $j = 1, 2, \cdots, s$ form the column basis (by reordering if necessary). Let $\hat{a}_i = (\alpha_{i1}, \cdots, \alpha_{is})$ and consider the equations:

$$\hat{a}_i.y = 0, i = 1, 2, \cdots, r.$$

r equations in s unknowns have a non-trivial solution $\bar{y} = (\eta_j)$; also given the fact that a_i forms a row basis for $i = 1, \cdots, r$, any other row a_k can be represented as a linear combination of the first r rows, that is, $a_k = \sum_{i=1}^{r} \mu_{ki} a_i$ for some numbers μ_{ki}; hence we shall also have $\hat{a}_k = \sum_{i=1}^{r} \mu_{ki} \hat{a}_i$; thus we must have $\hat{a}_k . \bar{y} = 0$ for all k and hence we have $\sum_{j=1}^{s} \eta_j a^j = 0$: which contradicts the fact that a^j, $j = 1, 2, \cdots, s$ forms a column basis. Hence $r \geq s$; applying this argument to A^T: the transpose of the matrix A by interchanging rows and columns, we have $s \geq r$ and hence $s = r$ as claimed. •

It is, therefore, meaningful to describe the common value of the row and column rank as the **rank** of a matrix.

7.2 Solution to Equations and Inequalities: Separating Hyperplane Theorem

Given a matrix A which is of order $m \times n$, we are interested in the question of solutions to the system of equations of the form

$$A.x = b$$

or in the system of inequalities of the forms

$$A.x \leq b.$$

Consider the equations first. Note that if there is a solution then we are able to express the vector b as a linear combination of the columns of the matrix A: this is only possible if the vector b lies in the column space of the matrix A, hence:

Claim 7.2.1 A necessary and sufficient condition for the existence of a solution to the equation $A.x = b$ is that

$$Rank\ of\ A = Rank\ of\ \{A, b\}.$$

The proof is left for the reader; note that if A is $m \times n$ then by the matrix $\{A, b\}$ we refer to the matrix of order $m \times (n + 1)$ made up of the first n columns of the matrix A and the last column being the vector b.

Once again the proof is left for the reader. An alternative statement of the previous result is contained in the following:

Claim 7.2.2 $A.x = b$ has a solution if and only $y.A = 0$, $y.b = 1$ have no solution.

Proof Note that if $A.x = b$ has a solution $\hat{x}$ then $y.A = 0$, $y.b = 1$ can have no solution: for if there is a solution $\bar{y}$, say then $0 = \bar{y}.A.\hat{x} = \bar{y}.b = 1$: a contradiction. So now suppose that $A.x = b$ has no solution, that is, $r = $ Rank $A < $ Rank $\{A, b\} = r + 1$; let a^j, $j = 1, \cdots, r$ denote the r independent columns of the matrix A and consider the matrix made up of these r columns and the column b; these $r + 1$ columns are linearly independent. Call this matrix of order $m \times (r + 1)$ by C; now the rank of C $= r + 1$; further there must be $r + 1$ linearly independent row vectors of C (row rank being equal to column rank); these vectors must form a basis of $\mathfrak{R}^{r+1}$; consequently any $(r+1)$–vector may be represented as a linear combination of these row vectors. In particular consider the $(r+1)-$ vector $\hat{c} = (0, 0, 0, \dots 0, 1)$; there is thus a solution $u = (\eta_i)$ such that $\sum_i \eta_i c_i = \hat{c}$, where c_i are the rows of the matrix C; thus $u.a^j = 0, j = 1, \cdots, r$ and $u.b = 1$; note that $u.a^k = 0 \forall k > r$; hence $u.A = 0, u.b = 1$. •

The above result has the following corollary:

Claim 7.2.3 A necessary and sufficient condition for the solution to $A.x = b$ being unique is that the n columns of the matrix be linearly independent.

We shall have an occasion to return to this result later.

Often we are not only interested in solutions to equations but to non-negative solutions of equations. There is one result in this connection which is of great significance. It is referred to as the Theorem of the Separating Hyperplane. This is the term used by David Gale in his famous book and we provide the statement according to Gale. First of all, we adopt Gale's convention to call $x \in \mathfrak{R}^n$ **non-negative** if $x = (\xi_j)$ and $\xi_j \geq 0 \forall j$ and we write $x \geq 0$; we call $x \in \mathfrak{R}^n$ **semi-positive** if $x = (\xi_j)$ and $\xi_j \geq 0 \forall j$ and for some j, $\xi_j > 0$ for some j; in this case we write $x \geq 0$; we call $x \in \mathfrak{R}^n$ **positive** if $x = (\xi_j)$ and $\xi_j > 0 \forall j$ and we write $x > 0$.

*Claim 7.2.4 Exactly one of the following alternatives **must** hold: **either** $A.x = b$ has a non-negative solution or $y.A \gneq 0$, $y.b < 0$ have a solution.*

Instead of providing a formal proof of the proposition, let us first consider what the statement says. Note it is trivial to note that

$$A.x = b \tag{7.1}$$

has a non-negative solution and

$$y.A \gneq 0, \quad y.b < 0 \tag{7.2}$$

have a solution cannot occur simultaneously. For, if to the contrary, there $\bar{x} \geq 0$ such that $A.\bar{x} = b$ and there is $\bar{y}$ such that $\bar{y}.A \geq 0$, $\bar{y}.b < 0$. Then note that $0 > \bar{y}.b = \bar{y}.A.\bar{x} \geq 0$: a contradiction. Thus, at most one of the above can hold. The Separating Hyperplane Theorem says that exactly one of them must hold. Figure 7.1 shows why this must be so and also provides an explanation of why it is so.

Figure 7.1 shows the geometry of the situation when A, the non-negative linear combination of two vectors a^1, a^2 does not contain the vector b; in this situation one may easily construct the dashed line which separates A from b; that is, A lies on one side of the dashed line and b on the other. The equation for the dashed line may be taken to be $y = mx$ or alternatively, there is some m such that all (x, y) on the line satisfy:

$$(m - 1).\begin{pmatrix} x \\ y \end{pmatrix} = 0;$$

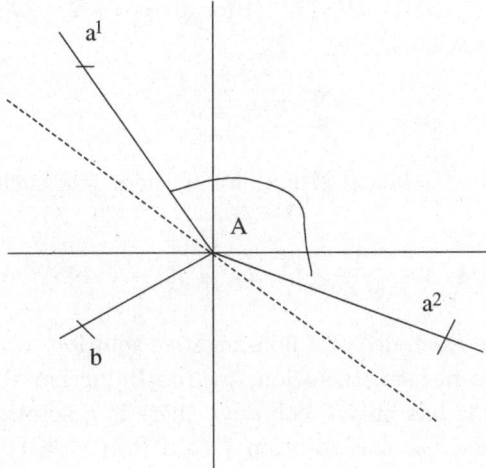

FIGURE 7.1   The Separating Hyperplane Theorem

Consequently,

$$\forall (x, y) \in A, \ \ (m - 1) . \begin{pmatrix} x \\ y \end{pmatrix} \geq 0$$

while

$$(m - 1) . \begin{pmatrix} b_1 \\ b_2 \end{pmatrix} < 0.$$

We provide a proof for the existence of such a vector for the general case.

Proof We have seen that (7.1) and (7.2) cannot both hold.

Suppose then that (7.1) has no non-negative solution. There are then two possibilities: that (7.1) has no solution or that there is a solution but that it is not non-negative. If the former, then by virtue of the result considered before, there must exist a solution $\bar{y}$ to $y.A = 0$, $y.b = 1$ and hence it follows that $-\bar{y}$ satisfies (7.2).

So now suppose that (7.2) has a solution but that it is not non-negative. The proof proceeds by induction on the number of columns of the matrix A. Suppose that there is only one column a^1; then there is a λ such that $a^1 \lambda = b$ and $\lambda < 0$ (there is a solution but it is not non-negative); consider $y = -b$; then $-b.a^1 \geq 0$, $-b.b < 0$: so the result is true when there is only one column.

We make the Induction Hypothesis that the result is true for k columns; consider the case when there are $k + 1$ columns; that is, if $\sum_{j=1}^{k+1} a^j \lambda_j = b$ has no non-negative solution then $\sum_{j=1}^{k} a^j \lambda_j = b$ has no non-negative solution as well; hence by the Induction Hypothesis, there is y^1 such that $y^1.\{a^j, j = 1, ...k\} \geq 0, y^1.b < 0$; if $y^1.a^{k+1} \geq 0$ then there is nothing more to show since then y^1 will solve (7.2) when there are $k + 1$ columns. So suppose that $y^1.a^{k+1} < 0$.

Consider $\bar{a}^j = a^{k+1}(y^1.a^j) - a^j(y^1.a^{k+1})$ for all $j = 1, 2, ..., k$ and $\bar{b} = (y^1.b)a^{k+1} - (y^1.a^{k+1}).b$; now the equation

$$\sum_{j=1}^{k} \bar{a}^j \xi_j = \bar{b} \tag{7.3}$$

can have no non-negative solution either; for if there was such a solution, then we would have:

$$\sum_{j=1}^{k} a^j \xi_j + \frac{1}{-y^1.a^{k+1}} [\sum_{j=1}^{k} (y^1.a^j)\xi_j - y^1.b]a^{k+1} = b$$

which demonstrates the existence of a non-negative solution to (7.1): a contradiction; hence (7.3) has no non-negative solution; by the Induction Hypothesis, applicable since the matrix in (7.3) has only k columns, there is a solution $\bar{y}$ to $\bar{y}\bar{a}^j \geq 0$, $j = 1, 2, .., k$ and $\bar{y}.\bar{b} < 0$ and use this solution $\bar{y}$ to define: $y' = (\bar{y}.a^{k+1})y^1 - (y^1.a^{k+1})\bar{y}$. Check that $y'.a^j = \bar{y}.\bar{a}^j \geq 0$, $j = 1, 2, ..., k$; $y'.b = \bar{y}.\bar{b} < 0$; and $y'.a^{k+1} = 0$. We thus have a solution to (7.2) as claimed. This proves the result. •

The above result may be used to identify conditions when inequalities may yield non-negative solutions, for example:

*Claim 7.2.5   Exactly one of the following alternatives hold: **either** A.x ≤ b has a non-negative solution or y.A ≥ 0, b.y < 0 have a non-negative solution.*

This follows immediately from the theorem of the Separating Hyperplane if one notes that the existence of a non-negative solution to $A.x \leq b$ is equivalent to the existence of a non-negative solution to $(A, I). \begin{pmatrix} x \\ z \end{pmatrix} = b$ where I has the same number of rows as A, say m and the number of columns of I is also m; in addition the elements of I are as follows $I = (b_{ij})$, $b_{jj} = 1$, $b_{ij} = 0$ if $i \neq j$. Now see that the existence of a solution to $y.(A, I) \geq 0$, $y.b < 0$ will imply that $y.A \geq 0$, $y.I \geq 0$, $y.b < 0$ so that the claim follows.

Several related results follow by using the theorem of the Separating Hyperplane. For details, the reader may like to look at Gale's *Theory of Linear Economic Models*, Chapter 2. The exercises of that chapter are an excellent collection (some of them may appear quite tough though!).

Two implications of the above for homogenous equations and inequalities are noted without proof, for future use. It would be instructive for the reader to construct proofs of these propositions, by converting them to one of the above noted forms.

*Claim 7.2.6   Exactly one of the following alternatives hold: **either** A.x = 0 has a semi-positive solution or y.A > 0 has a solution.*

We also have:

*Claim 7.2.7   Exactly one of the following alternatives hold: **either** A.x ≤ 0 has a semi-positive solution or y.A > 0 has a non-negative solution.*

7.3 Determinants

We have considered matrices in the last section; we shall confine attention to square matrices in this section. If the number of rows and columns is equal to say, n, then we say the matrix is of the order n; we can define a real-valued function $d(A)$ for any square matrix A with the following properties:

1. If the matrix B is obtained from A by replacing the i-th row a_i, by αa_i then $d(B) = \alpha d(A)$;
2. $d(A) = 0$ whenever two rows of A are identical; and
3. $d(I) = 1$.

Recall that the square matrix I was defined in the last section. The three properties are enough to specify that given any matrix $A = (a_{ij})$, of order $n \times n$,

$$d(A) = \sum \pm a_{1i_1} a_{2i_2} \cdots a_{ni_n} \qquad (7.4)$$

where the sum extends over all permutations of $1, 2, \cdots, n$ and where for each such permutation $i_1, i_2, \cdots, i_n$, the term $a_{1i_1} a_{2i_2} \cdots a_{ni_n}$ appears with either a plus sign or a minus sign depending on whether the permutation is even or odd, that is, whether the number of transpositions required to pass from $1, 2, \cdots, n$ to $i_1, i_2, \cdots, i_n$ is even or odd.

For instance, consider the natural order 1 2 3 4; then 3 4 2 1 is a permutation of the natural order and one may get back the natural order from 3 2 4 1 by shifting 1 three places to before 3, to get 1 3 2 4 and then shift 2 once to before 3 to get the natural order 1 2 3 4—all together 4 transpositions; so the original permutation 3 2 4 1 will be classified as even; note that we could have gone from 3 2 4 1 by shifting 2 two steps to get 3 4 1 2; then shift 3 three steps to get 4 1 2 3 and then shift 4 three steps to get back the natural order 1 2 3 4; this method took $2 + 3 + 3 = 8$ transpositions, but again an even number. There is no unique way of going from 3 2 4 1 to 1 2 3 4 but all of them will require an even number of transpositions, which is why 3 2 4 1 is called an even permutation. The fundamental fact is that *if a permutation of the natural order requires an even (odd) number of transpositions to revert to the natural order by a series of changes, then any other method of reverting to the natural order from the given permutation will remain even (odd, respectively).*

$d(A)$ is the **determinant** of the matrix A ('an unmitigated nuisance', according to Dorfman, Samuelson, and Solow (1958)).

To clarify matters, consider the following square matrix of order 2×2:

$$A = \begin{pmatrix} a_{11} & a_{12} \\ a_{21} & a_{22} \end{pmatrix}$$

Note that by virtue of the defining properties we must have $d(a_{11}, a_{12}; a_{21}, a_{22}) = A a_{11} + B a_{12}$ where A, B depend only on the elements of the second row; in particular, we must have $A = C_1 a_{21} + D_1 a_{22}$ and $B = C_2 a_{21} + D_2 a_{22}$ and the quantities C_1, C_2, D_1, D_2

must be independent of the elements of the rows of A. In other words, we must have

$$d(a_{11}, a_{12}; a_{21}, a_{22}) = (C_1 a_{21} + D_1 a_{22})a_{11} + (C_2 a_{21} + D_2 a_{22})a_{12}. \qquad (7.5)$$

We have this on the basis of the first property; we use the second property to obtain

$$0 = d(a_{11} + a_{21}, a_{12} + a_{22}; a_{21} + a_{11}, a_{22} + a_{12})$$

by virtue of the two rows of the matrix being identical. The right hand side (rhs) may be expanded by means of (7.5) to obtain:

$0 = (C_1 a_{21} + D_1 a_{22})a_{11} + (C_1 a_{11} + D_1 a_{12})a_{11} + (C_2 a_{21} + D_2 a_{22})a_{12} + (C_2 a_{11} + D_2 a_{12})a_{12} +$
$(C_1 a_{21} + D_1 a_{22})a_{21} + (C_1 a_{11} + D_1 a_{12})a_{21} + (C_2 a_{21} + D_2 a_{22})a_{22} + (C_2 a_{11} + D_2 a_{12})a_{22}$

or collecting terms and using the basic (7.5), we have:

$0 = d(a_{11}, a_{12}; a_{21}, a_{22}) + d(a_{11}, a_{12}; a_{11}, a_{12}) + d(a_{21}, a_{22}; a_{21}, a_{22}) + d(a_{21}, a_{22}; a_{11}, a_{12})$

or that

$$0 = d(a_{11}, a_{12}; a_{21}, a_{22}) + d(a_{21}, a_{22}; a_{11}, a_{12})$$

since the other terms are zero by property 2 as they involve determinants of matrices with identical rows. We have derived a very important property, that is, interchanging rows leads to a change in signs of the determinant. We thus have

$$(C_1 a_{21} + D_1 a_{22})a_{11} + (C_2 a_{21} + D_2 a_{22})a_{12} = -[(C_1 a_{11} + D_1 a_{12})a_{21} + (C_2 a_{11} + D_2 a_{12})a_{22}]$$

Comparing coefficients, we have: $C_1 = 0$, $D_2 = 0$, $D_1 = -C_2$; consequently, we have

$$d(a_{11}, a_{12}; a_{21}, a_{22}) = D_1(a_{11}a_{22} - a_{12}a_{21})$$

Finally $D_1 = 1$ from property 3. We have thus established the unique defining properties of a determinant; for determinants of higher orders, the same three properties will be sufficient to pin down the determinant value as specified in (7.4).

One may deduce further that:

1. If B is obtained from A by an interchange of rows then $d(B) = -d(A)$.
2. If B is obtained from A by changing rows into columns, then $d(B) = d(A)$; such a B is called the transpose of A and written as A^T.
3. If the word 'rows' is replaced by 'columns' and 'columns' by 'rows', all claims made will remain valid.
4. As we have seen, it is expected that $d(A) = a_{i1}A_{i1} + \cdots + a_{in}A_{in}$ for some numbers A_{ij} which do not depend on the elements of row i; the numbers A_{ij} are called the *co-factors* of the ij-th element and are defined by $A_{ij} = (-1)^{i+j}d(D_{ij})$ where D_{ij} is the matrix obtained by deleting the i-th row and j-th column in A; D_{ij} also called a *minor* of order $n-1$ of the original matrix.
5. In case A, B are both square matrices of the same order, then $d(A.B) = d(A).d(B)$ while $d(A + B) \neq d(A) + d(B)$.

Clearly, by deleting several rows and equal number of columns of a square matrix, we can get minors of different orders. If a minor (which is also a square matrix) has a diagonal which is part of the diagonal of the original matrix, then it is called a principal **minor** of the original matrix.

Square matrices whose determinants are not zero are called non-singular. Let A be of order n. One may show that $d(A) = 0 \Leftrightarrow$ rank of A is n. In this case we say that the rank is full.

If $d(A) = 0$ we call A **singular**. Singular matrices cannot have full rank and the columns (rows) are linearly dependent; in such cases, there exist vectors $x, y \neq 0$ such that $y.A = 0, A.x = 0$.

For any non-singular matrix A, we can construct $B = (b_{ij})$ such that

$$b_{ij} = \frac{A_{ji}}{d(A)}$$

with the property $A.B = B.A = I$, where I is the identity matrix; B is called the 'inverse' of A and written A^{-1}. Note that if there is an inverse (that is, the matrix is non-singular) it is unique. This is very helpful in solving equations of the form $A.x = b$; for then, $x = A^{-1}.b$ is the unique solution and this yields the well-known Cramer's Rule for solving equations of this type:

$$x_i = \frac{1}{det A}\{\sum_{j=1}^{n} A_{ji}b_i\}$$

We note that in case $A.B$ is non-singular then $(A.B)^{-1} = B^{-1}.A^{-1}$.

7.4 Characteristic Roots and Vectors

We continue our investigation of square matrices in this section. Given a square matrix A, *the characteristic equation* of A is given by $d(A - \lambda I) = 0$; if A is of order n then this equation is a polynomial of order n and has n roots λ_j, $j = 1, 2, \cdots, n$. These are called characteristic roots of the matrix A; note that for each λ_j there is $x, y \neq 0$ such that $x.(A - \lambda_j I) = 0$ and $(A - \lambda_j I).y = 0$: x, y are the characteristic vectors corresponding to λ_j.

Even if all the entries of a square matrix are real numbers, characteristic roots may be complex and since $x.A = \lambda_j x$ or $A.y = \lambda_j y$, the characteristic vectors too may have complex elements. Characteristic vectors corresponding to distinct characteristic roots have an important property:

Claim 7.4.1 Let $x.A = \lambda x$ and $A.y = \mu y$ where $\lambda \neq \mu$; then $x.y = 0$.

Proof Note that from the first, we have by multiplying by y, $x.A.y = \lambda x.y$; while by multiplying the second equation by x we have, $x.A.y = \mu x.y$ and hence $\lambda x.y = \mu x.y$ and since $\lambda \neq \mu$ we must have $x.y = 0$. •

For symmetric matrices, that is, $A = A^T$, we have the following property:

Claim 7.4.2 Characteristic roots for symmetric matrices are real.

Proof Let A be symmetric and let $\lambda = \alpha + \iota\beta$ be a characteristic root of A; we shall show that $\beta = 0$ so that the root has to be real. Consider the corresponding characteristic vectors x so that $A.x = \lambda x$; given that λ is complex, x must be complex too and we shall write $x = u + \iota v$; then from the equation $A.(u+\iota v) = (\alpha+\iota\beta)(u+\iota v)$, equating the real parts and the complex parts, we have:

$$A.u = \alpha u - \beta v, \quad A.v = \beta u + \alpha v$$

Multiplying the first by v and the second by u, we have

$$v.A.u = \alpha v.u - \beta v.v, \quad u.A.v = \beta u.u + \alpha u.v$$

Now note that $v.A.u = u.A.v$ since A is transitive; hence $\beta(u.u + v.v) = 0$ and since $x \neq 0$, $u.u + v.v > 0$; thus $\beta = 0$ as claimed. •

In addition, for symmetric matrices, *column and row characteristic vectors may be taken to be the transpose of one another*: for if $A.x = \lambda x$ then $x^T.A^T = \lambda x^T$ or $x^T.A = \lambda x^T$; hence using the result deduced earlier, if x, y are characteristic vectors corresponding to characteristic roots (λ, μ respectively, $\lambda \neq \mu$) then $x.y = 0$.

Consider a situation when a symmetric matrix A of order $n \times n$ has **distinct** characteristic roots $\lambda_1, \lambda_2, \cdots, \lambda_n$ with characteristic vectors $x^1, x^2, \cdots, x^n$, $x^i \neq 0$; let Λ denote the diagonal matrix (θ_{ij}) with $\theta_{ii} = \lambda_i, \theta_{ij} = 0, i \neq j$. We then have:

Claim 7.4.3 There is a matrix X such that $X^T.X = I$ and $X^T.A.X = \Lambda$

Proof Since $A.x^i = \lambda_i x^i$, we choose x^i such that $x^i.x^i = 1$ for all i and consider the $n \times n$ matrix X such that its i-th column is x^i; note then that we have $A.X = X.\Lambda$.

First, note that X must be non-singular, for if it is singular, then some column must be linearly dependent on the remaining, that is, say $x^1 = \sum_{i=2}^{n} \alpha_i x^i$; then we must have $x^1.x^1 = 0$: a contradiction. Hence X^{-1} exists and we have $X^{-1}.A.X = \Lambda$.

Denote the i-th row of X^{-1} by y_i and note that $X^{-1}.A = \Lambda X^{-1}$; hence $y_i.A = \lambda_i y_i$ and since $X^{-1}.X = I$, we have that $y_i.x^j = 0$ $i \neq j$, $y_i.x^i = 1$; thus one may demonstrate that the collection $\{x^j, j \neq i, y_i^T\}$ forms a linearly independent set of vectors. Since this is a collection of n n-vectors, they form a basis of $\Re^n$ and consequently, we should be able to write $x^i = \alpha_i y_i^T + \sum_{j\neq i} \alpha_j x^j$. Since $x^i.x^j = 0, y_i^T.x^j = 0, j \neq i$, it follows that $x^i = \alpha_i y_i^T$ and since $y_i.x^i = 1$, it follows that $\alpha_i = 1$ and $y_i^T = x^i$ so that $X^{-1} = X^T$ as claimed. •

In general, of course, a square matrix may be symmetrical but the characteristic roots may be repeated or the matrix may not be symmetric. In such cases, the transformation of A to the matrix of characteristic roots becomes more complicated. Interested readers may consult more advanced texts for this topic.

7.5 Quadratic Forms

Given a symmetric square matrix A, the expression $x^T.A.x$ is called a quadratic form; note that it is customary to take the matrix in a quadratic form to be symmetric since if it is not so, then $x^T.A.x = 1/2x^T.(A + A^T).x$ and $A + A^T$ is symmetric. Given the above, there is no loss of generality in taking the matrix defining the quadratic form, A, to be symmetric.

The quadratic form $x^T.A.x$ is said to be **positive (negative) definite** if for any $x \neq 0$, $x^T.A.x > 0 (< 0,$ respectively). Given the results noted above, $x^T.A.x = x^T.(X.\Lambda.X^T).x = y^T.\Lambda.y$ where $y = X^T.x$. Thus $x^T.A.x = \sum_i \lambda_i y_i^2 > 0$ for all x iff $\lambda_i > 0 \forall i$. In such situations, we shall call the matrix A **positive definite** as well. Note then that A is positive definite iff $-A$ is negative definite.

The quadratic form $x^T.A.x$ is said to be **positive (negative) semi-definite** if $x^T.A.x \geq (\leq$ respectively$)$ 0 for all x. The matrix of the quadratic form is accordingly referred to as being positive or negative semi-definite depending on whether the quadratic form is positive or negative semi-definite. The crucial difference between positive definite and positive semi-definite forms lies in the possibility of the existence of a vector $x \neq 0$ such that $x^T.A.x = 0$; note that in case there is no such x and the form is positive semi-definite, then one may conclude that the form is actually positive definite. It turns out that if such an x exists then the matrix A must be singular:

Claim 7.5.1 *In case A is positive semi-definite and if there exists $x \neq 0$ such that $x^T.A.x = 0$ then $A.x = 0$.*

Proof Suppose $A.x \neq 0$; then there is y such that $x^T.A.y > 0$; consider a scalar α and construct $z = x + \alpha y$; consider then:

$$0 \leq z^T.A.z = 2\alpha x^T.A.y + \alpha^2 y^T.A.y = \alpha\{2x^T.A.y + \alpha y^T.A.y\}$$

By choosing $\alpha < 0$ and small, the last term becomes strictly negative: a contradiction. Hence the claim follows. •

Thus, a non-singular positive semi-definite matrix is positive definite. The positive definiteness of a matrix has some connection with the determinants of principal minors of the matrix. Given a matrix $A = (a_{ij})$ denoted by:

$$D_k = \begin{pmatrix} a_{11} & \cdots & a_{1k} \\ a_{21} & \cdots & a_{2k} \\ \cdots & \cdots & \cdots \\ a_{k1} & \cdots & a_{kk} \end{pmatrix}$$

The matrices D_k are nested principal minors of the matrix A and if A is of order n then there are n such nested principal minors; if A is positive definite, each D_k is positive definite too. Clearly $D_n = A$; we claim:

Claim 7.5.2 *A is positive (negative) definite iff $d(D_k) > 0 \forall k$ (respectively, $(-1)^k d(D_k) > 0 \forall k$).*

Proof When $n = 2$, A is positive definite iff $a_{11}x_1^2 + a_{22}x_2^2 + 2a_{12}x_1.x_2 > 0$ for all $(x_1, x_2) \neq (0, 0)$, that is, if $a_{11} + a_{22}x_2^2 + 2a_{12}x_2 > 0$ for all x_2, obtained by choosing $x_1 = 1$. The quadratic expression is always positive if there is no **real** solution to the quadratic $a_{11}+a_{22}x_2^2+2a_{12}x_2 = 0$ and if $a_{11} > 0$, that is, if $a_{11} > 0$ and $a_{12}^2 < a_{11}a_{22}$ as claimed. The completion of the proof is left to the interested readers; however, they should be warned that it is long and tedious. •

Next, define the principal minors:

$$D_{ij}^k = \begin{pmatrix} a_{ii} & \cdots & a_{ij} \\ \cdots & \cdots & \cdots \\ a_{ji} & \cdots & a_{jj} \end{pmatrix} \quad i \leq j$$

k signifies that there are k rows and columns in the matrix D_{ij}^k, $k = 1, 2, \cdots n$. Then we have the following result which is included without proof.

Claim 7.5.3 *A is positive (negative) semi-definite, iff $d(D_{ij}^k) \geq 0 \forall k, i, j, i \leq j$ (respectively, $(-1)^k d(D_{ij}^k) \leq 0 \forall k, i, j, i \leq j$).*

7.6 Dominant Diagonal Matrices

Consider the square matrix A of order n; construct the matrix $B_A = (b_{ij})$ such that $b_{jj} = |a_{jj}|$, $b_{ij} = -|a_{ij}|$, $i \neq j$. The matrix A is said to have a **column dominant diagonal** if there is $d \geq 0$ such that $d^T.B_A > 0$; the matrix A is said to have a **row dominant diagonal** if A^T has a column dominant diagonal or alternatively, if there exists $c \geq 0$ such that $B_A.c > 0$. We note:

Claim 7.6.1 *A square matrix A has a column dominant diagonal iff it has a row dominant diagonal.*

Proof Let the matrix A have a column dominant diagonal; that is, there is $d^\star \geq 0$ such that $d^{\star^T}.B_A > 0$. Now suppose to the contrary that there is no non-negative solution to $B_A.c > 0$; then a direct application of Claim 7.2.7, yields the existence of $x^\star$ semi-positive satisfying $x^{\star^T}.B_A \leq 0$. Thus, for any scalar $\alpha > 0$, we must have $(d^\star - \alpha x^\star)^T.B_A > 0$; in particular, choose $\alpha = \min_{j \ni x_j^\star > 0} d_j^\star/x_j^\star$; for such a choice one of the components of the vector $d^\star - \alpha x^\star$ must be zero and consequently, $(d^\star - \alpha x^\star)^T.B_A > 0$ cannot hold. Hence, there must be a non-negative solution to $B_A.c > 0$ or A must have a row dominant diagonal. To complete the proof of the claim, apply the argument to A^T. •

The above claim allows us to refer to matrices with dominant diagonals (dd). We do not need to distinguish between column dominant diagonal and row dominant diagonal. Thus A is a matrix with dominant diagonal if there exist non-negative vectors $d, c \geq 0$ such that $d^T.B_A > 0$ and $B_A.c > 0$. Note that there is no necessity for $d = c$; this may be ensured if we know that the matrix A is symmetric, for example.

A crucial property of matrices with dominant diagonal is presented next.

Claim 7.6.2 *If A has a dd, then A is non-singular.*

Proof Since by definition, $d.B_A > 0$ for some non-negative d, by virtue of Claim 7.2.7, $B_A.y \leq 0$ has no semi-positive solution. However, if A is singular, there is $z \neq 0$ such that $A.z = 0$; that is,

$$-a_{ii}.z_i = \sum_{j \neq i} a_{ij} z_j \text{ for all } i$$

or

$$|a_{ii}|.|z_i| = |\sum_{j \neq i} a_{ij} z_j| \leq \sum_{j \neq i} |a_{ij}|.|z_j| \text{ for all } i$$

or writing $y_i = |z_i|$, $y = (y_i)$ we note that $y \geq 0$ and $B_A.y \leq 0$: which is a contradiction. Hence the claim. •

We shall now focus attention on matrices with the sign pattern of the matrix B_A; that is, **with all off diagonals non-positive**; we shall call such matrices B-**matrices**. For B-matrices, then a dominant diagonal exists with $b_{ii} > 0 \forall i$ **iff** there is $d \geq 0$ such that $B.d > 0$. In such a case, we say that B has a +ve dd; B−matrices with a +**ve dd** have very nice properties and we turn our attention to them next.

Claim 7.6.3 *If B is a B-matrix then $B.x = c$ has a unique non-negative solution for any $c \geq 0$, iff B has a +ve dd.*

Proof Suppose B has a +ve dd; then B is non-singular and the equation $B.x = c$ has a unique solution for any $c \geq 0$. Consider y such that $y^T.B \geq 0$; if possible, let $J = \{j : y_j < 0\}$ and suppose $J \neq \emptyset$. Then writing $y^J = (y_i : i \in J)$, $B^J = (b_{ij}), i, j \in J$, $y^{J^T}.B^J \geq 0$; thus $z^T.B^J \leq 0$ has a strictly positive solution; hence by virtue of Claim 7.2.7, $B^J.x > 0$ has no non-negative solution. This contradicts the fact that B has a +ve dd. Hence $J = \emptyset$; thus $y^T.B \geq 0 \Rightarrow y \geq 0$; hence for any $c \geq 0$, $y^T.B \geq 0$, $y^T.c < 0$ can have no solution and hence by the Separating Hyperplane Theorem, $B.x = c$ has a non-negative solution for any $c \geq 0$. The solution is unique as we have said. The converse is trivial. •

A result proved in the above proof is noted separately for future reference:

Claim 7.6.4 *If B is a B-matrix such that $B.x = c$ has a non-negative solution for any $c \geq 0$ then $y^T.B \geq 0 \Rightarrow y \geq 0$.*

An immediate implication of the result, Claim 7.6.3, is contained in the following:

Claim 7.6.5 *For any B-matrix, B, $B^{-1} \geq 0 \Leftrightarrow B$ has a +ve dd.*

And finally, the Hawkins-Simon Condition:

Claim 7.6.6 *For any B-matrix, B has a +ve dd iff*

$$\det B_k = \begin{pmatrix} b_{11} & \cdots & b_{1k} \\ \cdots & \cdots & \cdots \\ b_{k1} & \cdots & b_{kk} \end{pmatrix} > 0$$

for all $k = 1, 2, \cdots, n$.

Proof Suppose B has a +ve dd. We shall show that $\det B_k > 0$ for all $k = 1, 2, \cdots, n$. For the case when $n = 1$, $\det B_1 = b_{11} > 0$ and the claim is trivially true. We make the Induction Hypothesis that the claim is true for all $n < K$ and now consider the case when $n = K$. We may write the matrix B as:

$$\begin{pmatrix} B_{K-1} & b_{\sim K} \\ b_{K\sim}^T & b_{KK} \end{pmatrix}$$

where $b_{\sim K} = (b_{1K}, \cdots, b_{K-1K})^T$, $b_{K\sim}^T = (b_{K1}, \cdots, b_{KK-1})$; note that B_{K-1} is a B-matrix of order $< K$ and hence the Induction Hypothesis implies $\det B_k > 0$, $k = 1, 2, \cdots, K-1$. Our claim will be established if $\det B = \det B_K > 0$. Since B has a +ve dd, B is non-singular and hence $\det B_K \neq 0$; if possible let $\det B_K < 0$. Now consider, $\theta \in \Re$, $0 \leq \theta \leq 1$:

$$f(\theta) = \det B(\theta) = \det \begin{pmatrix} B_{K-1} & \theta b_{\sim K} \\ \theta b_{K\sim}^T & b_{KK} \end{pmatrix}$$

Note, $f(1) < 0$ by assumption; also $f(0) = b_{KK} \det B_{K-1} > 0$; moreover, $f(\theta)$ is a continuous function and hence by the Intermediate Value Theorem (Section 4.2), there is $\bar{\theta}, 0 < \bar{\theta} < 1$ such that $f(\bar{\theta}) = 0$; or $B(\bar{\theta})$ is singular. Since $B(1)$ has a +ve dd, it follows that $B(\theta)$ has a +ve dd for all $\theta < 1$: a contradiction. Hence $f(1) > 0$ and our assertion is true for $n = K$ and hence for all values of n.

For the converse, assume that $\det B_k > 0$ for all $k =, 2, .., n$; we shall show that this implies that $B^{-1} \geq 0$ and hence use one of our earlier claims to conclude that B must have a +ve dd. Assume then that $n = 2$; it is straightforward to check that for this case, $b_{11} > 0$, $\det B_2 = \det B > 0 \Rightarrow B^{-1} \geq 0$. Thus our assertion is true for $n = 2$; next, we make the Induction Hypothesis that the assertion is true for all $n < K$ and consider the situation when $n = K$. Since $\det B_K = \det B > 0$, we know that B^{-1} exists. Writing B, as before:

$$B = \begin{pmatrix} B_{K-1} & b_{\sim K} \\ b_{K\sim}^T & b_{KK} \end{pmatrix}$$

we write $B^{-1} = C$ in a conformable form as:

$$C = \begin{pmatrix} C_{K-1} & c_{\sim K} \\ c_{K\sim}^T & c_{KK} \end{pmatrix}$$

and noting that $C.B = I = B.C$, we note that:

$$c_{KK} = \frac{\det B_{K-1}}{\det B_K} > 0$$

by hypothesis; further, we have the following:

$$B_{K-1}.c_{\sim K} + c_{KK} b_{\sim K} = 0.$$

By the Induction Hypothesis, since $B_{K-1}^{-1} \geq 0$, and $b_{\sim K} \leq 0$, from the property of a B-matrix, we obtain:

$$c_{\sim K} = c_{KK}.B_{K-1}^{-1}(-b_{\sim K}) \geq 0.$$

Again since $C.B = I$, we have:

$$c_{K\sim}^T.B_{K-1} + c_{KK}.b_{K\sim}^T = 0$$

and arguing as above, we obtain:

$$c_{K\sim}^T \geq 0$$

and finally, we also have:

$$B_{K-1}.C_{K-1} + b_{\sim K}.c_{K\sim}^T = I_{K-1},$$

where I_{K-1} is the identity matrix of order $K - 1$. This yields $C_{K-1} = B_{K-1}^{-1}[I_{K-1} - b_{\sim K}.c_{K\sim}^T] \geq 0$ (recall that $-b_{\sim K} \leq 0$ from the properties of a B-matrix; thus $C = B^{-1} \geq 0$ and the assertion is true for $n = K$. Hence the assertion is true for all values of n. This completes the proof for the claim. •

The above set of results have very wide applicability in the theory of linear economic models and we shall have a chance to investigate some of these properties in subsequent sections. This set of results may also be used to establish the special properties of square matrices which are non-negative as well. We turn to these matters, next.

7.6.1 Non-negative square matrices

A square matrix A of order n is said to be non-negative if all entries are non-negative and we shall write $A \geq 0$. Characteristic roots of such matrices have some special properties and we investigate them in this section. The results of the previous section, particularly the properties of B-matrices have a crucial role to play.

Let $A \geq 0$ be a square matrix; unless otherwise stated A will be always taken to be non-negative; define $F(A) = \{\pi \in \Re : \pi I - A \text{ has a +ve dd } \}$. Note that for any π, $\pi I - A$ is a B-matrix; also note that for any A, $F(A) \neq \emptyset$: since any π greater than the largest column sum of A must be an element of $F(A)$. Further, $\pi \in F(A) \Rightarrow \pi \geq 0$. Thus

$$\pi_A^\star = \inf_{\pi \in F(A)} \pi$$

exists (recall the properties of real numbers). We shall drop the subscript A whenever the context makes it clear which matrix we are referring to. We note then:

Claim 7.6.7 $\pi^\star \notin F(A)$, $\pi^\star \geq 0$ and $\pi > \pi^\star \Rightarrow \pi \in F(A)$. Thus $F(A) = (\pi^\star, +\infty)$

Proof First of all note that if $\pi^\star \in F(A)$ then, by definition, there is $d \geq 0$ such that $(\pi^\star I - A)d > 0$; thus for some $\epsilon > 0$, $((\pi^\star - \epsilon)I - A)d > 0$ and hence $\pi^\star - \epsilon \in F(A)$: a contradiction. Hence $\pi^\star \notin F(A)$; then the next two assertions follow directly from the property of the infimum. First, we have already seen that 0 is a lower bound for $F(A)$, hence $\pi^* \geq 0$; next since π^* is the infimum, it follows that for any $\pi' > \pi^*$ there is $\pi < \pi'$ such that $\pi \in F(A)$ that is, there is $d \geq 0$ such that $(\pi I - A).d > 0$; but then $(\pi' I - A).d > 0$ as well and hence $\pi' \in F(A)$. Hence the fact that $F(A)$ is an open interval is established. •

We show next:

Claim 7.6.8 π^* is a characteristic root (cr) of A and has an associated semi-positive characteristic vector (cv); further, for any other c.r. α, $|\alpha| \leqq \pi^*$.

Proof Let c be a positive column vector $(n \times 1)$; consider the matrix $(n \times (n+1))$

$$D = (\pi^* I - A \ -c)$$

where the first n columns are the columns of $\pi^* I - A$ and the last column is $-c$. Note that $y^T.D > 0$ can have no solution since otherwise, if a solution $\bar{y}$ exists, then $\bar{y}^T.(\pi^* I - A) > 0$ and $\bar{y}^T.c < 0$ or for some small $\epsilon > 0$, $\bar{y}^T.((\pi^* + \epsilon)I - A) \geq 0$ and $\bar{y}^T.c < 0$, so that by the Separating Hyperplane Theorem, there can be no non-negative solution to $((\pi^* + \epsilon)I - A).x = c$ and consequently by virtue of Claim 7.6.3, $\pi^* + \epsilon \notin F(A)$: a contradiction. Hence no such $\bar{y}$ exists. Thus by virtue of Claim 7.2.6, there is a semi-positive solution to:

$$\pi^* I - A \ -c \cdot \begin{pmatrix} x \\ \alpha \end{pmatrix} = 0$$

that is, there exists $\begin{pmatrix} x^* \\ \alpha^* \end{pmatrix} \geq 0$ such that $(\pi^* I - A).x^* = \alpha^* c$; in case $\alpha^* > 0$ this means that $(\pi^* I - A)$, a B−matrix, has a +ve dd and hence $\pi^* \in F(A)$: a contradiction. Hence, $\alpha^* = 0$; then $x^* \geq 0$ and π^* is a c.r. with x^* as its associated c.v., as claimed.

Let α be any other cr and z an associated cv; that is, $A.z = \alpha z$; hence, we have, for all i:

$$|\alpha|.|z_i| = |\sum_j a_{ij} z_j| \leqq \sum_j a_{ij} |z_j|$$

or writing $y_i = |z_i|$, $y = (y_i) \geq 0$ and $(|\alpha|I - A).y \leq 0$; in case $|\alpha| > \pi^*$, $|\alpha| \in F(A)$ and hence $(|\alpha|I - A)^{-1} \geqq 0$ and we obtain $y \leq 0$: a contradiction; hence $|\alpha| \leq \pi^*$, as claimed. •

We have the following results as well:

Claim 7.6.9 $(\pi I - A)^{-1} \geqq 0 \Leftrightarrow \pi > \pi^*$; further $L_{ij}(\pi) \geqq 0$ for all $\pi \geqq \pi^*$, where $L_{ij}(\pi)$ denotes the co-factor the j–i-th element in $(\pi I - A)$.

Proof $(\pi I - A)^{-1} \geqq 0 \Leftrightarrow (\pi I - A)^{-1}$ has a +ve dd $\Leftrightarrow \pi \in F(A)$ and the assertion follows. Writing $L(\pi)$ for the determinant of $(\pi I - A)$, we have $L(\pi) > 0$ for $\pi \in F(A)$, from Claim 7.6.6; thus since $L_{ij}(\pi)/L(\pi) \geq 0$ for $\pi > \pi^*$, we must have $L_{ij}(\pi) \geqq 0$ for all $\pi > \pi^*$; hence $L_{ij}(\pi^*) \geqq 0$ as well, from continuity. •

Claim 7.6.10 If $A \geq C \geq 0$ then $\pi_A^\star \geq \pi_C^\star \geq 0$; if C is a principal minor of A, then $\pi_A^\star \geq \pi_C^\star \geq 0$.

Proof First, note that if $A \geq C \geq 0$ then $F(A) \subseteq F(C)$; this follows since $\pi \in F(A)$, that is, $(\pi I - A)$ has a dominant diagonal, that is, there is a non-negative d such that $(\pi I - A)d > 0$ or $\pi d > A.d \geq C.d \Rightarrow (\pi I - C).d > 0 \Rightarrow \pi \in F(C)$. Hence, $\pi_A^\star \geq \pi_C^\star \geq 0$. When C is a principal minor of A, fill C with 0s to obtain D which is of the same dimension as A; note then that $A \geq D$ and consequently, $\pi_A^\star \geq \pi_D^\star \geq 0$, as we have just shown and further $\pi_D^\star = \pi_C^\star$. •

Thus for any non-negative matrix A, the following may be said:

(i) There is a cr $\pi^\star \geq 0$ such that $\pi^\star \geq |\alpha|$ for any other cr α and further there are semi-positive vectors x, p such that $(\pi^\star I - A).x > 0$ and $p^T.(\pi^\star I - A) > 0$.

(ii) $(\pi I - A)^{-1} \geq 0$ iff $\pi > \pi^\star$.

(iii) If $A.y \geq |\alpha| y$ and $y \geq 0$ then $|\alpha| \leq \pi^\star$.

This collection of results constitutes the **Perron-Frobenius Theorem** on Non-negative matrices; $\pi^\star$ is often called the **dominant** root. If we impose the property of indecomposability on the matrix A then some sharper results are possible. We turn to these matters next.

Let $\mathcal{I} = \{1, 2, \cdots, n\}$, the index set. A square matrix of order n, $A = (a_{ij})$, is said to be **decomposable**, if there is a **non-empty, proper** subset $J \subseteq \mathcal{I}$ such that $a_{ij} = 0, i \notin J, j \in J$. If no such subset exists then the matrix is said to be **indecomposable**. With this definition, we can present the following:

Claim 7.6.11 If $A \geq 0$ is also indecomposable then (using the notation introduced in the earlier claims):

(i) $\pi^\star > 0$

(ii) $x^\star > 0$ (and $p^\star > 0$).

(iii) any other cv corresponding to $\pi^\star$ is a scalar multiple of $x^\star$ (or $p^\star$).

(iv) No other cr of A has an associated non-negative c.v.

(v) If C is such that $A \geq C$ and $C \neq A$ then $\pi_A^\star > \pi_C^\star$.

(vi) If C is a principal minor of A and $C \neq A$ then $\pi_A^\star > \pi_C^\star$.

(vii) $(\pi I - A).x \geq 0$ for $x \geq 0 \Rightarrow (\pi - A)^{-1} > 0$.

(viii) $\pi^\star$ is a simple root of the characteristic equation of A: $\det(\pi I - A) = 0$.

Proof In case $\pi^\star = 0$, we have $A.x^\star = 0$; Let $J = \{j : x_j^\star > 0\}$; then $J \neq \emptyset$; note then that $j \in J \Rightarrow a_{ij} = 0 \forall i$ which contradicts indecomposability. Hence the claim. The positivity of $x^\star$ follows from a similar argument.

Now suppose that there is another cr α with an associated non-negative cv y; in fact $A.y = \alpha, y \geq 0$. Let $J = \{j : y_j > 0\}$. If possible let $J \neq \mathcal{I}$; then we have $\sum_j a_{ij} y_j = 0 \forall i \notin J$; hence $a_{ij} = 0 i \notin J, j \in J$: a contradiction. Hence $J = \mathcal{I}$ and

$y > 0$. Thus for any $\epsilon > 0$, $((\alpha + \epsilon)I - A).y = \epsilon.y > 0$ so that $\alpha + \epsilon \in F(A)$ for any $\epsilon > 0$ so that $\alpha = \pi^\star$.

The other claims are refinements of the previous arguments and are not repeated. We consider only the last claim. Note that α is a c.r. of $B = (\pi^\star I - A)$ iff only if $\pi^\star - \alpha$ is a c.r. of A. We shall show that 0 is a simple root of the characteristic equation of B. The characteristic equation of $B = (b_{ij})$ is $\pi^n - \pi^{n-1}.(\sum_j b_{jj}) + \cdots + (-1)^{n-1}(\text{ sum of all principal minors of B of order } n - 1) + (-1)^n \det B = 0$; note that by virtue of the previous claim, $L_{jj}(\pi^\star) > 0$ consequently, the last but one term in the characteristic equation cannot vanish and hence 0 cannot be a multiple root. •

7.6.2 Stable matrices

A square matrix A is said to be **stable**, if all its characteristic roots have real parts negative; thus, for example, a negative definite matrix is a stable matrix.

The conditions under which the stability of a matrix may be ensured will form the focus of attention in this section. The basic propositions about such matrices are contained in the following, which we shall state without proof:

Proposition 7.1   (Liapunov's Theorem): A necessary and sufficient condition for a square matrix A to be stable is that there exist a positive definite matrix $\mathcal{B}$ such that $\mathcal{B}.A + A^T.\mathcal{B}$ is negative definite.

Proposition 7.2   (Routh-Hurwitz Theorem): A necessary and sufficient condition for the square matrix A to be stable is that $s_i > 0$ for all i where

$$s_i = (-1)^i (\textit{sum of the determinant of all principal minors of order } i)$$

and in addition:

$$\det \begin{pmatrix} s_1 & 1 \\ s_3 & s_2 \end{pmatrix} > 0, \quad \det \begin{pmatrix} s_1 & 1 & 0 \\ s_3 & s_2 & s_1 \\ s_5 & s_4 & s_3 \end{pmatrix} > 0 \cdots$$

As would be evident, the above two results provide different characterizations of the same property. Here, we provide a sufficient condition, which may be useful in some contexts.

Claim 7.6.12   If A has a +ve dd then $-A$ is a stable matrix.

Proof Since A has a +ve dd, it follows that there is a non-negative d such that $B_A.d > 0$. Also if α is a cr for A then $-\alpha$ is a cr for $-A$; so if we can show that all crs for A have a positive real part then the claim will be established. Consider the fact that in case $z = a + \imath b$ is a cr and $a \leq 0$ then $|z - a_{ii}| \geq |a_{ii}|$ for all i; now consider the matrix $C = (z.I - A)$ and consider B_C then $B_C.d \geq B_A.d > 0$. Thus, C has a +ve dd and must be non-singular. Hence z cannot be a cr unless the real part is positive. •

We shall show below the importance of these results to economic theory.

EXERCISES

1. Let $a_1, a_2, \ldots\ldots, a_m$ be m- vectors. Prove that the equations $a_i y = \beta_i$ $i = 1, \ldots, m$ have a unique solution y iff the equations $a_i y = 0$ $i = 1, \ldots, m$ have no non-zero solution.

2. Consider the following collection of 3-vectors: $a_1 = (0, 1, -2)$, $a_2 = (1, 1, 1)$, $a_3 = (1, 2, 3)$, $a_4 = (2, 0, 3)$. If they are linearly dependent, show their linear dependence by expressing one of them as a linear combination of the others.

3. A function f from $\Re^m$ into $\Re$ is called a *linear function* if
$f(x + y) = f(x) + f(y)$ for all x, $y \epsilon \Re^m$ ......(1);
$f(\lambda x) = \lambda f(x)$ for all $x \epsilon \Re^m$, $\lambda \epsilon \Re$.....(2).
Show that if f is a linear function, there exists a vector a in $\Re^m$ such that $f(x) = xa$ for all x in F^m.

4. Give an example of a system of two linear equations in two unknowns which has no solution.

5. Let W_1 *and* W_2 be sub-spaces of a vector space V such that $W_1 + W_2 = V$ and $W_1 \cap W_2 = \{0\}$. Prove that for each vector α in V there are unique vectors $\alpha_1 \in W_1$ and $\alpha_2 \in W_2$ such that $\alpha = \alpha_1 + \alpha_2$.

6. Are the vectors
$\alpha_1 = (1, 1, 2, 4)$, $\alpha_2 = (2, -1, -5, 2)$,
$\alpha_3 = (1, -1, -4, 0)$, $\alpha_4 = (2, 1, 1, 6)$
linearly independent in R^4?

7. Show that a necessary and sufficient condition that
$$A = \begin{bmatrix} a & b \\ c & d \end{bmatrix}$$
has two distinct, real characteristic roots u and v is that there exists some invertible matrix C, such that
$$CAC^{-1} = \begin{bmatrix} u & 0 \\ 0 & v \end{bmatrix}.$$

8. Determine the values for α (if any) such that vectors $\begin{bmatrix} 1 + \alpha \\ 1 - \alpha \end{bmatrix}$ and $\begin{bmatrix} 1 - \alpha \\ 3 \end{bmatrix}$ are orthogonal.

9. Describe the set of vectors $\begin{bmatrix} \alpha \\ \beta \end{bmatrix}$ which are orthogonal to $\begin{bmatrix} 3 \\ 4 \end{bmatrix}$ as well as to $\begin{bmatrix} 1 \\ 4 \end{bmatrix}$.

10. Let A be a square matrix of order n and let (A_{ij}) denote the square matrix of order n where A_{ij} denotes the determinant of the co-factor corresponding to the $i-j$-th element in A. Show that $\det(A_{ij}) = (\det A)^{n-1}$.

9. Let A be a square matrix of order n and rank $n - 1$; further let there be $p \neq 0$ such that $p'.A = A.p = 0$. Writing the determinant of the co-factor corresponding to the $i-j$-th element in A as A_{ij}, show that there exist constants λ_i such that $A_{ij} = \lambda_i.p_j = \lambda_j p_i$.

KEY TERMS

Characteristic Roots	Non-singular matrix
Characteristic Vectors	Perron-Frobenius Theorem
Cramer's Rule	Quadratic Forms
Determinants	Rank of a Sub Space
Dominant Diagonal Matrices	Routh-Hurwitz Theorem
Induction Hypothesis	Scalar Multiple
Liapunov's Theorem	Scalar Vector Sum
Linearly Dependent	Separating Hyperplane Theorem
Linearly Independent	Singular Matrix
Matrices	Stable Matrix
Non-negative Square Matrix	Vector Space

USEFUL WEB LINKS

http://www.cds.caltech.edu/~murray/courses/primer-f01/mls-lyap.pdf (accessed on 2 August 2010): Liapunov's Theorem—relevant to differential equations and stability

http://www.cems.uvm.edu/~lakobati/08_fall/EE_295/Routh_Hurwitz_Criterion. pdf (accessed on 2 August 2010): Routh-Hurwitz Theorem—a compact statement of the conditions

http://www.prenhall.com/divisions/esm/app/ph-linear/leon/html/perron.html (accessed on 2 August 2010);

http://www2.warwick.ac.uk/fac/sci/maths/people/staff/oleg_zaboronski/fm/pf_ theory.pdf (accessed on 2 August 2010): Perron-Frobenius Theorem

CHAPTER EIGHT

Functions of Several Variables

8.1 Differentiability

We shall consider functions of the type $z = f(x, y)$ or $z = f(x_1, x_2, ...x_n)$, that is, $f : D \subset \Re^n \to \Re$; first, we need to define when such functions will be called continuous. It is straightforward to extend the earlier definition to say that the function $f : D \subset \Re^n \to R$ is continuous at a point $a \in D$ if a is a limit point of the domain of definition D of the function and if whenever $x^n \in D, x^n \to a$ as $n \to \infty \Rightarrow f(x^n) \to f(a)$.

While attempting to extend the notion of a derivative, we run into some difficulties. We shall consider in this section, functions f whose domain of definition S is an open subset of R^n; then for every $x \in S$, there is a neighbourhood $N(x) \subset S$. Note also that any $y \in N(x)$ may be represented as $x + \lambda u$ where u is some unit vector, that is, $|u| = 1$. For any unit vector u, we may define the **directional derivative of f at x in the direction u** by

$$D_u f(x) = \lim_{\lambda \to 0} \frac{f(x + \lambda u) - f(x)}{\lambda}.$$

In the one variable case, note that this reduces to the usual definition of a derivative (taking $u = 1$). The more interesting situation is when we take u to be such that $u_j = 0 \forall j \neq i, u_i = 1$. In other words, we are considering only the change in the i-th variable. This is the partial derivative with respect to i. We define this to be the **partial derivative of f with respect to x_i**; or alternatively, the partial derivative at the point $x = a$ with respect to the variable i is defined by the following:

$$\frac{\partial f(a)}{\partial x_i} \equiv f_i(a) = \lim_{h \to 0} \frac{f(a_1, a_2, ..., a_i + h, a_{i+1}, ..., a_n) - f(a)}{h}$$

Thus, while evaluating partial derivatives with respect to a particular variable, all other variables are being treated as fixed or constants and this allows us to use the definition of a derivative which is used for a single variable as the definition of the partial derivative. We shall also use the notation

$$\nabla f(x) = (f_1(x), f_2(x), ..., f_n(x))$$

as the vector of partial derivatives of the function $f(x)$, provided, of course, that all the partial derivatives exist.

The first important point of departure arises in this connection as the function may not be continuous at $x = a$, even if $\nabla f(x)$ is well defined at $x = a$. For example, consider the function:

$$f(x, y) = \begin{cases} x + y & \text{if } x = 0 \text{ or } y = 0 \\ 1 & \text{otherwise} \end{cases} \tag{8.1}$$

First, check that the function is not continuous at $(0, 0)$; but check that $f_1(0, 0)$, $f_2(0, 0)$ are both well defined. Thus, the existence of partial derivatives at a point does not imply that the function is continuous at that point. (Recall that in the case of functions of a single variable, existence of a derivative does imply continuity.) We need to strengthen the existence of partial derivatives by adding some more conditions to deduce continuity in the case of several variables. This relates to the existence of a **differential**:

Let $f : S \to R$ where S is an open subset of R^n and let $x \in S$. We shall say that f **has a differential at** x if there is another function $g(x; t)$ defined for $x \in S$ and for every $t \in R^n$ satisfying the following:

1. $g(x; \alpha t + \alpha' t') = \alpha g(x; t) + \alpha' g(x; t')$ for any scalars α, α';
2. For every $\epsilon > 0$ there is a neighbourhood $N(x)$ such that $y \in N'(x)$ implies
   $|f(x) - f(y) - g(x; y - x)| < \epsilon|y - x|$.

The function $g(x; t)$ satisfying the above properties is called the differential of the function $f(.)$ at x.

Claim 8.1.1 In case the differential exists, it is clear that the differential must be of the form:
$g(x; t) = \sum_{i=1}^{n} \alpha_i(x) t_i$
where $\alpha_i(x)$ depend on x but not on t.

Claim 8.1.2 In addition, if the differential exists, one may show that it is unique and is given by $g(x; t) = \nabla f(x).t = \sum_{i=1}^{n} f_i(x) t_i$. In addition, the directional derivative $D_u f(x) = \nabla f(x).u$ whenever the differential at x exists.

It is the existence of the differential at a point which implies the continuity of the function at that point.

Claim 8.1.3 A sufficient condition for the existence of the differential at a point $x \in S$ is that all partial derivatives exist and are continuous at that point.

The knowledge of the sign of partial derivatives may be used to sketch diagrams of functions of several variables. Consider, for example, $y = f(x_1, x_2)$; to sketch a diagram we need three axes. However, we can consider the family of level curves of the function f, that is, $\{(x_1, x_2) : f(x_1, x_2) = c\}$ where c is some constant. Now we can easily plot this family of curves in the $x_1 - x_2$ plane by considering different values of the function c. To plot these curves accurately, we need to have a clear idea about the slopes of these curves. This information is contained in the following:

$$\frac{dx_1}{dx_2}|_{y=c} = -\frac{f_2}{f_1} \text{ provided } f_1 \neq 0.$$

Partial derivatives of a higher order are computed exactly as before. Consider $f(x, y)$; we have $\nabla f(x) = (f_1(x), f_2(x))$; we then define $f_{ij}(x) = \frac{\partial f_i(x)}{\partial x_j}$. Most of the time, we shall have $f_{ij}(x) = f_{ji}(x)$, that is, the order of taking derivatives will not matter. But then this is NOT the general situation; for instance, consider:

$$f(x, y) = \begin{cases} \dfrac{xy(x^2 - y^2)}{x^2 + y^2} & \text{if } (x, y) \neq (0, 0) \\ 0 & \text{otherwise} \end{cases} \tag{8.2}$$

It should be useful to know what is required to remove this problem:

Claim 8.1.4 *In case $f_1(x)$, $f_2(x)$, $f_{21}(x)$ are continuous in a neighbourhood of the point x^o then $f_{12}(x^o)$ exists and $f_{12}(x^o) = f_{21}(x^o)$.*

(In the statement, f_{12}, f_{21} may be interchanged.) Thus given $f(x_1, x_2)$, there are four second order partial derivatives of the function: $f_{11}(x)$, $f_{22}(x)$, $f_{12}(x)$, $f_{21}(x)$. Under suitable restrictions only three of these are distinct. We usually write them in the form:

$$\nabla^2 f(x) = \begin{pmatrix} f_{11}(x) & f_{12}(x) \\ f_{21}(x) & f_{22}(x) \end{pmatrix}$$

This matrix is symmetric when the function is restricted and is known as the Hessian of the function f. Note that in case the function f is defined over some subset of R^n, $\nabla^2 f(x)$ will be an $n \times n$ symmetric matrix whose i–j-th element will be $f_{ij}(x)$.

In some situations, we may have $z = f(x_1, x_2)$ and in turn, $x_i = g_i(y_1, y_2)$ so that actually z is a function of the variables y_1, y_2. To compute partial derivatives we note the following rules to be followed in such situations, we follow the following chain rule:

$$\frac{\partial f(.)}{\partial y_i} = \sum_{k=1}^{2} \frac{\partial f(.)}{\partial x_k} \frac{\partial g_k(.)}{\partial y_i}$$

In case we have the functions $x_i = g_i(y)$ where $y \in R$; indirectly, the function f is a function of a single variable and one may be interested in computing the derivative of f with respect to the variable y, in case it exists. This is given by the following:

$$\frac{df(.)}{dy} = \sum_{k=1}^{2} \frac{\partial f(.)}{\partial x_k} \frac{dg_k(.)}{dy}$$

Given two points $a, b \in R^n$, $L(a, b) = \{c_\lambda = \lambda a + (1 - \lambda)b : 0 \leq \lambda \leq 1\}$ denotes the line segment joining the two points. Note that for any real valued function, $f(x)$ along the line segment $L(a, b)$ is given by $f(c_\lambda) = h(\lambda)$, say, which is just a function of the variable λ alone; consequently, $h'(\lambda) = \nabla f(c_\lambda).(b - a)$.

We may use this to express the Mean Value Theorem $(h(1) = h(0) + h'(\lambda),$ $0 < \lambda < 1)$ thus:

$$f(b) = f(a) + \nabla f(c_\lambda).(b - a)$$

for some $c_\lambda \in L(a, b), 0 < \lambda < 1$; or we may expand it thus:

$$f(b) = f(a) + \nabla f(a).(b - a) + \frac{1}{2}(b - a)^t.\nabla^2 f(c_\lambda).(b - a)$$

for some $c_\lambda \in L(a, b)$. Thus, as before, approximate values for functions may be written as

$$f(b) \simeq f(a) + \nabla f(a).(b - a)$$

or

$$f(b) \simeq f(a) + \nabla f(a).(b - a) + \frac{1}{2}(b - a)^t.\nabla^2 f(a).(b - a)$$

provided $|b - a|$ is small enough.

Finally consider $f : S \subset R^n \to R^n, n > 1$. We then usually consider it to be of the form $f = (f_1, ... f_n)$ where each $f_i : S \to R$. (Here each f_i is a function of the type we have discussed so far and not to be confused with the partial derivative; the notation will have to be understood from the context.) Associated with such maps or functions from R^n to R^n, we have the notion of the Jacobian $J(f) = (\frac{\partial f_i(x)}{\partial x_j})$ which is an $n \times n$ matrix, each row of which is the $\nabla f_i(x)$. Sometimes the word Jacobian is used only in connection with the determinant of the $n \times n$ matrix described above. Continuity of such functions will be ensured by the continuity of each of the functions f_i and the continuity of all the partial derivatives will ensure the existence of a differential for the function $f : R^n \to R^n$. The function f will then be said to be **continuously differentiable**; this is sometimes symbolically represented as $f \in C^1$. We note some consequences of the Jacobian having a non-zero determinant at some point $x^o \in S$. First, we have the following result on the existence of a local inverse which is continuously differentiable.

Claim 8.1.5 If $f = (f_1, ..., f_n) \in C^1$ on an open set $S \subset R^n$ and suppose that for some $x^o \in S$, $\det J(f(x^o)) \neq 0$; let $T = f(S)$. Then there is a uniquely determined function g and two open sets $X \subset S, Y \subset T$ such that:

$x^o \in S$ and $f(x^o) \in Y$,
$Y = f(X)$,
f is one to one on X,
g is defined on Y, $g(Y) = X$ and $g[f(x)] = x \forall x \in X$, and
$g \in C^1$ on Y.

Consider next $f : S \subset R^{n+k} \to R^n$; we shall write a typical element of S as $(x; t)$ where $x \in R^n$ and $t \in R^k$; thus $f(x; t) = y \in R^n$. Also assume that $(x^o; t^o) \in S$ is such that $f(x^o; t^o) = 0$ and that $\det(f_{ij}(x^o; t^o)) \neq 0, (i, j = 1, 2, ..., n)$; then we have the Implicit Function Theorem:

Claim 8.1.6 There exists a k-dimensional neighbourhood T^o of t^o and a unique function g defined on T^o having values in R^n such that

$$g \in C^1 \text{ on } T^o,$$
$$g(t^o) = x^o,$$
$$f(g(t); t) = 0 \text{ for every } t \in T^o.$$

8.2 Some Special Functions

A real valued function $f: S \subset R^n \to R$ where S is such that $x \in S \Rightarrow \lambda x \in S \forall \lambda \in R$, is said to be **homogenous of degree** r if $f(\lambda x) = \lambda^r f(x)$ for any $\lambda \in R, x \in S$. One may note that if partial derivatives exist, then given that $f(x)$ is homogenous of degree r, $f_i(x)$, for each i is homogenous of degree $r - 1$ and further, $\nabla f(x).x = r.f(x)$. A homogeneous function of degree one is often referred to as linear function. A set $S \subset R^n$ is said to be **convex** if for any points $a, b \in S$ we have $L(a, b) \subset S$; consider a function $f: S \to R$ where $S \subset R^n$ is an open convex subset. If for any $a, b \in S$, $f(\lambda a + (1 - \lambda)b) \geq \lambda f(a) + (1 - \lambda)f(b) \forall \lambda, 0 \leq \lambda \leq 1$ then the function is said to be **concave** on S; $f(x)$ is said to be convex on S if $-f(x)$ is concave on S. If the inequality is strict for all $\lambda, 0 < \lambda < 1$, we shall say that $f(x)$ is strictly concave.

For a function of a single variable, a concave differentiable function $f(x)$ amounts to the restriction that $f''(x) \leq 0$. Sometimes it is wrongly claimed that strict concavity implies that this inequality is strict; that this is not so may be seen from the strictly concave function $f(x) = -x^4$ whose second derivative is zero at the origin; however, it is true that should the second derivative be strictly negative, then the function is strictly concave. For the case of a function of several variables, the corresponding restriction for concavity is that the Hessian $\nabla^2 f(x)$ is negative semi-definite. (A square symmetric matrix A is said to be negative (positive) semi-definite, if for all x, $x^t A x \leq$ (respectively $\geq$) 0; the matrix is said to be positive (negative) definite if $x^t.Ax > 0$ (respectively < 0) $\forall x \neq 0$.) Similarly, for convexity, the restriction is that $\nabla^2 f(x)$ be positive semi-definite.

Another property for concave and convex functions is the fact that at any point the tangent (line or plane, as the case may be) to the function keeps the function on one side. Drawing a diagram of a concave (convex) function of a single variable should convince one of this geometric fact. For differentiable concave functions of several variables, it follows that for any two points x, y in the domain of the definition of the function, we must have

$$f(x) - f(y) \leq \nabla f(y).(x - y) \tag{8.3}$$

For convex functions, the inequality is reversed. This relationship follows immediately on account of the Mean Value Theorem and due to the property of the Hessian noted above.

These restrictions will have an important role to play when we consider questions of maxima and minima.

Another type of function which is often used in economics is a **quasi-concave** function. Consider a function $f : S \to R$ where $S \subset R^n$ is an open convex subset. If for any $a, b \in S$, $f(a) \geq f(b) \Rightarrow f(\lambda a + (1 - \lambda)b) \geq f(b) \forall \lambda, 0 \leq \lambda \leq 1$ then the function is said to be **quasi-concave** on S; this definition may be used to yield the following implication of quasi-concavity. Let f, S be as above and in addition, let f be differentiable on S then:

$$f(a) \geq f(b) \Rightarrow \nabla f(b).(a - b) \geq 0 \tag{8.4}$$

Note that all concave functions are quasi-concave (however, the converse is not true; for instance $y = x^2$ is quasi-concave but NOT concave). Equivalently for quasi-concave functions, the set $\{x : f(x) \geq \alpha\}$ for any number α is a convex set.

8.3 Maps and Fixed Points

Consider a function $f : S \to S$, where $S \subset \Re^n$; a point $x^* \in S$ is said to be a **fixed point** of the map f if $f(x^*) = x^*$. We shall see that identifying conditions under which fixed points are ensured will be useful in many contexts. A **Fixed Point Theorem** does precisely this.

Brouwer's Fixed Point Theorem states that *a continuous function $f : S \to S \subset \Re^n$, admits a fixed point provided S is convex and compact.*

If $S \subset \Re$, that is, $n = 1$, an easy demonstration is available. Consider, for example $S = [0, 1]$; note that S is convex and compact; let $f : [0, 1] \to [0, 1]$ be any continuous function; then consider $f(0)$. We know $f(0) \in [0, 1]$, by definition; if $f(0) = 0$: then we have a fixed point and there is nothing left to show. So consider $f(0) > 0$; similarly $f(1) = 1$ then too we have a fixed point; so suppose that $f(1) \neq 1$; then since $f(1) \in [0, 1]$, we must have $f(1) < 1$. Next consider $g(x) = f(x) - x$; note that $g(x)$ is continuous on $[0, 1]$; moreover, $g(0) > 0$ and $g(1) < 0$; using the continuity of the function g, it follows that $g(\bar{x}) = 0$ for some $\bar{x} \in (0, 1)$ and we have our fixed point. Thus either there is a fixed point at the end points of the interval $[0, 1]$ or there must a fixed point inside the interval. For higher dimensions, the proof is beyond the scope of our considerations.

From the above argument, also note the importance of the domain of definition S being convex and compact; that is, if S had been the open set $(0, 1)$ then we would not be able to argue about the fixed points at the end points, since there are none; and if S had been made up of several sub-intervals, say $S = [0, 1/4] \cup [2/3, 1]$; S is not convex and the argument to establish a fixed point in S need not go through. The cross-over from negative to positive values for the function g may occur precisely in the middle which is outside the set S.

There are some instances where we may need to employ a fixed point argument where the function f may not be a function, that is, $f(x) \subset S$ or f maps S into subsets of the set S: $f: S \to 2^S$ (2^S denotes the family of all subsets of a set S); such a map is called a **correspondence**. In such a situation, we say that $X^* \in S$ is a fixed point of the correspondence f if $x^* \in f(x^*)$. To handle the existence of such a fixed point, we need to take the help of the following:

Kakutani Fixed Point Theorem *Let $f: S \to 2^S$, where S is a convex and compact subset of $\Re^n$; if F is closed at every point of S and if $f(x)$ is a non-empty, convex subset of S for every $x \in S$ then there is fixed point.*

We need to define the notion of the correspondence being closed at some point. f is closed at $x^o \in S$, if for any sequence $x^n \in S, n = 1, 2, \cdots, x^n \to x^o$, as $n \to \infty$, $y^n \in f(x^n)$, $y^n \to y^o \to y^o \in f(x^o)$; note that for functions, this is identical to continuity at a point; for correspondences, however, this definition prevents the correspondence from only 'collapsing' at a point. It must also be pointed out that these definitions have been made with respect to $\Re^n$; for more general spaces, we need to make some adjustments in the definition.

8.4 Separation Theorems

In the earlier chapter on Real Linear Algebra, the principal mathematical tool was the theorem of the Separating Hyperplane: a result which was stated in terms of existence of non-negative solutions to equations. A reconsideration of the result in terms of a slightly different perspective will convince readers that the result had to do with two sets: the set of non-negative solutions to the simultaneous linear equations $Ax = b$ and the set of solutions to the inequalities $yA \geq 0, y.b < 0$ and the elementary fact that both cannot be non-empty simultaneously. The fundamental truth is that if one of these sets is empty, the other must be non-empty. That is, if the vector b does not lie in the set generated by the non-negative linear combinations of the columns of A (a convex set) then one could construct a (hyper -) plane (with equation $y.x = 0$ for some non-zero y such that the convex set is on one side and the vector b on the other. This is a special case of a more general set of results called Separation Theorems. The specialization use in the fact that the relevant disjoint sets, which are convex, are capable of being represented through linear inequalities. General convex sets with no points in common can also be separated. These results play a fundamental role in economic theory and we turn to some basic results concerning them.

First, recall from the chapter on linear algebra the following fact: A real valued function $f(x)$, $f : \Re^n \to \Re$, is linear iff there is some $a \in \Re^n$ such that $f(x) = a.x$. Consider a linear function $f(.)$ which is not identically zero: that is, $f(x) = a.x, a \neq 0$. We can now define, with such a linear function f, and for any real number α, a **half-space** $H(f, \alpha)_+$: $\{x \in \Re^n : f(x) \geq \alpha\}$; we can similarly define the half-space $H(f, \alpha)_-$ with the inequality sign reversed: $\{x \in \Re^n : f(x) \leq \alpha\}$; clearly the whole space is made up of these two spaces. We shall also describe $\{x \in \Re^n : f(x) = \alpha\}$ as the **plane**

$H(f, \alpha)$ separating these two half-spaces. Formally, we shall say that the plane $H(f, \alpha)$ separates two sets A, B if $\sup_{x \in A} f(x) \leq \alpha \leq \inf_{x \in B} f(x)$.

We then have:

Claim 8.4.1 If $C \subset \Re^n$, is a non-empty, closed and convex set not containing the origin, then there is a real valued linear function $f(x)$ and a positive number α such that $x \in C \Rightarrow f(x) > \alpha$.

Proof Define $B(0, \lambda) = \{x \in \Re^n : ||x|| \leq \lambda\}$ where $||x|| = \sqrt{(\sum_i x_i^2)}$ and λ is so chosen that $\hat{C} = B(0, \lambda) \cap C \neq \phi$; now $\hat{C}$ is compact and consequently, $||x||$ attains a minimum over $\hat{C}$: say at $\bar{x} \in \hat{C}$. Also $\bar{x} \neq 0$ and $||x|| \geq ||\bar{x}|| \ \forall x \in \hat{C}$; consequently, $||x|| \geq ||\bar{x}|| \ \forall x \in C$. We claim that $f(x) = x.\bar{x}$ serves the purpose of the assertion since it may be shown that $x \in C \rightarrow x.\bar{x} \geq ||\bar{x}||^2 > 0$. Suppose there is $y \in C$ such that $y.\bar{x} = ||\bar{x}||^2 - \epsilon$ for some $\epsilon > 0$; then note that $z = ty + (1-t)\bar{x} \in C$, for t, $0 < t < 1$ and further that $||z||^2 = \sum_i (t^2 y_i^2 + (1-t)^2 \bar{x}_i^2 + 2t(1-t)y_i \bar{x}_i) = t^2 ||y||^2 + (1-t)^2 ||\bar{x}||^2 + 2t(1-t)y.\bar{x} = t^2 ||y||^2 + (1-t)^2 ||\bar{x}||^2 + 2t(1-t)(||\bar{x}||^2 - \epsilon) = ||\bar{x}||^2 + t[t(||y||^2 - ||\bar{x}||^2) - 2(1-t)\epsilon] < ||\bar{x}||^2$ for a small enough ϵ: this contradicts the fact that $z \in C$; hence no such y can exist and the claim is established for $\alpha = ||\bar{x}||^2 > 0$ and $f(x) = \bar{x}.x$. •

We show next what we can get if the set C is known to be only convex (but not necessarily closed). We then have the following:

Claim 8.4.2 If C is a non-empty convex set not containing the origin, then there is a linear function $f(x)$, not identically zero, such that $x \in C \Rightarrow f(x) \geq 0$.

Proof For any $x \in C$, consider $A_x = \{y \in \Re^n : y.x \geq 0, y/||y|| = 1\}$; for any finite number of point $x_i \in C, i = 1, 2, \cdots, k$, consider $C_{x_1, \ldots, x_k} = \{x = \sum_{i=1}^k x_i p_i,$ where $p_i \geq 0, \sum_{i=1}^k p_i = 1\}$; now $C_{x_1, \ldots, x_k}$ is a non-empty closed and convex set not containing the origin. Hence, there is, by the previous result a $y \neq 0$ and a $\alpha > 0$ such that $x \in C_{x_1, \ldots, x_k} \Rightarrow y.x > \alpha$; hence $y \in \cap_{i=1}^k A_{x_i}$; thus for any finite number of points $x_i \in C$, $\cap A_{x_i} \neq \phi$; since $\{A_x | x \in C\}$ is a family of compact subsets, any finite number of which have a non-empty intersection, it follows that $\cap_{x \in C} A_x \neq \phi$; let $y \in \cap_{x \in C} A_x$; then the linear function $y.x$ has the desired property. •

The above results may now be used to provide the separation between convex sets.

Proposition 8.1 If $C, C' \subset \Re^n$ are non-empty disjoint convex sets then there is a plane $H(f, \alpha)$ which separates them.

Proof Consider $D = C - C' = \{z : z = x - x', x \in C, x' \in C'\}$; then D is a convex set not containing the origin. Hence by the previous result there is a linear function f, not identically zero such that $z \in D \Rightarrow f(z) \geq 0$; note that $f(z) = f(x - x') = f(x) - f(x') \geq 0$ (recall the fact that the function f is linear); hence it follows that:

$$\inf_{x \in C} f(x) \geq \sup_{x' \in C'} f(x')$$

Thus, for a suitable choice of α say $\inf_{x \in C} f(x)$, $H(f, \alpha)$ separates the two sets. •

Finally:

Proposition 8.2   If $C, C' \subset \mathfrak{R}^n$ are non-empty disjoint convex sets and C is compact and C' is closed, then there is a plane $H(f, \alpha)$ which separates them strictly, that is, $\inf_{x \in C} f(x) > \alpha > \sup_{x' \in C'} f(x')$.

Proof Notice that the set D constructed as in the proof of the previous proposition is convex and closed and does not contain the origin. Hence there is a linear function f and a number $\gamma > 0$ such that $z \in D \Rightarrow f(z) > \gamma > 0$, that is, $f(x) - f(x') > \gamma$; hence it follows that:

$$\inf_{x \in C} f(x) \geq \sup_{x' \in C'} f(x') + \gamma > \sup_{x' \in C'} f(x')$$

Therefore, there is a number α such that:

$$\inf_{x \in C} f(x) > \alpha > \sup_{x' \in C'} f(x')$$

as claimed. ●

KEY TERMS

Brouwer's Fixed Point Theorem

Concave Function

Continuously Differentiable Function

Convex Function

Differentiability

Fixed Point Theorem

Kakutani Fixed Point Theorem

Partial Derivative

Quasi-concave Function

Separation Theorems

USEFUL WEB LINKS

http://homepages.nyu.edu/~caw1/UMath/Handouts/ums06h23convexsetsand-functions.pdf (accessed on 2 August 2010): collection of properties and results concerning quasi-concave, quasi-convex, concave and convex functions

http://www.hss.caltech.edu/~kcb/Notes/SeparatingHyperplane.pdf (accessed on 2 August 2010): Separating Hyperplane Theorem

http://www.sjsu.edu/faculty/watkins/kakutani.htm (accessed on 2 August 2010): Kakutani Fixed Point Theorem with illustrations

http://www.math.hmc.edu/funfacts (accessed on 2 August 2010): Brouwer's Fixed Point Theorem

CHAPTER NINE

Static Optimization

We shall consider problems of maximization or minimization of functions $f : S \to R$ where S is a subset of R^n. In all cases, the function will be assumed to have continuous second order partial derivatives. We shall provide definitions and conditions for the existence of a maximum.

9.1 Unconstrained Optimization

$x^o \in S$ provides a **local maximum** for the function $f(x)$ if there is a neighbourhood $N(x^o)$ such that $x \in N(x^o) \cap S \Rightarrow f(x) \leq f(x^o)$; if $f(x) \leq f(x^o) \forall x \in S$ then x^o provides a **global maximum**. Local and global minimum are defined similarly by noting that a local (global) minimum for the function $f(x)$ is actually a local (global) maximum for the function $-f(x)$. If x^o is an interior point of S, then a local maximum at x^o is termed to be a local interior maximum and we have:

Claim 9.1.1 A local interior maximum (or minimum) at x^o implies that $\nabla f(x^o) = 0$ and $\nabla^2 f(x^o)$ is negative (positive) semi-definite.

In case we have a local interior maximum at x^o, there is a neighbourhood $N(x^o) \subset S$, $x \in N(x^o) \Rightarrow f(x) \leq f(x^o)$. Note that in case $\nabla f(x^o) \neq 0$ there would be some $t \neq 0, t \in R^n$ such that $\nabla f(x^o).t > 0$ and hence, choosing some scalar $\alpha > 0$ sufficiently small, $x^o + \alpha t \in N(x^o)$, $f(x^o + \alpha t) \simeq f(x^o) + \alpha \nabla f(x^o).t > f(x^o)$: a contradiction. The claim relating to $\nabla^2 f(x^o)$ follows from the approximation introduced earlier: suppose this is not the case; that is, for some $y \neq 0$, $y^t.\nabla^2 f(x).y > 0$; next note that since we have shown that $\nabla f(x) = 0$ it follows that for all $x \in N(x^o)$ $f(x) \simeq \frac{1}{2}(x - x^o)^t.\nabla^2 f(x^o).(x - x^o)$; now choose α positive and small enough so that $x^o + \alpha y \in N(x^o)$ and establish a contradiction as before.

The above result contains the **necessary** conditions for a local interior maximum. Next, we report some conditions which are **sufficient** to ensure that x^o is a local maximum (minimum):

Claim 9.1.2 If x^o is an interior point of S and if $\nabla f(x^o) = 0$ and $\nabla^2 f(x^o)$ is negative (positive) definite, then we may conclude that x^o is a local interior maximum (minimum) for the function $f(x)$.

9.2 Constrained Optimization

The general class of problems we shall consider is given by:

$$\max f(x) \text{ subject to } x \in X$$

where we shall take $f : S \to R$, S an open subset of R^n and $f \in C^2$: that is, continuous partial derivatives of the second order exist. We shall call X the constraint set and usually this set is represented by either

$$\text{equality constraints } g_j(x) = 0, j = 1, .., m$$

or by

$$\text{inequality constraints } g_j(x) \geq 0 , \; j = 1, .., m, \text{ and } x \geq 0$$

We shall consider them separately. In the representation of the constraint set, once again we shall take each function $g_j(x) \in C^1$.

We shall say that $x^o \in X$ is a **local maximum (LM)** if there is a neighbourhood $N(x^o)$ such that $f(x) \leq f(x^o) \forall x \in N(x^o) \cap X$. We shall say that $x^o \in X$ **solves the problem (M)** if $\forall x \in X, f(x) \leq f(x^o)$. Although, we are interested in obtaining a solution to the specified problem it may not be possible to obtain such a solution and we need to be satisfied with a LM. We shall see that the following function, the Lagrangean, will be crucial for solving the above problems:

$$\phi(x, \lambda) = f(x) + \sum_{i=1}^{m} \lambda_j g_j(x)$$

Note that $\phi : R^{n+m} \to R$.

9.3 Equality Constraints

The constraint set is now specified by $X = \{x \in R^n : g_j(x) = 0, j = 1, 2, .., m\}$. The functions $f, g_j(.)$ will be assumed to be of class C^2, that is, with continuous second order partial derivatives.

First, we have the following **necessary condition** for a LM at x^o:

Claim 9.3.1 If x^o provides a local maximum for $f(x)$ in X and if the rank of $G^o = (\frac{\partial g_k(x^o)}{\partial x_j})$, which is an $m \times n$ matrix is m then there is $\lambda^o \in R^m$ such that

$$\nabla f(x^o) + \sum_{i=1}^{m} \lambda_i^o \nabla g_i(x^o) = 0. \tag{9.5}$$

It should be clear that this places a restriction on m, that is, $m \leq n$. Also note that the condition (9.5) may be represented as:

$$\phi_x(x^o, \lambda^o) = 0$$

where the subscript refers to the variables with respect to which partial derivatives are being considered and $\phi()$ is the Lagrangean function introduced above. To ensure that we have a local maximum, which is a **sufficient condition**, we have:

Claim 9.3.2 If (x^o, λ^o) satisfy (9.5), $x^o \in X$ and if

$$x^t[\nabla^2 f(x^o) + \sum_{j=1}^{m} \lambda_j^o \nabla^2 g_j(x^o)]x < 0 \ \forall x \neq 0, \ \nabla g_j(x^o).x = 0 \ \forall j \qquad (9.6)$$

then x^o provides a local maximum for the function $f(x)$.

We note that the matrices $A = (\nabla^2 f(x^o)) = (a_{ij})$ (say) is an $n \times n$ matrix whereas $B = (\nabla g_j(x^o)) = (b_{ij})$ is an $m \times n$ matrix. We write $A_r = (a_{ij})$ $i, j = 1, 2, ..., r$; $r = 1, 2, ..., n$ and $B_{mr} = (b_{ij})$, $i = 1, ..., m$, $j = 1, 2, ..r$, $r = 1, 2, ..n$. Using these notations, we write

$$C_r = \begin{pmatrix} 0 & B_{mr} \\ B_{mr}^t & A_r \end{pmatrix} \ r = m+1, .., n$$

In the above, note that 0 is a block $m \times m$; the total dimension of the matrix C_r is $(m + r) \times (m + r)$. A more common representation of the condition (9.6), using the above notation is contained in the following:

$$(-1)^r \det C_r > 0 \ \forall r = m+1, ..., n$$

The matrix C_r is known as the **bordered Hessian**. The above version using the bordered Hessian, is often referred to as the **second order condition** for a local maximum. Note that we are still far from obtaining a maximum for the problem in the whole constraint set X. Towards this end, we need some further restrictions on the nature of the functions. We have:

Claim 9.3.3 If for some $x^o \in X$, there is λ^o satisfying (9.5) and if $\phi(x, \lambda^o)$ is concave in x then x^o provides a maximum (M) for $f(x)$ in X.

Recall (8.3) for differentiable concave functions. Consequently $\phi(x, \lambda^o)$ is concave in x; this implies that we must have:

$$\phi(x, \lambda^o) - \phi(x^o, \lambda^o) \leq \phi_x(x^o, \lambda^o).(x - x^o)$$

for any x; hence using (9.5) (that is, $\phi_x(x^o, \lambda^o) = 0$) we have for any x:

$$\phi(x, \lambda^o) \leq \phi(x^o, \lambda^o) = f(x^o) \text{ since } x^o \in X$$

that is, we have $f(x) + \sum_i \lambda_i^o g_i(x) \leq f(x^o)$; thus, we must have $f(x) \leq f(x^o) \forall x \in X$ since $\lambda_i^o \geq 0$, $g_i(x) = 0 \forall i$ when $x \in X$. This proves the claim.

We also have the following:

*Claim 9.3.4 If $f(x), g_j(x), \forall j$ are **quasi-concave functions** and if (x^o, λ^o) satisfy (9.5), $x^o \in X, \nabla f(x^o) \neq 0, \lambda^o \geq 0$ then x^o provides a maximum for $f(x)$ in X.*

Recall (8.4) for differentiable quasi-concave functions. Suppose then that (9.5) holds at $x^o \in X, \lambda^o \geq 0$. Consider any $x \in X$; then since $g_j(x) = g_j(x^o) = 0 \forall j$, using the quasi-concavity property referred to above, we have $\nabla g_j(x^o).(x - x^o) \geq 0 \forall j$ for all $x \in X$; further by virtue of (9.5), we have

$$\nabla f(x^o).(x - x^o) = -\sum_j \lambda_j^o g_j(x^o).(x - x^o) \leq 0$$

Next, if possible, let there be $\hat{x} \in X$ such that $f(\hat{x}) > f(x^o)$; by quasi-concavity of f, we have $\nabla f(x^o).(\hat{x} - x^o) \geq 0$; so that we may conclude that $\hat{x} \in X$ and $f(\hat{x}) > f(x^o)$ implies $\nabla f(x^o).(\hat{x} - x^o) = 0$. Consider such a $\hat{x}$; by continuity of the function f, there is a neighbourhood $N(\hat{x})$ such that $x \in N(\hat{x}) \Rightarrow f(x) > f(\hat{x})$. Now since $\nabla f(x^o) \neq 0$ there is $\tilde{x} \in N(\hat{x})$ such that $\nabla f(x^o).\tilde{x} < \nabla f(x^o).\hat{x}$; but then $\nabla f(x^o).(\tilde{x} - x^o) < 0$ so that we may conclude that $f(\tilde{x}) \leq f(\hat{x})$: a contradiction. Hence, no such $\hat{x}$ can exist and the claim is proved.

We may also replace the requirement that $g_j(.)$ be quasi-concave by the requirement that the set X be convex. In this situation, we have:

Claim 9.3.5 If $f(x)$ is a quasi-concave function, X is a convex set and if (x^o, λ^o) satisfy (9.5), $x^o \in X, \nabla f(x^o) \neq 0$ then x^o provides a maximum for $f(x)$ in X.

Thus the non-negativity of the multipliers is no longer required. Consider any $x \in X$ and $x^o \in X, \lambda^o$ satisfying (9.5) with $\nabla f(x^o) \neq 0$. Since X is convex, $(\lambda x + (1 - \lambda)x^o) = (x^o + \lambda(x - x^o)) \in X$ for any λ satisfying $0 \leq \lambda \leq 1$. Hence, it follows that for all j, $g_j(x^o + \lambda(x - x^o)) = 0$, $\lambda \in [0, 1]$. Thus, for each j, $\nabla g_j(x^o).(x - x^o) = 0, x \in X$. Consequently, (9.5) implies that $\nabla f(x^o).(x - x^o) = 0$ if $x \in X$ and $f(x) > f(x^o)$. The rest of the demonstration follows the proof of the previous claim.

9.4 Inequality Constraints

We consider next the situation when the constraint set X is specified by:

$$g_j(x) \geq 0 \ , \ j = 1, .., m, \text{ and } x \geq 0$$

As earlier we shall assume that all functions appearing have continuous second order partial derivatives.

We shall say that (x^o, λ^o) provides a **saddle point** provided $\phi(x, \lambda^o) \leq \phi(x^o, \lambda^o) \leq \phi(x^o, \lambda)$ for all $x \geq 0, \lambda \geq 0$. In the case when the constraints set is specified by inequality constraints, we shall see that the obtaining M or LM is intimately related to obtaining an SP. First, we have:

Claim 9.4.1 Kuhn-Tucker Conditions (KTC) If (x^o, λ^o) constitutes an SP then

$(i)\ \ \phi_x(x^o, \lambda^o) \leq 0,\ \ x^o.\phi_x(x^o, \lambda^o) = 0,\ \ x^o \geq 0$

$(ii)\ \phi_\lambda(x^o, \lambda^o) \geq 0,\ \ \lambda^o.\phi_\lambda(x^o, \lambda^o) = 0,\ \ \lambda^o \geq 0$

These conditions follow, once we note that in the definition of the SP, we seek to find (x^o, λ^o) such that $\phi(x, \lambda)$ is maximized with respect to x (at x^o) given $\lambda = \lambda^0$ and is minimized with respect λ at λ^0 given that $x = x^o$.

Conversely, we have:

Claim 9.4.2 If (x^o, λ^o) satisfy KTC and if $\phi(x, \lambda)$ is concave in x given λ, then (x^o, λ^o) constitutes an SP.

We shall write $g(x) = (g_j(x))$.

By virtue of concavity, we have $\phi(x, \lambda^o) \leq \phi(x^o, \lambda^o) + \phi_x(x^o, \lambda^o).(x - x^o) \leq \phi(x^o, \lambda^o)$, the last step following for all $x \geq 0$ on account of KTC. Also $\phi(x^o, \lambda^o) - \phi(x^o, \lambda) = (\lambda^o - \lambda).g(x^o) = -\lambda g(x^o) \leq 0 \forall \lambda \geq 0$, using KTC. Thus the string of inequalities valid at SP are established.

The connection with solving the maximum problem (M) is obtained through the following:

If (x^o, λ^o) solves SP then x^o provides a maximum for $f(x)$ in X.

This follows since by virtue of SP, $\phi(x, \lambda^o) \leq \phi(x^o, \lambda^o) \forall x \geq 0$; so that using KTC, we have $f(x) + \lambda^o g(x) \leq f(x^o) + \lambda^o.g(x^o) = f(x^o)$ and hence $f(x) \leq f(x^o) \forall x$ such that $g(x) \geq 0$ since $\lambda^0 \geq 0$

In addition, by virtue of the two previous results: if $\phi(x, \lambda)$ is concave in x given λ and if for some λ^o, (x^o, λ^o) satisfies KTC then x^o provides a maximum for the function $f(x)$ in X.

It should be pointed out that it is not necessary that at a maximum for the function $f(x)$ there should be some λ^o satisfying KTC. For example, consider the problem:

$$\max\ x \text{ subject to } -x^2 \geq 0, x \geq 0$$

By observation we know that $x^o = 0$ solves the problem; yet considering $\phi_x(x, \lambda) = 1 - 2\lambda x$, we conclude that there cannot be any configuration satisfying KTC if $x^o = 0$. To remove problems of this type, we need to impose **constraint qualification conditions**. There are several such conditions and all of them serve to make KTC necessary at a maximum. We mention some of these conditions:

1. $m = 1$ with $g(x^o) = 0$ (the constraint is binding) and $\nabla g(x^o) \neq 0$, then x^o solves the maximum problem (M) implies that there is $\lambda^o \geq 0$ such that (x^o, λ^o) satisfy KTC.

2. **Slater's Condition**: If $f(x), g_j(x)$ are all concave functions and there is $\bar{x} \geq 0$ such that $g_j(\bar{x}) > 0 \forall j$, then x^o solves the maximum problem (M) implies that there is $\lambda^o \geq 0$ such that (x^o, λ^o) satisfy KTC.

3. If $g_j(x) = a_j.x + b_j \forall j$ (linear affine constraints), then x^o solves the maximum problem (M) implies that there is $\lambda^o \geq 0$ such that (x^o, λ^o) satisfy KTC.

4. **Kuhn-Tucker Constraint Qualification (KTCQ)**: Let $X = \{x : g_j(x) \geq 0, \forall j, j = 1, 2, .., m, x \geq 0\}$, the constraint set and let $\bar{x} \in X$ such that $E = \{j : g_j(\bar{x}) = 0\} \neq \emptyset$ (that is, there are binding constraints at $\bar{x}$). Let $\hat{x} \geq 0$ such that $\nabla g_j(\bar{x}).(\hat{x} - \bar{x}) \geq 0 \ \forall j \in E$, then there is a differentiable function $h : [0, 1] \to R^n_+$, differentiable at 0 such that $h(0) = \bar{x}, h(t) \in X \forall t \in [0, 1]$, $h'(0) = \alpha(\hat{x} - \bar{x})$, $\alpha > 0$. If x^o solve the maximum problem (M) and if a function such as h exists at x^o (that is, x^o replacing $\bar{x}$ in the above), then there is $\lambda^o \geq 0$ such that (x^o, λ^o) satisfy KTC.

5. **Non-degenerate Constraint Qualification**: Let x^o solves the maximum problem (M); assume that out of the m constraints, only k constraints are binding at x^o, $g_j(x^o) = 0$ for $j = j_1, ..., j_k$ and k = rank of $(\frac{\partial g_j(x^o)}{\partial x_i})$, $j = j_1, ..., j_k; i = 1, 2, ... n$ then there is $\lambda^o \geq 0$ such that (x^o, λ^o) satisfy KTC.

As in the case for equality constraints, it is possible to extend coverage to situations where the functions are differentiable quasi-concave functions. We mention here the results we may obtain in this connection. First, we have:

Claim 9.4.3 If $\hat{x}$ is an absolute maximum $f(.)$ in X, if Slater's condition holds and $\nabla g_j(x) \neq 0$ for all j and $x \in X$ then there is $\hat{\lambda} \geq 0$ such that $(\hat{x}, \hat{\lambda})$ forms an SP for $\phi(x, \lambda) = f(x) + \sum_j \lambda_j g_j(x)$; in other words, KTC must hold.

Claim 9.4.4 If for some $\hat{x} \in X$ there is $\hat{\lambda}$ satisfying KTC, Slater's condition is met and $\nabla f(\hat{x}) \neq 0$ then $\hat{x}$ provides an absolute maximum for $f(x)$ in X.

The first result establishes the necessity of the KTC at a maximum while the second provides conditions under which KTC are also sufficient. It should be recalled that the above results are for **quasi-concave differentiable** functions. Proofs, being similar to the cases considered above, are omitted.

9.5 A Duality Theorem

In the literature on mathematical programming theory, duality theorems refer to results which specify a relationship between the solutions of two optimization exercises; a standard example of such a relationship between two linear programming exercises is known as the Fundamental Theorem of Linear Programming. We shall present in this section, a somewhat more general statement of this relationship. There are three crucial aspects of this relationship:

1. Among the two problems, one is a constrained maximization problem (M), while the other is a constrained minimization problem (N).

2. The existence of a solution to any one of the problems implies that the other has a solution too and the two extremum values are equal.
3. If the constraints in any one of the problems is inconsistent while the other is consistent then there is a sequence of feasible points in the consistent problem on which its objective function becomes infinitely large in absolute value.

The Fundamental Theorem of Linear Programming satisfies all three. We shall provide a duality theorem in the same spirit. Consider $f(x)$ to be a real valued differentiable **concave function** and let $g_j(x) = a_j.x - b_j$, $a_j, x \in \Re^n$, $b_j \in \Re$. Let A denote the $m \times n$ matrix whose j-th row is a_j and $b \in \Re^n$ denote n-vector whose j-th component is b_j; the problem (M) is then specified by:

$$\max f(x) \text{ subject to } A.x \geq b$$

Let $\phi(x, u) = f(x) + u'(A.x - b)$, where $u \in \Re^m$. Note then that $\phi_x(x, u) = \nabla f(x) + u'.A$. Consider the problem (N) defined by:

$$\min \phi(x, u) \text{ subject to } \phi_x(x, u) = 0 \text{ , } u \geqq 0$$

The duality theorem will consist of the relationship between the two problems (M) and (N). Let $\mathcal{X}, \mathcal{Y}$ be defined as follows:

$$\mathcal{X} = \{x \in \Re^n : A.x \geq b\} \text{ ; } \mathcal{Y} = \{(x, u) \in \Re^n \times \Re^m_+ : \phi_x(x, u) = 0\}$$

and further, let $v = \sup_{x \in \mathcal{X}} f(x)$, $V = \inf_{(x,u) \in \mathcal{Y}} \phi(x, u)$. We have then:

Claim 9.5.1 $\mathcal{X}, \mathcal{Y} \neq \emptyset \Rightarrow v \leq V$.

Consider $(\bar{x}, \bar{u}) \in \mathcal{Y}$; then $\phi_x(\bar{x}, \bar{u}) = 0$ which implies that $\phi(x, \bar{u})$ is maximized at $x = \bar{x}$; hence $\phi(x, \bar{u}) \leq \phi(\bar{x}, \bar{u}) \forall x$, that is, we have

$$f(x) + \bar{u}'(A.x - b) \leq f(\bar{x}) + \bar{u}'(A.\bar{x} - b) \forall x \tag{9.7}$$

Consider next $\hat{x} \in \mathcal{X}$; then, $A.\hat{x} \geq b$. Hence using (9.7), we obtain the following (since $\bar{u} \geqq 0$)

$$f(\hat{x}) \leq f(\hat{x}) + \bar{u}'(A.\hat{x} - b) \leq \phi(\bar{x}, \bar{u}) \tag{9.8}$$

Since (9.8) holds for any $\hat{x} \in \mathcal{X}$ and any $(\bar{x}, \bar{u}) \in \mathcal{Y}$ the claim follows.

Claim 9.5.2 $\mathcal{X} = \emptyset \Rightarrow$ *The problem (N) has no solution.*

In case $\mathcal{X} = \emptyset$, $A.x \geq b$ has no solution and hence it follows that $u'.A = 0$, $u'.b = 1$ have a solution say $\tilde{u} \geq 0$. Suppose next that the problem (N) has a solution $(\bar{x}, \bar{u})$. Then we have $-\nabla f(\bar{x}) = \bar{u}'.A$. Note then that $(\bar{x}, \bar{u} + \theta \tilde{u}) \in \mathcal{Y}$ for any $\theta > 0$ since $(\bar{u} + \theta \tilde{u}) \geq 0$ and $(\bar{u} + \theta.\tilde{u})'.A = \bar{u}'.A = -\nabla f(\bar{x})$. Further it follows that $\phi(\bar{x}, \bar{u} + \theta \tilde{u}) = f(\bar{x}) + \bar{u}'(A.\bar{x} - b) - \theta$ since $\tilde{u}'.A = 0$, $\tilde{u}'.b = 1$. Thus $\phi(\bar{x}, \bar{u} + \theta \tilde{u})$ can be made as small as possible by choosing θ as large as possible. The claim follows.

One may similarly show that:

Claim 9.5.3 $\mathcal{Y} = \emptyset \Rightarrow$ *The problem (M) has no solution.*

Suppose $\bar{x}$ solves (M). Now it follows that the system $u'.A = -\nabla f(\bar{x})$ can have no non-negative solution since $\mathcal{Y}$ is empty. Hence there exists z such that $A.z \geq 0$, $-\nabla f(\bar{x}).z < 0$. Also note that for any scalar $\theta > 0$, $A.(\bar{x}+\theta.z) \geq b$ so that $\bar{x}+\theta z \in \mathcal{X}$; however $f(\bar{x} + \theta z) \approx f(\bar{x}) + \theta.\nabla f(\bar{x}).z > f(\bar{x})$ and this completes the proof.

Next, we may claim:

Claim 9.5.4 *If* x^* *solves the problem (M) then there is some* $u^* \geq 0$ *such that* (x^*, u^*) *solves (N) and* $f(x^*) = \phi(x^*, u^*)$.

If x^* solves the problem (M), and the nature of the constraints of the problem (M) (linear affine), there must be u^* such that (x^*, u^*) forms an SP for the function $\phi(x, u) = f(x) + u'(A.x - b)$; that is, $\phi(x, u^*) \leq \phi(x^*, u^*) \leq \phi(x^*, u) \forall x$ and $\forall u \geq 0$. Since $\phi(x, u)$ is differentiable this means

$$\phi_x(x^*, u^*) = 0 \text{ and } \phi_u(x^*, u^*) = A.x^* - b \geq 0, u^{*'}.\phi_u(x^*, u^*) = 0, u^* \geq 0;$$

hence, (x^*, u^*) is feasible for (N) and $f(x^*) = \phi(x^*, u^*)$. By virtue of (9.8), for any $(\bar{x}, \bar{u})$ feasible for (N), we must have $\phi(\bar{x}, \bar{u}) \geq f(x^*) = \phi(x^*, u^*)$ and the claim follows.

Now if (x^*, u^*) solves the problem (N), it is not clear whether it is necessary that there be v^* such that (x^*, u^*, v^*) satisfies KTC for the function

$$\psi(x, u, v) = -\phi(x, u) + \phi_x(x, u).v = -(f(x) + u'[A.x - b]) + [\nabla f(x) + u'A].v \quad (9.9)$$

where $\nabla f(x) = (\frac{\partial f(x)}{\partial x_i})$; recall that the necessity of the KTC follows from the constraint qualification condition. For the problem (N) the nature of the constraints is unclear (unlike, the problem (M)) since we do not know what the properties of the function $\nabla f(x)$ are; we only know that $f(x)$ is concave; it may even be that the constraints of (N) are linear affine as well, but we cannot appeal to the necessity. Hence we state the following:

Claim 9.5.5 *If* (x^*, u^*) *solves (N) and if there is a* v^* *such that* (x^*, u^*, v^*) *forms an SP for the function* $\psi(x, u, v)$ *defined above, that is,* $\psi(x, u, v^*) \leq \psi(x^*, u^*, v^*) \leq \psi(x^*, u^*, v) \forall x$ *and* $\forall u \geq 0, \forall v$, *and if further the rank of* $(\frac{\partial^2 f(x^*)}{\partial x_i \partial x_j})$ *is full, then* x^* *solves (M) and* $f(x^*) = \phi(x^*, u^*)$.

One may note that the necessary conditions for the SP are given by the following:

$$\psi_x(x^*, u^*, v^*) = 0 \ ; \ \psi_u(x^*, u^*, v^*) \leq 0, u^{*'}.\psi_u(x^*, u^*, v^*) = 0, u^* \geq 0; \psi_v(x^*, u^*, v^*) = 0$$

that is, we have:

$$\psi_x(x^*, u^*, v^*) = -[\nabla f(x^*) + u^{*'}A] + \nabla^2 f(x^*).v^* = 0 \quad (9.10)$$

where $\nabla^2 f(x^*) = (\frac{\partial^2 f(x^*)}{\partial x_i \partial x_j})$. Further we have:

$$\psi_u(x^*, u^*, v^*) = -(A.x^* + b) + A.v^* \leqq 0, \; u^{*\prime}[-(A.x^* + b) + A.v^*] = 0 \; u^* \geq 0 \quad (9.11)$$

and finally,

$$\nabla f(x^*) + u^{*\prime}.A = 0 \qquad (9.12)$$

Using (9.12) in (9.10), we have $\nabla^2 f(x^*).v^* = 0$ and hence using the full rank property of $\nabla^2 f(x^*)$ we have, $v^* = 0$; thus from (9.11), we obtain $-[A.x^* + b] \leq 0$ so that x^* is feasible for the problem (M). Further since $f(x^*) = f(x^*) + u^{*\prime}[-(A.x^* + b) + A.v^*]$, we can conclude that x^* solves the problem (M).

In case the function $f(x) = c'.x$, a linear function, the duality theorem takes on a very simple form. Note that first, while all the other results hold, the proof of the last result becomes inapplicable since $(\frac{\partial^2 f(x^*)}{\partial x_i \partial x_j})$ is a null matrix. However, the special linear form allows a direct demonstration. We begin by noting that the maximum problem (M) may be written as:

$$\max c'.x \text{ subject to } -A.x \leq -b$$

Hence, now $\phi(x, u) = c'.x + u'.(A.x - b)$ and hence $\phi_x(x, u) = c' + u'.A$ so that the problem (N) reduces to

$$\min u'.(-b) \text{ subject to } u'(-A) = c' , u \geq 0$$

Suppose now that u^* solves (N); then there is x^* such that (u^*, x^*) constitute an SP for the function $\psi(u, x) = u'.b + [u'.(-A) - c'].x$: given the linearity of the constraints (and the objective function) it is necessary that there be such a x^*. Thus we must have

$$\psi(u, x^*) \leqq \psi(u^*, x^*) \leqq \psi(u^*, x) \forall x, \; \forall u \geq 0$$

The first inequality leads to $u'(b - A.x^*) \leqq u^{*\prime}(b - A.x^*) \forall u \geq 0$; this is possible only if $A.x^* \geq b$ and hence leads to the conclusion that x^* is feasible for the problem (M). Thus we must have $c'.x^* \leq u^{*\prime}.(-b)$.

Further since $u^{*\prime}(-A) = c'$ it follows that $\psi(u^*, x^{star}) = u^{*\prime}.b$ and hence $\psi(u, x^*) \leqq u^{*\prime}.b \; \forall u \geq 0 \Rightarrow -c'.x^* \leq u^{*\prime}.b$.

Combining the above we have $c'.x^* = u^*.(-b)$ and hence we may conclude that x^* solves the problem (M) and the values of the two problems are equal.

The collection of results in this section for the linear case constitutes the **Fundamental Duality Theorem of Linear Programming**. The 'standard' form of the maximum problem for the linear case is usually represented as:

$$\max c'.x \text{ subject to } A.x \leq b, x \leq 0$$

and the corresponding 'dual' minimum problem then is:

$$\min u'.b \text{ subject to } u'.A \geq c', u \geqq 0$$

To convert the standard from into the form we have been considering, we may write this problem as

$$\max c'.x \text{ subject to } \begin{pmatrix} A \\ I \end{pmatrix}.x \leq \begin{pmatrix} b \\ 0 \end{pmatrix}$$

where I is $n \times n$ identity matrix. Now applying our result, the corresponding minimum problem is:

$$\min u'.b + v'.0 \text{ subject to } (u, v)'. \begin{pmatrix} A \\ I \end{pmatrix} = c', (u, v) \geq 0$$

Note that one may remove the variable v so that this problem is equivalent to:

$$\min u'.b \text{ subject to } u'.A = c' - v' \geq c', u \geq 0.$$

EXERCISES

1. Consider the functions $f(x, y) = y^2 + x^2 y + x^4$ and $f(x, y) = y^2 - x^3$ and in each case, find the extremum points and classify them (whether maxima or minima etc.).

2. Find the shortest distance from $(0, b)$ to the parabola $x^2 - 4y = 0$.

3. Solve the following:

   (a) $\max x^{1/2} y^{1/4}$ subject to $3x + 4y \leq 600$, $x \geq 0, y \geq 0$.
   (b) $\max x^2 + y^2$ subject to $x + 2y \leq 100$, $x \geq 0, y \geq 0$.
   (c) $\min 2x + 3y$ subject to $x^{1/2}.y^{1/2} \geq 1$, $x \geq 0, y \geq 0$.

KEY TERMS

Constrained Optimization
Constraint Qualification Conditions
Equality Constraints
Fundamental Duality Theorem of
   Linear Programming
Inequality Constraints
Kuhn-Tucker Constraint Qualification

Kuhn-Tucker Conditions
Local Maximum
Necessary and Sufficient Condition
Saddle Point
Second Order Conditions
Slater's Condition
Unconstrained Optimization

USEFUL WEB LINKS

http://are.berkeley.edu/courses/ARE210/fall2005/lecture_notes/Kuhn-Tucker.pdf
   (accessed on 2 August 2010): Kuhn-Tucker Conditions—constrained optimization

http://www.economics.utoronto.ca/osborne/MathTutorial/KTC.HTM (accessed
   on 2 August 2010): optimization with inequality constraints: Kuhn-Tucker
   Conditions

http://web.mit.edu/15.053/www/AMP-Chapter-04.pdf (accessed on 2 August
   2010): Duality in Linear Programming

CHAPTER TEN

Economic Applications II

10.1 Static Optimization I: Consumer Behaviour

The modern theory of consumer behaviour revolves around the relationship between two optimization problems; the first is the more traditional **Utility Maximization Problem (UMP)**:

$$Max\ U(x)\ s.t.\ p^t.x \leq M$$

where the decision-maker chooses a consumption bundle x subject to the budget constraint given by the price vector p and the income per period M. In other words, the data for UMP are $U(.), p, M$; with these being in place, a consumption bundle x is chosen so as to solve UMP. Related to this is the more recent **Expenditure Minimizing Problem (EMP)**:

$$Min\ p^t.x\ s.t. U(x) \geq U^*$$

where once again the choice variable is another consumption bundle x; but this time the choice is made so as to minimize the level of expenditure and attain a given level of satisfaction U^*; thus the data for EMP are $U(.), p, U^*$. The only difference between the two lies in the specification of the money income M in UMP as opposed to the level of real income U^* in EMP. For the first problem, a family of indifference curves is specified together with a budget line; and we need to pick the 'best' point on the line; for the second, a single indifference curve has been specified and along with a family of iso-cost lines; we need to choose the lowest iso-cost line which attains the given indifference curve. We shall investigate the relationship between these two problems below.

The following assumptions are involved:

1. The consumption possibility set $\mathcal{C}$ is *non-empty, closed, convex, and bounded below subset of* $\mathfrak{R}^n$. (Sometimes $\mathcal{C} = \mathfrak{R}^n_+$: the non-negative orthant.) In addition, we have $\mathfrak{R}^n_{++} \subseteq \mathcal{C}$.
2. There is an *ordering $\mathcal{R}$ on $\mathcal{C}$; ($\mathcal{R}$ is reflexive, complete, and transitive on $\mathcal{C}$). Further, the ordering $\mathcal{R}$ is monotone (that is, $x, y \in \mathcal{C}, x \geq y \Rightarrow x\mathcal{R}y$) and*

continuous (that is, the subsets $\mathcal{R}_x = \{y \in \mathcal{C} : y\mathcal{R}x\}, \mathcal{R}^x = \{y \in \mathcal{C} : x\mathcal{R}y\}$ are closed subsets of $\mathcal{C}$ for every $x \in \mathcal{C}$).

It may be recalled that given the above, there exists a **continuous real-valued utility indicator function** $U : \mathcal{C} \to \mathfrak{R}$. For the demonstration, we shall consider $\mathcal{C} = \mathfrak{R}^n_+$. The proof proceeds by first identifying some $x^* \in \mathfrak{R}^n_{++}$ and setting $U(x^*) = 1$; next, consider any point on the ray $X^* = \{y \in \mathcal{C} : y = \lambda x^*, \lambda \in \mathfrak{R}\}$; for all such points define $U(y) = \lambda$ where λ is such that $y = \lambda x^*$; note that such a definition must necessarily agree with the ranking $\mathcal{R}$ on the ray, on account of monotonicity. Next consider some $x \in \mathcal{C}$ but $\notin X^*$. Note that there is $y^h \in X^*, y^h \geq x \Rightarrow y^h \in \mathcal{R}_x$; in addition, there is $y^\ell (= 0, \text{possibly}) \in X^*, x \geq y^\ell \Rightarrow y^\ell \in \mathcal{R}^x$, that is, $x\mathcal{R}y^\ell$. Now consider $X^*_1 = \{\lambda \in \mathfrak{R} : \lambda x^* \mathcal{R}x\}, X^*_2 = \{\lambda \in \mathfrak{R} : x\mathcal{R}\lambda x^*\}$. We have just shown that these sets are non-empty; note also that $\lambda_1 \in X^*_1, \lambda_2 \in X^*_2 \Rightarrow \lambda_1 \geq \lambda_2$; otherwise $\lambda_1 < \lambda_2 \Rightarrow \lambda_1 x^* < \lambda_2 x^* \Rightarrow \lambda_2 x^* \mathcal{P}\lambda_1 x^* \mathcal{R}x\mathcal{R}\lambda_2 x^*$: a contradiction. Define $\lambda^* = \text{glb}\{\lambda \in X^*_1\}; \lambda_* = \text{lub}\{\lambda \in X^*_2\}$; show that $\lambda^* = \lambda_* = \lambda(x)$ say; further $\lambda(x)x^*\mathcal{I}x$; hence define $U(x) = \lambda(x)$ and the demonstration is complete.

We use this utility indicator function in the definition of the problems UMP and EMP; note also that there is an extra constraint in both the problems which we have not mentioned explicitly, that is, $x \in \mathcal{C}$. Generally, the image of the function U is taken to be $\mathfrak{R}_+$: the non-negative portion of the real line but this, of course, is not necessary; the image is say a set $B \subseteq \mathfrak{R}$; hereafter whenever we speak of a utility level U, we shall mean some point in the set B.

We first show that these problems have solutions under some meaningful conditions.

- Consider the problem UMP, first. Note that we are maximizing a continuous function over the budget set; if the budget set is compact (closed and bounded) then we are assured of there being a solution. The compactness of the budget set is assured if the price vector p is strictly positive (check this carefully). For the EMP, the objective function is linear in x and hence continuous; once again, note that when the price vector p is strictly positive the constraint set is bounded below by say some b; it is closed by definition. Using the continuity of the utility function, note that the minimum will be attained over the set and the problem will be solvable. Henceforth, we shall assume that $p > 0$ or that for all components $j, p_j > 0$. Thus, we have the following result: *Whenever $p > 0$ both UMP and EMP have solutions.*

- **The Hicks-Allen Theory:** The classical approach was concerned with locating the solution to the problem UMP given the data p, M and obtaining the behaviour of the solution when the data changed. To do so, we need to impose additional restrictions on the utility indicator function $U(.)$. Such a set of restrictions and the development along these lines is considered separately.

- *If $p > 0$ and there is $x \in C$ such that $p^t.x \le M$ let x^* solve UMP; define EMP for the same p with $U^* = U(x^*)$. Then x^* solves EMP.*

Note that x^* is feasible for EMP; consequently, if x^* does not solve, there must be another bundle $y \in C$, $U(y) \ge U^*$, $p^t.y < p^t.x^* (\le M)$. Thus there exists $y' > y$, $p^t.y' \le M$; since $U(y') > U(y) \ge U(x^*) = U^*$, it follows that x^* could not have solved UMP: this proves the assertion.

- *If $p > 0$ let x^* solve EMP for some U^*; define UMP for the same p, with $M = p^t.x^*$; then x^* solves UMP.*

Note that the bundle x^* is feasible for the problem UMP; so if it does not solve the problem it means that there must be some $y \in C$, $U(y) > U(x^*)$, $p^t.y \le M = p^t.x^*$; from the continuity of $U(.)$, $\exists y' < y$, $U(y') > U(x^*)$, $p^t.y' < p^t.x^*$: which means that x^* could not have solved the problem EMP. This proves the assertion.

The two results mentioned above constitute the basic relationship between the two problems.

We next proceed to analyse the solutions to each of the problems and the value of the objective functions when the problems are solved. In this connection, we need to introduce some further notation; given p, $U(.)$, M, U^*, let $x^*, \hat{x}$ solve UMP, *EMP* respectively; note that there may be many other solutions to the problems. But the value of the objective function at any such solution is well defined. Thus $U(x^*)$ is the unique maximum level of utility that may be attained given the data and we shall denote this by writing $U(x^*) = V(p, M)$ (the indirect utility function); similarly, $p^t.\hat{x} = E(p, U^*)$ (the expenditure function). In case $x^*, \hat{x}$ solve their respective problems uniquely, we shall write $x^* = F(p, M)$, $\hat{x} = H(p, U^*)$; the first is called the Marshallian Demand Function while the second is called the Hicksian Demand Function. All of consumer behaviour is about the properties of these functions.

Note that we have demonstrated the validity of the following:

$$E(p, V(p, M)) = M \tag{10.1}$$

$$V(p, E(p, U)) = U \tag{10.2}$$

And in case the functions $F(p, M), H(p, U)$ are well defined then we have shown that:

$$H(p, U) = F(p, E(p, U)) \tag{10.3}$$

and

$$H(p, V(p, M)) = F(p, M) \tag{10.4}$$

(10.3) also identifies $H(p, U)$ as the **compensated demand function** whenever it is well defined: since along with price change, the income is changed so as to maintain the person at a pre-assigned level of utility.

Next, we have:

- $E(p, U)$ *is a concave function of p for each U; $E(p, U)$ is a continuous function of $p, U, \forall p > 0, U \in B$; $E(p, U)$ is homogenous of degree one in prices for any $U \in B$.*

Let $p', p'' > 0$ and let x', x'' solve EMP given p', p''; U is kept fixed at some level. Define $p = \theta p' + (1 - \theta)p'', 0 \leq \theta \leq 1$; $p > 0$; also let x solve EMP given p, U being kept at the same level. This allows us to conclude that $p''.x \geq p''.x'$ and $p'''.x \geq p'''.x''$; multiplying the first inequality by θ and the second by $1 - \theta$ and adding yields: $(\theta p' + (1 - \theta)p'')'.x \geq \theta E(p', U) + (1 - \theta)E(p'', U)$ or $E(p, U) \geq \theta E(p', U) + (1 - \theta)E(p'', U)$, which proves concavity.

For the remaining part of the claim, consider a sequence $(p^s, U^s) \to (\overline{p}, \overline{U})$; where $\overline{p} > 0, \overline{U} \in B$; let x^s solve EMP for each s given p^s, U^s so that $E(p^s, U^s) = p^{st}.x^s$. Note also that for each s, $U(x^s) \geq U^s$. Since $\overline{U} \in B$, there is $\tilde{x} \in C$ such that $U(\tilde{x}) = \overline{U}$; further since $\overline{p} > 0, \overline{p}^t.\tilde{x}$ is finite and $E(\overline{p}, \overline{U}) \leq \overline{p}^t.\tilde{x}$. Now the sequence x^s must be bounded; if not, since $x^s \in C$, which is bounded below, some component of $x^s, x_k^s \to +\infty$ with s and hence $p^{st}.x^s \to +\infty$; in addition, given the properties of the ordering, it follows that increasing each component of $\tilde{x}$ by a positive amount, say $\varepsilon, U(\tilde{x} + \Xi) > \overline{U}$ where Ξ denotes a vector with each component ε and hence it follows that for all s large enough, $U(\tilde{x} + \Xi) > U^s$ and since $p^{st}.(\tilde{x} + \Xi) < p^{st}.x^s$ for s large enough, since the left hand side (lhs) is finite while the rhs is becoming infinitely large; this implies that the rhs cannot be $E(p^s, U^s)$: which is a contradiction. Hence the sequence x^s must be bounded and must contain limit points; let $\overline{x}$ be any one such; note that $\overline{x} \in C$ (which was assumed to be closed) and since $U(x^s) \geq U^s, U(\overline{x}) \geq \overline{U}$ and we claim that it solves EMP at $(\overline{p}, \overline{U})$: for if not, $\exists x', U(x') > \overline{U}, \overline{p}^t.x' < \overline{p}^t.\overline{x}$. Note this means that there exist elements of the sequence p^k, U^k such that x^k solves EMP at (p^k, U^k) and x^k is arbitrarily close to $\overline{x}$; but then $U(x') \geq U^k, p^{kt}.x' < p^{kt}.x^k$: a contradiction. Hence $\overline{p}^t.\overline{x} = E(\overline{p}, \overline{U})$. Note that this was shown for any arbitrary limit point of the sequence and consequently $E(p^s, U^s) \to E(\overline{p}, \overline{U})$ which proves continuity. Finally, note that changing p to λp does not change the solution to EMP. Thus if x solved under p, it continues to do so under λp and this proves the claim.

- $p \geq p' \Rightarrow E(p, U) \geq E(p', U) \forall p, p' > 0, U \in B; U \geq U' \Rightarrow E(p, U) \geq E(p, U'), \forall p > 0, U, U' \in B$. The proofs of these claims are straightforward and follow from the definition.
- *If $H(p, U), E(p, U)$ are differentiable at some $(p, U), p > 0, U \in B$ then the following hold:*

$$(i) \sum_{j=1}^{n} p_j \frac{\partial H_j(p, U)}{\partial p_k} = 0 \forall k$$

$$(ii) \frac{\partial E(p, U)}{\partial p_k} = H_k(p, U) \forall k$$

Consider $\overline{p} > 0, \overline{U} \in B$ such that $H(\overline{p}, \overline{U})$ exists and partial derivatives of the function $H(p, \overline{U})$ exist too at $p = \overline{p}$; this would mean that in some small neighbourhood $N_\delta(\overline{p})$ of $\overline{p}$, $H(p, \overline{U})$ exists $\forall p \in N_\delta(\overline{p})$. Consequently, from the definition of the problem EMP, it follows that we must have

$$\overline{p}^t.H(p, \overline{U}) \geq \overline{p}^t.H(\overline{p}, \overline{U}) \forall p \in N_\delta(\overline{p})$$

In other words, the function $g(p) = \overline{p}^t.H(p, \overline{U})$ attains a local interior minimum at $p = \overline{p}$. Hence it follows that derivatives of the function $g(p)$ at $p = \overline{p}$ must vanish or that:

$$\sum_{j=1}^n \overline{p}_j \frac{\partial H_j(\overline{p}, \overline{U})}{\partial p_k} = 0 \forall k$$

which is the first part of the claim. The second part follows from the fact that, by definition, we have $E(p, U) = p^t.H(p, U)$ whenever the function $H(p, U)$ is defined; now taking derivatives of both sides with respect to p_k and by using the first part of the claim, the second part is immediate.

- *An immediate consequence of the above is the fact that whenever derivatives exist the matrix*

$$(\frac{\partial H_j(p, U)}{\partial p_k})$$

  is negative semi-definite. This follows from (ii) and the fact that the function $E(p, U)$ is concave and the matrix above is the Hessian of this function.

Further note from (10.3) that $H(p^*, U^*) = F(p^*, E(p^*, U^*))$ so that differentiating both sides with respect to the k-th price, we have (*denotes evaluation at the * values).

$$\frac{\partial H_i(*)}{\partial p_k} = \frac{\partial F_i(*)}{\partial p_k} + \frac{\partial F_i(*)}{\partial M}\frac{\partial E(*)}{\partial p_k}$$

Rearranging terms and using our earlier results, we obtain:

$$\frac{\partial F_i(*)}{\partial p_k} = \frac{\partial H_i(*)}{\partial p_k} - \frac{\partial F_i(*)}{\partial M}H_k(*) = \frac{\partial F_i(*)}{\partial p_k}|_{U=U^*} - \frac{\partial F_i(*)}{\partial M}F_k(p^*, M^*)$$

which is the well-known Slutsky Equation or what Hicks calls the Fundamental Equation of the Theory of Value; the derivatives of the compensated demand function or the substitution terms we know form a negative semi-definite matrix and hence the own price substitution effect is always non-positive.

- Finally, recall that we have shown that $V(p, E(p, U)) = U$ for any $p > 0, U \in B$. Differentiating with respect to prices we have

$$\frac{\partial V(p, E(p, U))}{\partial p_i} + \frac{\partial V(p, E(p, U))}{\partial M}.\frac{\partial E(p, U)}{\partial p_i} = 0$$

so that we have

$$H_i(p, U) = -\frac{\dfrac{\partial V(p, E(p, U))}{\partial p_i}}{\dfrac{\partial V(p, E(p, U))}{\partial M}}$$

and hence, writing $E(p, U) = M, H_i(p, U) = F_i(p, M)$, *we have Roy's identity*

$$F_i(p, M) = -\frac{\dfrac{\partial V(p, M)}{\partial p_i}}{\dfrac{\partial V(p, M)}{\partial M}}$$

10.2 The Hicks-Allen Theory

We provide a brief introduction to these developments. As mentioned earlier, we need to impose the following in addition to the assumptions made earlier:

$U(.)$ is a continuously differentiable function over $\mathcal{C} = \Re_+^n$ and is strictly quasi-concave over the interior of its domain of definition.

Strict quasi-concavity means the following: if $x, y \in \mathcal{C}$ and $U(x) \geq U(y)$ then $U(\lambda x + (1 - \lambda)y) > U(y)$ for all $\lambda, 0 < \lambda < 1$. Note that under this condition:

1. The problem UMP has a unique solution whenever $p > 0$. In other words, the function $F(p, M)$ is well defined for all $p > 0, M > 0$.

    To see this, note that if given $p, M, x \neq x'$ both solve UMP, then $U(x) = U(x')$ and hence by strict quasi-concavity, $U(\frac{x + x'}{2}) > U(x)$; further since $\mathcal{C}$ is convex $\frac{x + x'}{2} \in \mathcal{C}$ and $p\{\frac{x + x'}{2}\} \leq M$ so that x and hence x' could not have solved UMP.

    Accordingly, we shall assume that $p > 0, M > 0$ henceforth.

2. Thus given the fact that $\nabla U(x) > \forall x \in \mathcal{C}$, $p = \nabla(M - p^t.x) > 0$ and since $M > 0$, there is $x \geq 0, p^t.x < M$ (Slater's condition), a necessary and sufficient condition for $x^\star$ to solve UMP is that there should be a $\lambda^\star$ such that KTC hold:

    (i) $\nabla U(x^\star) \leq \lambda^\star p, [\nabla U(x^\star) - \lambda^\star p].x^\star = 0 , x^\star \geq 0$
    (ii) $M - p^t.x^\star \geq 0 , [M - p^t.x^\star].\lambda^\star = 0 , \lambda^\star \geq 0$

These results follow from our investigation of constrained maximization problems where the objective function and the function defining the constraint is quasi-concave. Finally, the Hicks-Allen Theory relate to **interior maxima** only and we shall assume henceforth that we have interior maxima, that is, $x^\star > 0$. This means that by virtue of KTC (i), $\nabla U(x^\star) = \lambda^\star p$. By virtue of monotonicity, of course, we have that **all income is spent** namely from KTC (ii), $M = p^t.x^\star$. These equations then define the solution for UMP: The Marshallian Demand Function $F(p, M)$.

In addition, we shall impose that the maxima under consideration be **regular**, that is, the second order conditions will be assumed to be satisfied. To introduce these, consider the matrix $(n + 1) \times (n + 1)$:

$$A = \begin{pmatrix} 0 & -p^t \\ -p & \nabla^2 U(x^\star) \end{pmatrix}$$

where $(\nabla^2 U(x)) = (\frac{\partial^2 U(x)}{\partial x_i \partial x_j})$: the Hessian of the function $U(.)$. A is called the bordered Hessian. We have seen that the second order conditions are given by:

$$(-1)^r \det A_r > 0 \ , \ r = 2, 3, ... n + 1$$

and

$$A_r = \begin{pmatrix} 0 & -p_1 & \cdots & -p_r \\ -p_1 & U_{11}(x^\star) & \cdots & U_{1r}(x^\star) \\ \cdots & \cdots & \cdots & \cdots \\ -p_r & U_{r1}(x^\star) & \cdots & U_{rr}(x^\star) \end{pmatrix}$$

To demonstrate the role of second order conditions, consider the case **when there are only two goods**. In this case, the second order condition reduces to a single determinant condition:

$$\det A_2 = \det \begin{pmatrix} 0 & -p_1 & -p_2 \\ -p_1 & U_{11}(x^\star) & U_{12}(x^\star) \\ -p_2 & U_{21}(x^\star) & U_{22}(x^\star) \end{pmatrix} > 0 \qquad (10.5)$$

We have moreover, from considerations above the first order conditions (recall, we insist on an interior maximum):

$$\frac{\partial U_i(x^\star)}{\partial x_i} = \lambda^\star p_i \ , \ \forall i = 1, 2 \qquad (10.6)$$

and we also have

$$M - p_1 x_1^\star - p_2 x_2^\star = 0 \qquad (10.7)$$

We are interested in how the demand bundle changes when there is a change in the data of the system (that is, prices and income). Note that changes in the data lead to changes in the above set of equations (10.6, 10.7) but the equations must be maintained after the changes have taken place (slight or small changes, shift points of tangency *and* do not introduce other second order problems); thus whatever changes there are, must balance. To put it differently, we need to consider total differentials of both sides of the equations and these must match. Note that these considerations lead to the system of equations given by

$$
\begin{pmatrix}
0 & -p_1 & -p_2 \\
-p_1 & U_{11}(x^\star) & U_{12}(x^\star) \\
-p_2 & U_{21}(x^\star) & U_{22}(x^\star)
\end{pmatrix}
\cdot
\begin{pmatrix}
d\lambda \\
dx_1 \\
dx_2
\end{pmatrix}
=
\begin{pmatrix}
\sum_i x_i^\star dp_i - dM \\
\lambda^\star dp_1 \\
\lambda^\star p_2
\end{pmatrix}
\tag{10.8}
$$

Consequently, that the above system has a unique solution given the second order condition (10.5), since then the square matrix on the left of the equation is non-singular has an inverse and hence the system has an unique solution given by:

$$
\begin{pmatrix}
d\lambda \\
dx_1 \\
dx_2
\end{pmatrix}
=
\begin{pmatrix}
0 & -p_1 & -p_2 \\
-p_1 & U_{11}(x^\star) & U_{12}(x^\star) \\
-p_2 & U_{21}(x^\star) & U_{22}(x^\star)
\end{pmatrix}^{-1}
\cdot
\begin{pmatrix}
\sum_i x_i^\star dp_i - dM \\
\lambda^\star dp_1 \\
\lambda^\star dp_2
\end{pmatrix}
$$

Now, writing out the solutions given above, say for dx_1, we have by virtue of Cramer's Rule:

$$
dx_1 = \dfrac{\det \begin{pmatrix}
0 & \sum_i x_i^\star dp_i - dM & -p_2 \\
-p_1 & \lambda^\star dp_1 & U_{12}(x^\star) \\
-p_2 & \lambda^\star dp_2 & U_{22}(x^\star)
\end{pmatrix}}{\det A_2}
\tag{10.9}
$$

The above relates changes in consumption of the first good to changes in prices and income. To simplify matters we shall consider the variation in only one price, say p_1; then $dp_2 = dM = 0$ and then the basic equation (10.9) reduces to:

$$
dx_1 \cdot \det A_2 = \det \begin{pmatrix}
0 & x_1^\star dp_1 & -p_2 \\
-p_1 & \lambda^\star dp_1 & U_{12}(x^\star) \\
-p_2 & 0 & U_{22}(x^\star)
\end{pmatrix}
\tag{10.10}
$$

or evaluating the determinant, we have

$$
\left(\dfrac{\partial x_1}{\partial p_1}\right)_{dp_2 = dM = 0} \cdot \det A_2 = -x_1^\star[-p_1 U_{22}(x^\star) + p_2 U_{12}(x^\star)] - p_2^2 \lambda^\star
\tag{10.11}
$$

To interpret the terms on the rhs, consider a change in M alone with no change in the prices; setting $dp_1 = dp_2 = 0$. Then the basic equation (10.9) reduces to:

$$
dx_1 \cdot \det A_2 = dM[-p_1 U_{22} + p_2 U_{12}] \Rightarrow (\dfrac{\partial x_1}{\partial M})_{dp_i = 0} \cdot \det A_2 = [-p_1 U_{22} + p_2 U_{12}]
$$

thus rewriting (10.11) we have:

$$
\left(\dfrac{\partial x_1}{\partial p_1}\right)_{dp_2 = dM = 0} \cdot \det A_2 = -x_1^\star \cdot (\dfrac{\partial x_1}{\partial M})_{dp_i = 0} - p_2^2 \lambda^\star
$$

Finally, we wish to find out when together with a change in prices, money income is changed so as to maintain the person at the same level of utility, that is, $dU(.) =$

$0 \Rightarrow \nabla U(x^\star).dx = 0 \Rightarrow \lambda^\star p.dx = 0 \Rightarrow p.dx = 0$ since $\lambda^\star > 0$ from the assumption of monotonicity. This means from the equation (10.7) that $x^\star.dp = dM$. Substitution into the basic equation (10.9) leads to the following:

$$(\frac{\partial x_1}{\partial p_1})_{dU=0}. \det A_2 = -\lambda^\star p_2^2$$

So now collecting terms and substituting into (10.11), we have

$$(\frac{\partial x_1}{\partial p_1})_{dp_2=dM=0} = (\frac{\partial x_1}{\partial p_1})_{dU=0} - x_1^\star.(\frac{\partial x_1}{\partial M})_{dp_i=0}$$

which is the well-known Slutsky Equation; note that the **second order condition did imply the existence of derivatives for demand functions**.

10.3 Producer Behaviour

Consider a decision-maker who has access to methods of converting some goods and services (inputs) to other goods and services (outputs) which are then marketed; usually this activity is called a **production activity** and the decision-maker is called a producer. We shall analyse the behaviour of such a decision-maker in this section. Traditionally, this analysis closely follows the analysis of the behaviour of the consumer discussed earlier. We shall consider moreover, the behaviour of a **competitive** producer; and as before, this will imply that the decision-maker cannot control the outcome to his advantage and passively reacts to market prices.

Thus, the producer has access to a set of **activity** vectors $a^j \in \Re^n$; each component of the vector a_{kj}, $k = 1, 2\cdots, n$ represents the usage of commodity or service k in the j-th activity; the sign of the component indicates whether it is an input or an output; a negative sign indicates an input while a positive sign indicates an output; a zero entry indicates that the commodity does not figure in the production process. The collection of all such activities that a producer has is the **technology** for the producer and denoted by $\mathcal{I}$; this is the case when a producer may be producing a wide variety of goods and services as output.

To simplify matters, consider the case when a producer produces only a single good as output, say the first; that is, $a \in \mathcal{I} \Rightarrow a_1 > 0$; $a_k \leq 0 \forall k \neq 1$. Thus the technology for the producer then consists of various ways of producing various quantities of good 1 using the other goods as inputs. We shall then write a typical element of $\mathcal{I}$ as $(x, -a)$; thus we shall use x for output and a for inputs. With such a structure, the first problem that the producer may need to resolve is what is the cheapest method of producing the output given the prices of all other goods. If $(x, -a) \in \mathcal{I}$ we shall sometimes write $a = (a_2, a_3, \cdots, a_n) \in V(x)$ and interpret it to mean that it is possible to produce the output quantity x by using as inputs the quantities $(a_2, a_3, \cdots, a_n) \in \Re_+^{n-1}$ of the various other goods since we do not need to distinguish between inputs and outputs. $V(x)$ is the **input requirement set**. Sometimes, we may represent the technology through a **production function**: $x = f(a)$ when we write x to be the maximum possible output given the inputs a; in such a situation, $V(x) = \{a : x \leq f(a)\}$. We shall generally use the input requirement sets to represent the technology.

10.3.1 Cost and profit functions

The first problem that confronts the decision-maker is the cheapest way of producing a given level of output, given the technology and the factor prices. Formally, the problem may be written as:

$$\min \sum_{k=2}^{n} w_k a_k \text{ subject to } (a_2, a_3, \cdots, a_n) \in V(x); \textbf{CM}^*$$

solving this problem, the producer will determine the cheapest way of producing x units of output; if the problem is solvable, then the values of the objective function will be denoted by $C(x, w_2, \cdots, w_n)$: the **cost function**; the function C maps $\mathfrak{R}^n_+$ to $\mathfrak{R}_+$ that is for each output configuration x and each configuration of input prices $w = (w_2, w_3, \cdots, w_n) \in \mathfrak{R}^{n-1}_+$, a level of cost of producing x units is determined. Our first set of results concerns the nature of this cost function.

We first consider the function $C_{\overline{w}}(x) = C(x, \overline{w})$ where $\overline{w} \in \mathfrak{R}^{n-1}_+$ and $x \in \mathfrak{R}_+$ that is with factor prices fixed at $\overline{w}$. Before we proceed, we need to set out the nature of the technology $\mathcal{I}$ or the sets $V(x)$ for all $x \in \mathfrak{R}_+$.

We shall assume that the technology is **regular**, that is, $V(x)$ is a **closed, non-empty subset** of $\mathfrak{R}^{n-1}_+$ for every $x > 0$; we shall also assume that the technology is monotone that is, $a' \geq a, \ a \in V(x) \Rightarrow a' \in V(x)$.

We may claim:

Claim 10.3.1 $C(x, w)$ is well defined for every $x \geq 0, w > 0$ and there is $a \in V(x)$ such that $C(x, w) = w.a$.

Note that $\inf_{a \in V(x)} w.a$ is well defined since $V(x)$ is non-empty and $a \in V(x) \Rightarrow w.a \geq 0$; let this infimum be α; then there is a sequence $a^s \in V(x)$ such that $w.a^s \to \alpha$ as $s \to \infty$. It follows that a^s is bounded and possesses a limit point a^o; $a^o \in V(x)$ since $V(x)$ is closed and hence $\alpha = w.a^o = C(x, w)$ as claimed.

We shall henceforth take $w > 0$. Further,

Claim 10.3.2 $x \geq y \Rightarrow C_{\overline{w}}(x) \geq C_{\overline{w}}(y)$.

This follows since otherwise $C(x, \overline{w}) < C(y, \overline{w})$ and hence $\overline{w}.a_x < \overline{w}.a_y$ for some $a_x \in V(x)$ and $a_y \in V(y)$; but by virtue of monotonicity, $a_x \in V(y)$ which contradicts the definition of $C(y, \overline{w})$.

Traditional producer behaviour is developed along the properties of $C_{\overline{w}}(x)/x$ **the average cost** and the **marginal cost**, $C'_{\overline{w}}(x)$ whenever the derivative exists.

---

* The problem is one of the forms of cost minimization.

Whenever the set $V(x)$ is represented by means of a production function $f(a)$, the cost minimization problem may be written as:

$$\min w.a \text{ subject to } x \leq f(a)(\mathbf{CM});$$

whenever the optimization techniques introduced earlier are applicable, the value of the Lagrangian multiplier determined at the optimum may be shown to be the marginal cost.

Some examples of production functions are provided below:

1. Cobb-Douglas $x = A.a_1^\alpha.a_2^{1-\alpha}$, $0 < \alpha < 1$.
2. Leontief $x = \min[a_1/\alpha, a_2/\beta]$, $\alpha, \beta > 0$.
3. CES $x = [a_1^\rho + a_2^\rho]^{1/\rho}$

The reader may investigate the form of the cost function in each case. The main properties of cost functions $C_x(w) = C(w, x)$, for a given output level x are mentioned below.

Claim 10.3.3 (i) $C_x(\lambda.w) = \lambda.C_x(w)$ *(homogeneity of degree one in factor prices)*
(ii) $C_x(w)$ *is a concave function of w, for each given value of x, when $w > 0$, that is,*
$\theta C_x(w^1) + (1 - \theta)C_x(w^2) \leq C_x(w)$ *where* $w = \theta w^1 + (1 - \theta)w^2$ *for any θ, $0 \leq \theta \leq 1$.*
(iii) Whenever partial derivatives exist,

$$\frac{\partial C_x(w)}{\partial w_j} = a_j(w) \ \forall j$$

where $a_j(w)$ is the j-th component of $a(w)$ which solves the cost minimization problem, that is, (CM) or (CM) (**Shephard's Lemma**).*
(iv) Whenever partial derivatives exist:

$$\text{the matrix } \left(\frac{\partial a_j(w)}{\partial w_k}\right) = \nabla^2 C_x(w) = \left(\frac{\partial^2 C_x(w)}{\partial w_k \partial w_r}\right)$$

and is symmetric and negative semi-definite.

The proofs follow directly from the definition of the cost function. To see the validity of (i), let a^1 solve

$$\min \lambda w.a \text{ subject to } x \leq f(a)$$

that is, $C_x(\lambda w) = \lambda w.a^1$; similarly let a^2 solve

$$\min w.a \text{ subject to } x \leq f(a)$$

that is, $C_x(w) = w.a^2$. It follows then that $\lambda w.a^1 \leq \lambda w.a^2$ and that $w.a^2 \leq w.a^1$ just from the definition of cost minimization and hence it follows that $C_x(\lambda w) = \lambda w.a^1 = \lambda w.a^2 = \lambda C_x(w)$ as claimed. For (ii), consider the problems

$$\min w^i.a \text{ subject to } x \leq f(a) \ i = 1, 2$$

and let a^i denote their respective solutions; now consider $w = \theta w^1 + (1 - \theta)w^2$, $0 \le \theta \le 1$; and let a' solve

$$\min w.a \text{ suject to } x \le f(a);$$

since $w^i > 0$, $w > 0$ and the solutions exist in each case. Then note that $w^i.a^i \le w^i a'$, $i = 1, 2$, that is, we have $C_x(w^i) \le w^i.a'$ and hence $\theta C_x(w^1) + (1 - \theta)C_x(w^2) \le (\theta w^1 + (1 - \theta)w^2).a' = C_x(w)$. Since this is true for any θ, and any $w^i > 0$, the claim follows. Consider the cost minimization problem at $\overline{w} > 0$:

$$\min \overline{w}.a \text{ subject to } x \le f(a);$$

suppose that the solution is unique (this is necessary for the partial derivative to exist: the reader may provide a proof for this assertion.) and denote it by $a(\overline{w})$; consider a neighbourhood N of $\overline{w}$, over which the function $a(w)$ is defined then note that from the definition of cost minimization, we have, for all $w \in N$, $\overline{w}.a(\overline{w}) \le \overline{w}.a(w) = \phi(w)$ say; that is, the function $\phi(w)$ attains a local interior minimum at $w = \overline{w}$: this follows immediately as a necessary condition for local interior extrema, the following, for any $\overline{w} > 0$

$$\sum_i \overline{w}_i \frac{\partial a_i(w)}{\partial w_j}\big|_{w=\overline{w}} = 0 \ \forall j \tag{10.12}$$

Consequently, we have:

$$\frac{\partial C_x(w)}{\partial w_j} = \frac{\partial(w.a(w))}{\partial w_j} = a_j(w) + \sum_i w_i \frac{\partial a_i(w)}{\partial w_j} = a_j(w)$$

by virtue of (10.12). This concludes the demonstration for (iii). Property (iv) follows from (iii) and from (ii), the concavity of the cost function. Thus the properties of the cost function are similar to those of the Expenditure function for the theory of consumer behaviour deduced earlier and as demonstrated there, will also play a central role in the analysis of producer behaviour. Thus, we know that if the firm has a technology specified by means of a production function $x \le f(a)$ or by the input requirement set $V(x)$, then the producer knows the cheapest method of producing the level of output level x given the factor prices w; and we have investigated how this choice, varies with changes in factor prices w with constant x and with changes in x at constant w. We move on to the determination of the output level x. We consider the price of the output x to be p; then by choosing a level of output x and a method of production $a \in V(x)$, the producer is assured of returns $p.x - w.a$ and the objective of the producer then is to maximize these returns for all feasible choices of x, a (that is, $a \in V(x)$). This is the so-called **profit maximization problem**. A competitive firm then must decide on the solution to the optimization problem described below:

$$\max_{x,a} p.x - w.a \text{ subject to } a \in V(x) \text{ OR } (x, -a) \in \mathcal{I}$$

If $(\overline{x}, \overline{a})$ solves the problem then $\pi(p, w) = p.\overline{x} - w.\overline{a}$ denotes the **profits** of the firm at (p, w); note that the value of the objective function $\pi(p, w)$ is uniquely determined whenever the problem has a solution and hence $\pi(p, w)$ qualifies to be termed a function whenever the problem has a solution; we call this function the **profit function** whenever the solution to the profit maximization problem exists. First, we note:

Claim 10.3.4 Whenever $\pi(p, w)$ is well defined, $\pi(p, w) = p.x - C_x(w)$ for some x: thus profit maximization implies cost minimization.

The proof follows by noting that if this were to be false then there would be some (x, a) solving the profit maximization problem such that $\pi(p, w) = p.x - w.a$ and $w.a > C_x(w) = w.a'$ say , $a' \in V(x)$ but then $\pi(p, w) = p.x - w.a < p.x - w.a'$: a contradiction. •

Remark 10.1 Hence note that the profit maximization problem may be reduced to just the problem of choosing the level of output: $\pi(p, w) = \max_x\{p.x - C(x, w)\}$; if the cost function is differentiable and a positive level of output solves the profit maximizing problem, we must have

$$p = \frac{\partial C(x, w)}{\partial x} \tag{10.13}$$

which is the well-known condition of profit maximization under perfect competition: price must match marginal cost; the second order condition that this is surely a profit maximizing condition provided the marginal cost is not decreasing at that point follows trivially from the above.

It may be next shown that,

Claim 10.3.5 Whenever $\pi(p, w)$ is well defined we have:

(i) $\pi(p, w)$ *is homogenous of degree one in (p, w).*
(ii) $\pi(p, w)$ *is a convex function of (p, w), that is, $\pi(p, w) \geq \lambda\pi(\overline{p}, \overline{w}) + (1-\lambda)$*
    $\pi(\hat{p}, \hat{w})$, *for any λ, $0 \leq \lambda \leq 1$ and $(p, w) = \lambda(\overline{p}, \overline{w}) + (1-\lambda)(\hat{p}.\hat{w})$.*
(iii) *If derivatives exist, then*

$$\frac{\partial\pi(p, w)}{\partial p} = x(p, w) \text{ and } \frac{\partial\pi(p, w)}{\partial w_i} = -a_i(p, w)$$

where $(x(p, w), -a(p, w)) \in \mathcal{I}$ solves the profit maximizing problem.

The proof follows the steps outlined in the case of the cost function and it is recommended that the reader goes through the proofs once again.

Recall the technology set $\mathcal{I}$; the conditions under which the profit function is well defined are contained in the following:

Claim 10.3.6 If $\mathcal{I}$ is non-empty, closed and bounded above, then $\pi(p, w)$ is well defined for all $p > 0, w > 0$.

The proof follows by considering the set $A(p, w) = \{z : z = p.x - w.a, (x, -a) \in \mathcal{I}\}$; since $\mathcal{I}$ is bounded above, there is some (u, v) such that $(w, -a) \in \mathcal{I} \Rightarrow (w, -a) \leqq$

(u, v) and hence $z \in A(p, w) \Rightarrow z \leq p.u + w.v$ if $p > 0, w > 0$; thus $A(p, w)$ is a non-empty set of real numbers bounded above and hence possesses a supremum $\pi(p, w)$; note that $A(p, w)$ must be closed (the reader may find it instructive to construct the proof of this assertion) and hence the supremum must be attained in $\mathcal{I}$ and this, of course, is the basis for the existence of the profit function.

We say that the technology is convex if $\mathcal{I}$ is a convex set in $\Re^n$ or if $V(x)$ is a convex subset in $\Re^{n-1}$ for each x; that is, $(x, -a), (x', -a') \in \mathcal{I} \Rightarrow (\lambda x + (1 - \lambda)x', \lambda(-a) + (1 - \lambda)(-a')) \in \mathcal{I}$ for any λ, $0 \leq \lambda \leq 1$. In case $(x, -a) \in \mathcal{I}$ and $(y, -b) \leq (x, -a) \Rightarrow (y, -b) \in \mathcal{I}$, we shall say that $\mathcal{I}$ satisfies the assumption of **free disposal**. In case $\mathcal{I}$ satisfies convexity and free disposal, it is said to satisfy **strict convexity** if $(x, -a), (x', -a') \in \mathcal{I} \Rightarrow (x_\lambda, -a_\lambda) = (\lambda x + (1 - \lambda)x', \lambda(-a) + (1 - \lambda)(-a')) \in \mathcal{I}$ and there is an open set $\mathcal{U} \in \mathcal{I}$ such that $(x_\lambda, -a_\lambda) \in \mathcal{U}, \forall \lambda$, $0 < \lambda < 1$. The advantage of imposing strict convexity is contained in the following:

Claim 10.3.7 In case $\mathcal{I}$ is strictly convex, the profit maximization problem is uniquely solved in $\mathcal{I}$. Thus given $(p, w) > (0, 0)$ the solution to the profit maximizing problem may be written as **functions***, $x(p, w), -a(p, w)$. $x(p, w)$ is the* **supply function** *and $a(p, w)$ is the* **derived demand function** *for factors. Further, if derivatives exist, then:*

$$\frac{\partial x(p, w)}{\partial p} \geq 0 \; and \; \frac{\partial a_i(p, w)}{\partial w_i} \leq 0$$

Thus supply curves do not slope downwards: they slope upwards if at all; and derived demand curves slope downwards if at all. The problems encountered in deriving slopes of demand curves for consumers vanish. For proofs, note that if under strict convexity, profit is maximized at two points $(x, -a)$ and $(x', -a')$ then $p.x - w.a = p.x' - w.a' = p.x_\lambda - w.a_\lambda$ where $(x_\lambda, -a_\lambda)$ are as defined above; but there exists $(\overline{x}, -\overline{a}) \in \mathcal{U}$ such that $(\overline{x}, -\overline{a}) > (x_\lambda, -a_\lambda)$ which contradicts profit maximization, since $p.\overline{x} - w.\overline{a} > p.x_\lambda - w.a_\lambda$; hence a unique solution. Recall that the profit function was seen to be a convex function of (p, w) and hence the Hessian (the matrix of second order partial derivatives) of the profit function is positive semi-definite. By virtue of Claim 10.3.5, the rest follows.

Note the role played by the assumption that $\mathcal{I}$ is bounded above, in deriving a profit maximizing response. For certain types of technology sets where this assumption does not hold, we need to proceed differently. Technology sets are often described by the implicit **returns to scale**. To examine the returns to scale, we need to examine the effect on output if inputs are changed proportionally (scale changes); thus if $a \in V(x)$, what about λa ? For example, does $\lambda a \in V(\lambda x)$? More commonly, such arguments are carried through in terms of the production function $f(a)$ and the degree of homogeneity of this function. Recall that $f(a)$ is said to be homogenous of degree r if $f(\lambda a) = \lambda^r f(a)$ for all a in the domain of definition of the function. Accordingly, for a homogenous production function, we shall say that it satisfies **constant returns to scale** (crs) if $r = 1$; increasing returns to scale in case $r > 1$ and **decreasing**

returns to scale if $r < 1$. Note that crs may be alternatively defined whenever we claim $a \in V(x) \Leftrightarrow \lambda a \in V(\lambda x)$ for any x. It may be shown that under crs, the structure of cost functions is very simple; linear in x given w.

Claim 10.3.8 If $a \in V(x) \Leftrightarrow \lambda a \in V(\lambda x)$ for any x, then $C(x, w) = C(1, w).x$.

The proof involves noting that in case $C(x, w) = w.a_x$ for some $a_x \in V(x)$ then since crs holds, assuming $x > 0$, it must be the case that $a_x.\frac{1}{x} \in V(1)$ and hence if $C_1(w) = w.a_1$ it follows that $C(1, w) = w.a_1 \leq w.a_x\frac{1}{x}$. Again using crs, we must have $x.a_1 \in V(x)$ since $a_1 \in V(1)$; hence cost minimization implies $C(x, w) = w.a_x \leq w.x.a_1$; combining these two, we get the claim. •

For the converse, we need to invoke the property of convexity and we take up this matter next. The route to be followed, however, will be somewhat circuitous as we shall address a completely different matter first. The technology is specified through either the set $\mathcal{I}$ or through sets such as $V(x)$. Given the technology, we deduce the cost function $C(x, w)$. The question we need to examine is the following: what information does function $C(x, w)$ contain about the technology? To answer this question, we define the set:

$$V^\star(x) = \{a \in \Re^{n-1} : w.a \geq C(w, x) \forall w \geq 0\}$$

Since $a \in V(x) \Rightarrow w.a \geq C(w, x) \forall w \geq 0$ by definition of cost minimization, it follows that $a \in V(x) \Rightarrow a \in V^\star(x) \Rightarrow V(x) \subseteq V^\star(x)$. This is straightforward. However,

Claim 10.3.9 In case $V(x)$ is regular and convex, and satisfies free disposal, $V^\star(x) = V(x) \forall x$

Proof We know from definition that $V(x) \subseteq V^\star(x)$; if possible, for some x, let $\bar{a} \in V^\star(x)$ and $\bar{a} \notin V(x)$; since $V(x)$ is convex and non-empty, we may use the Separation theorem (See Section 8.4 of this book) to assert the existence of $w \neq 0$ such that $w.\bar{a} < w.a \; \forall a \in V(x)$. Since the assumption of free disposal holds, note that we may conclude that $w \geq 0$: this is so because if $w_j < 0$ then since in $V(x)$, $a_j \to +\infty$, it follows that $w.a \to -\infty$ but on $V(x)$, $w.a$ is bounded below by $w.\bar{a}$: a contradiction. But now consider $C(x, w)$; it follows that $w.\bar{a} < C(x, w)$: this contradicts the definition of $V^\star(x)$. Hence, no such $\bar{a}$ can exist. Thus it follows that $a \in V^\star(x) \Rightarrow a \in V(x)$ or that $V^\star(x) \subseteq V(x)$ and hence the claim follows. •

We may now use the above result to characterize completely linear cost functions through the following result:

Claim 10.3.10 For convex, regular technologies satisfying free disposal, $C(x, w) = \alpha.x$ iff the technology satisfies crs.

Proof We already know that crs technology implies that $C(x, w) = C(1, w).x = \alpha.x$ where $\alpha = C(1, w)$. In case $C(x, w) = \alpha.x$ and since we know that the technology is convex regular and satisfies free disposal, $V^\star(x) = V(x)$; thus $a \in V^\star(x) = V(x) \Rightarrow w.a \geq C(x, w) = \alpha.x \Rightarrow w.\lambda a \geq \alpha \lambda x = C(\lambda x, w) \Rightarrow \lambda a \in V^\star(\lambda x) = V(\lambda x)$; hence crs obtains as claimed. •

Finally, even though in general $V^{\star}(x) \neq V(x)$, the set $V^{\star}(x) - V(x)$ consists of input vectors which are never picked up in cost minimization and consequently if $C^{\star}(x, w) = \min_{a \in V^{\star}(x)} w.a$ then $C^{\star}(x, w) \leq C(x, w)$; if the strict inequality holds, then there is $a \in V^{\star}(x)$ such that $w.a < C(x, w)$: a contradiction. Hence,

Claim 10.3.11 $C^{\star}(x, w) = C(x, w)$ *for all* $x \geq 0, w > 0$.

Thus, specifying a cost function amounts to specifying the technology completely.

10.4 Market Equilibria

The last section describes determination of individual demand; aggregate or market demand is obtained by aggregating over all individuals.

The treatment of a firm's decision-making is completely analogous to the above. A supply curve for the firm is obtained on the basis of profit maximization; the market supply curve is obtained by aggregating over firms. The equilibrium in the market, when it is competitive, is determined on the basis of these forces. We shall simplify matters somewhat by considering only the exchange situation, that is, where there is no production.

10.4.1 The excess demand function

We have seen how demand is determined for competitive agents. We briefly discuss how a collection of such agents determine how much to buy and sell.

Consider a collection of such agents, each with a stock of goods $w^i \in \Re^n_+$, $i = 1, 2, \cdots, M$: the **endowment** of agent i. Thus, the entire stock available for this economy of M agents is given by $W = \sum_{i=1}^{M} w^i$; we shall consider that there is no other addition to these stocks, thus there is no possibility of production. We call such an economy an exchange economy. Each agent i has a continuous utility indicator function $U^i(.)$ which is assumed to be in addition, strictly quasi-concave.

The essential question for this economy is whether agents transact among themselves and if so, what would they end up with. Since the agents are assumed to be competitive, they believe that they cannot control the market prices to their own advantage. Thus given $p \in \Re^n$, agent i arrives at demand $x^i(p, w^i)$ which solves $\max U^i(x)$ subject to $p.x \leq p.w^i$; given strict quasi-concavity, the problem has a unique solution for every $p > 0$, so that the notation $x^i(p, w^i)$ is valid. Market demand is naturally, $X(p\{w^i\}) = \sum_i x^i(p, w^i)$. We define the **excess demand** as **Aggregate Demand – Aggregate Supply**. For the exchange economy, thus $Z(p, \{w^i\}) = X(p, \{w^i\}) - W$ is the excess demand; since the distribution of endowments is taken to be fixed, the excess demand is written as usually represented as $Z(p)$ but one should always keep in mind the fact that the distribution $\{w^i\}$ is a crucial determinant of excess demand. We shall discuss the properties of the function $Z(p)$ first.

It will be appropriate to summarize properties of the demand function $x^i(p, w^i)$ (we drop the superscript i below):

1. $x(p, w)$ is homogenous of degree zero in p, that is, $x(tp, w) = x(p, w)$ for any scalar $t > 0$. (Earlier we had discussed the properties of $x(p, M)$; to compare, note that here $M = p.w$; making this adjustment the properties deduced earlier will have similar counterparts.)

2. $x(p, w)$, whenever well defined, is a continuous function of p, w at all points $p, w > 0$. Consider $\overline{p}, \overline{w} > 0$ and let $(p^s, w^s) \to (\overline{p}, \overline{w})$ as $s \to \infty$; let $x^s = x(p^s, w^s)$ and let $m_j = \overline{p}\overline{w}/\overline{p}_j$; we note that for any $\delta > 0$ there exists $S(\delta)$, such that $s > S(\delta) \to m_j - \delta \leq p^s.w^s/p_j^s \leq m_j + \delta$. Thus since $0 \leq x_j^s \leq p^s.w^s/p_j^s$ we may conclude that x^s is bounded and must have a limit point x^o; clearly $\overline{p}.x^o \leq \overline{p}.\overline{w}$; so $x^o \neq x(\overline{p}, \overline{w}) \Rightarrow$ there is $\overline{x}, \overline{p}.\overline{w} \geq \overline{p}.\overline{x}, U(\overline{x}) > U(x^o)$. Hence, it follows that $U(\overline{x}) > U(x^s)$ for all s sufficiently large. Now if $\overline{px} < \overline{pw}$ then for all large s we must have $p^s\overline{x} \leq p^sw^s$: this would contradict the definition of x^s as the utility maximization choice at p^s, w^s; hence we must have $\overline{px} = \overline{pw}$. Consider then $y \geq 0$ such that $\overline{p}y < \overline{pw}$: such a y exists since $\overline{p}.\overline{w} > 0$ and define $y^s = t^s\overline{x} + (1-t^s)y$ where $t^s = \max[t : 0 \leq t \leq 1, p^s.y^s \leq p^sw^s]$; note that $t^s < 1 \Rightarrow p^sy^s = p^sw^s$ since otherwise t^s is not the maximum. Thus t^s being a bounded sequence must have a limit point, say t^o. In case $t^o < 1$, then $t^{s_r} < 1$ along some subsequence and hence $p^{s_r}y^{s_r} = p^{s_r}w^{s_r}$ so that in the limit the rhs tends to $\overline{pw}$ whereas the lhs approaches $\overline{p}[t^o\overline{x} + (1-t^o)y] < \overline{pw}$: a contradiction, again. So no such $\overline{x}$ can exist and hence $x^o = x(\overline{p}, \overline{w})$: which completes the demonstration.

3. In case, $x \geq y, x \neq y \Rightarrow U(x) > U(y)$, (the utility function is increasing) demand must exhaust all income, that is, $p.x(p, w) = p.w$. For otherwise, one may spend more and increase utility.

These properties are central for much of what follows. We now can state the following properties of the excess demand function for the exchange economy; as should be clear, these follow immediately from the previous properties and from the definition of the excess demand function. (We shall write the excess demand function as $Z(p)$.)

1. $Z(tp) = Z(p) \forall t$ scalar $> 0, p > 0$ (homogeneity of degree zero in prices).
2. Assuming that all utility functions are increasing, $p.Z(p) = 0$ (Walras Law). (Note this must hold at every p in the domain of definition of the excess demand function).
3. $Z(p)$ is a continuous function of prices and the distribution of endowments whenever the demand functions are continuous functions of prices and endowment. Also note that $Z(p)$ is bounded below; in the case of exchange economies, $Z(p) \geq -W$, for example.

We define $p^\star$ to be a competitive equilibrium if $Z(p^\star) \leq 0$; since we have Walras Law, this means that if $p^\star$ is an equilibrium then $p_j^\star > 0 \Rightarrow Z_j(p^\star) = 0$. We need to provide an argument to establish that there is such an equilibrium: notice it is only at such a configuration that the plans, of all the different agents, are mutually compatible. We shall show that the existence of a competitive equilibrium is a non-trivial proposition.

Given the properties of the excess demand functions, we narrow down our search for equilibrium by adopting some suitable normalization. One such method involves the choice of a unit of account. Note that by virtue of Walras Law, if $Z_j(p) \leq 0$ for all $j \neq j_1$ then $Z_{j-1}(p) \leq 0$ if $p_{j_1} \neq 0$; this has usually led to the choice of a particular good as numeraire, one whose price is always positive and to use that good as the unit of account, with all other price being measured as being relative to that good. Let us designate good n to be the numeraire, then the price configuration is $p = (p_1, p_2, \cdots, p_{n-1}, 1)$. Homogeneity of degree zero in prices provides another manner of normalization; since this property ensures that excess demands are constant, along rays through the origin, we confine attention to the unit simplex $S_n = \{p : p_i \geq 0 \forall i, \sum_i p_i = 1\}$. If there is an equilibrium then there must be one in S_n. We shall adopt one such method in most situations.

When excess demand functions are differentiable, define

$$Z_{ij}(p) = \frac{\partial Z_i}{\partial p_j}$$

so that the matrix $\bar{J}(p) = (Z_{ij}(p)), i, j = 1, 2, ..., n$ represents the Jacobian of the excess demand functions. If there is a numeraire, say good n, we often confine attention to the numeraire without the row and column relating to good n, that is, to $J(P) = (Z_{ij}(p)), i, j = 1, 2 \cdots, n-1$. Note that $\bar{J}(p)$ is singular at every p in the domain of definition of $Z(p)$ where the function has derivatives; in particular, we consider $p > 0$. This claim follows from the homogeneity property of degree zero in the prices:

$$\sum_{j=1}^{n} Z_{ij}(p)p_j = 0 \forall i \rightarrow \bar{J}(p).p = 0 \qquad (10.14)$$

Also differentiating the expression for Walras Law, we have:

$$\sum_{i=1}^{n} p_i Z_{ij}(p) + Z_j(p) = 0 \ \forall j$$

that is,

$$p'.\bar{J}(p) = -Z(p)' \qquad (10.15)$$

Note that by virtue of the above the **Jacobian can be symmetric only at a competitive equilibrium**.

10.4.2 The Existence Theorem and the Fixed Point Theorem

Consider the case when $n = 2$ and let us say that the price vector is (p_1, p_2); if at $p_1 = 0, Z_1(0, p_2) \leq 0$, for some $p_2 > 0$, then we have and equilibrium at $(0, p_2)$:

according to the definition provided since by virtue of Walras Law, $Z_2(0, p_2) = 0$ when $p_1 = 0$. Next, by the same logic, for any $p_1 > 0$, if there is no equilibrium then $Z_2(p_1, 0) > 0$; since otherwise, for some $p_1 > 0$, $Z_2(p_1, 0) \leq 0 \Rightarrow Z_1(p_1, 0) = 0$ and hence $(p_1, 0)$ would be an equilibrium.

Consider $S = \{(p_1, p_2) : p_1 + p_2 = 1, p_1, p_2 \geq 0\}$; we assume that the excess demand functions are continuous over S; if there is no equilibrium, then by the property of homogeneity, there cannot be any equilibrium in S. Consequently, $Z_1(0, 1) > 0$ and $Z_2(1, 0) > 0$; further from the property of continuity of excess demand functions, it follows that for some small $\epsilon > 0$, $Z_2(1 - \epsilon, \epsilon) > 0$; by Walras Law, since $(1 - \epsilon).Z_1(1 - \epsilon, \epsilon) + \epsilon Z_2(1 - \epsilon, \epsilon) = 0$ it follows that $Z_1(1 - \epsilon, \epsilon) < 0$; let $f(x) = Z_1(1 - x, x)$; then $f(1) = Z_1(0, 1) > 0$; whereas $f(\epsilon) = Z_1(1 - \epsilon, \epsilon) < 0$ and hence by the Intermediate Value Theorem, there is $\overline{x} \in (\epsilon, 1)$ such that $f(\overline{x}) = 0$ that is $Z_1(1 - \overline{x}, \overline{x}) = 0$ and hence $(1 - \overline{x}, \overline{x})$ provide an equilibrium. Thus, under the assumptions, an equilibrium must always exist.

This argument is very much like the simple version of the Brouwer's Fixed Point Theorem presented above. There is a reason for that similarity. We shall investigate that reason next. To proceed formally, let us state the *Existence Theorem (ET): Consider a function $Z : S_n \to \Re^n$ which is continuous on S_n, and satisfies Walras Law (that is, $p'.Z(p) = 0 \forall p \in S_n$); then there is $p^* \in S_n$ such that $Z(p^*) \leq 0$.* Our first claim is contained in the following proposition:

Proposition 10.1   Brouwers's Fixed Point Theorem implies that ET is valid.

Proof Given $Z : S_n \to \Re^n$, with the properties noted in the statement of ET, construct $\phi : S_n \to S_n$ as follows:

$$\phi(p) = (\phi_i(p)) \text{ where } \phi_i(p) = \frac{\max(0, p_i + Z_i(p))}{\sum_j \max(0, p_j + Z_j(p))}$$

first, note that $g(p) = \sum_j \max(0, p_j + Z_j(p)) > 0 \forall p \in S_n$; since otherwise for some p, $\max(0, p_j + Z_j(p)) = 0 \forall j$ that is, $p_j + Z_j(p)) \leq 0 \forall j$; hence multiplying by p_j and summing, we have, using Walras Law, $\sum_j p_j^2 \leq 0$ hence $p_j = 0 \forall j \Rightarrow p \notin S_n$. Consequently, $\phi(p)$ is well defined on S_n and takes values in S_n too; also the continuity of the map $\phi(.)$ follows from the continuity of $Z(.)$ on S_n. Thus, from Brouwer's Fixed Point Theorem, there is a fixed point $p^* \in S_n$ such that $\phi(p^*) = p^* \Rightarrow \phi_i(p^*) = p_i^*$. Thus,

$$p_i^* g(p^*) = \max(0, p_i^* + Z_i(p^*)) \forall i$$

Hence, we must have

$$p_i^* g(p^*) = p_i^* + Z_i(p^*) \forall i \text{ such that } p_i^* > 0$$

Therefore, multiplying by p_i^* and summing over all such i:

$$g(p^*) \sum_{i \ni p_i^* > 0} (p_i^*)^2 = \sum_{i \ni p_i^* > 0} (p_i^*)^2$$

by virtue of Walras Law. Hence $g(p^\star) = 1$. Thus, it follows that

$$p_i^\star = \max(0, p_i^\star + Z_i(p^\star))\forall i \Rightarrow Z_i(p^\star) \le 0 \forall i;$$

Thus, ET is valid. •

Proposition 10.2   ET implies that any continuous function $f : S_n \to S_n$ admits a fixed point.

Proof Consider any continuous function $f : S_n \to S_n$; define $\xi : S_n \to \Re^n$ as follows:

$$\xi(p) = (\xi_j(p)) \text{ where } \xi_j(p) = f_j(p) - \lambda(p)p_j, \ p \in S_n \text{ and } \lambda(p) = \frac{\sum_k p_k f_k(p)}{\sum_k p_k^2}$$

Note that $\xi(p) : S_n \to \Re^n$, is continuous and $\xi(p).p = 0$; hence the ET applies and one may conclude that there is $p^\star \in S_n$ such that $\xi(p^\star) \le 0$; that means we have $0 \le f_j(p^\star) \le p_j^\star \lambda(p^\star) \forall j$; in fact, the strict sign cannot hold for any j; now summing over j and recalling that $f(p), p \in S_n$, it follows that $\lambda(p^\star) = 1$ whence it follows that $p^\star = f(p^\star)$: which proves the claim. •

Note then that ET and Brouwer's Fixed Point Theorem are at par with one another; without using a fixed point theorem, once cannot provide a comprehensive proof of the existence of a competitive equilibrium. Checking for the existence of a competitive equilibrium is central since that is the only configuration where all agents are able to carry out their plans. In fact, this is the reason why the check for existence may be thought of as the first consistency check for our model of a competitive economy. Several other checks need to be made and we shall come to them in due course.

10.5 Non-competitive Market Equilibria

When markets are not competitive, we need a somewhat different approach to the determination of equilibria. This section is devoted to such considerations.

Consider a market for an intermediate good X, produced by a single seller S and bought by a single buyer B.

The seller has a cost function $C(x)$ for x units of X; the cost function satisfies $C'(x) > 0$ and will be taken to be a strictly convex differentiable function, $C''(x) > 0$. Thus the profits of S, while selling x units of X at the unit price of w per unit is given by the profit function $\pi_S = w.x - C(x)$.

The buyer B, buys X to use it in the production of another product Y which may be sold at a constant unit price of p per unit; the production of y units of Y requires inputs of x units of X according to the production function $y = f(x)$ which will be assumed to be a differentiable strictly concave function, $f''(x) < 0$. The profits of B are given by $\pi_B = p.f(x) - w.x = g(x) - w.x$ assuming that Y is produced from X alone.

We wish to determine how many units of X are bought and sold and at what price. To proceed, we need to specify behavioural assumptions further.

10.6 Perfect Competition

Assume that B and S each believe that they would be unable to alter the price w to their own advantage. Under this behavioural rule, note that S would choose x to maximize π_S for each given w: thus S would choose that level of x at which $C'(x) = w$ (price = marginal cost: note that this level is uniquely determined and is profit maximizing: these are to be checked). One should also check that the curve is upward rising in the $w - x$ plane.

Similarly, B would maximize π_B to determine how much X should be purchased when w is known: that is, B would choose to purchase that level of X at which $g'(x) = w$: once again one should check that this is indeed profit maximizing; this is the demand curve for X and it should be easy to check that it is downward sloping.

Equilibrium demands that there be w_c, x_c at which $C'(x_c) = w_c = g'(x_c)$: assume that there is such an intersection between the downward sloping demand curve and the upward rising supply curve.

10.7 Monopoly and Monopsony

The seller now realizes that there are possibilities in extracting monopoly profits by setting price; assume that the buyer B still believes that it would not be possible to affect the price to his own advantage. The buyer B maximizes π_B as before and there is thus a downward sloping demand curve for the product X derived exactly as before: $g'(x) = w$. The seller S uses this information while formulating his own plans and now considers his profit function to be given by $\pi_S^m = g'(x).x - C(x)$ and chooses x to maximize this expression. In effect, the seller is choosing that point on the demand curve which maximizes his own profit π_S^m. This provides us with the monopoly solution: $C'(x) = g' + xg''(x)$ or note that the solution x^m must satisfy $x^m < x_c$ and hence $w^m > w_c$. Note that here the earlier supply curve becomes inapplicable since the supplier chooses a point on the demand curve. But note that the curve $C'(x)$ does play a role.

For the monopsony solution, note that it has to be the buyer B who sets the price. This is done such that B maximizes profit subject to the fact that the seller operates on the supply curve. Accordingly, the profit function $\pi_B^m = g(x) - xC'(x)$ is maximized by choosing an appropriate x: $g'(x) = xC''(x) + C'(x)$. The solution to this takes place at $x_m < x_c$ and it is easy to check that $w_c > w_m$. Note that here, in contrast, it is the demand curve which becomes inapplicable since the buyer chooses a point on the supply curve. However, note that the curve $g'(x)$ does play an important role.

10.8 Bilateral Monopoly

With both B and S recognizing their potential for controlling the price in the market, we need different tools entirely: demand and supply curves will not do. We consider the $w - x$ plane and find that each x, w combination implies a level of profits π_B, π_S

for the two participants. We shall consider iso-profit contours for each participant and thus classify points on the $w - x$ plane according to the level of profits implied by each.

Consider an iso-profit contour for B: $\pi_B =$ constant $\Rightarrow pf(x) - wx =$ constant; along such an iso-profit contour, we have:

$$p.f'(x)dx - wdx - xdw = 0 \Rightarrow \frac{dw}{dx}\big|_{\pi_B = \text{const}} = \frac{pf'(x) - w}{x}$$

Note that an iso-profit contour has a horizontal tangent along the curve $pf'(x) = w$ (this, of course, was the curve which was the demand curve under perfect competition).

Next consider an iso-profit contour for S: $\pi_S =$ constant $\Rightarrow w.x - C(x) =$ constant. Thus along an iso-profit contour for S, we have:

$$dwx + wdx - C'(x) = 0 \Rightarrow \frac{dw}{dx}\big|_{\pi_s = \text{const}} = \frac{C'(x) - w}{x}$$

Note that if a x, w point is such that iso-profit contours cross then this cannot be agreed to as a solution; since better (for both) alternatives exist. The only candidates for solution are where the iso-profit contours are tangential: slopes are the same; note that this is possible only if $x = x_c$. However, price w cannot be pinned down. To pin down the two limits, consider minimum acceptable profit levels for each firm to be 0. Now consider the iso-profit contours $\pi_B = 0$ and $\pi_S = 0$ along the line $x = x_c$; the former signifies the highest price acceptable $w_{\max}$ while the latter provides the lowest price acceptable $w_{\min}$; the former is the best price for S while the latter is the best price for B.

Finally, given that the profits of one of the participants, say S, is ensured not to fall below a level, say π_S^* what can the other participant B do? That is supposing that B is allowed to choose (w, x) subject to the constraint that $\pi_s \geq \pi_S^*$. What should B do? Clearly B should solve the following problem:

$$\max_{(w,x)} \pi_B = g(x) - wx \text{ subject to } wx - C(x) - \pi_S^* \geq 0, \ (w, x) \geq (0, 0)$$

It should be noted that the above is a concave problem (objective function and the constraint are both concave functions); so long as the level $\pi_S^* < \pi^o$, (which may be taken to be the maximum aggregate social surplus which is shown below to exist), there would be a (w, x) satisfying the inequality with a strict inequality (Slater's condition is satisfied). Thus the KTC are necessary and sufficient at a maximum. Note too that these conditions are given by:

(i) $g'(x^*) - w^* + \lambda^*[w^* - C'(x^*)] \leq 0$, $x^*\{g'(x^*) - w^* + \lambda^*[w^* - C'(x^*)]\} = 0, x^* \geq 0$

(ii) $-x^* + \lambda^* x^* \leq 0$, $w^*\{-x^* + \lambda^* x^*\} = 0$, $w^* \geq 0$

(iii) $w^* x^* - C(x^*) - \pi_S^* \geq 0$, $\lambda^*\{w^* x^* - C(x^*) - \pi_S^*\} = 0$, $\lambda^* \geq 0$

The maximum would involve $(w^*, x^*) > (0, 0)$; this would mean that $\lambda^* = 1$ and hence $g'(x^*) = C'(x^*) \Rightarrow x^* = x_c$. Also w^* would be chosen to satisfy the constraint exactly (binding). Thus it would mean that B would choose to be at x_c; this is what makes the solution robust in the sense that there should be no deviation from it.

10.9 Social Welfare Maximization

We may define social welfare as the sum of consumer's surplus and producer's surplus. Now consumer's surplus (for B) is given by $B_s = \int_0^x g'(s)ds - wx$ if B purchases x units at the price w: since the integral denotes the maximum that B would have paid for the right of consuming x units while the second term denotes what the consumer actually pays. Similarly, the producer's surplus for S is given by $S_s = wx - \int_0^x C'(s)ds$ if S sells x units at the price w per unit since, the first terms denote the revenue earned and the second the cost incurred in earning the revenue.

It should be noted that maximization of social welfare entails maximizing $B_s + S_s = \int_0^x g'(s)ds - \int_0^x C'(s)ds$; note that the first order condition for this is that $g'(x) = C'(x) \Rightarrow x = x_c$. Thus, it is the competitive level of output which maximizes the total surplus. The maximum social surplus is then $\{g(x_c) - g(0)\} - \{C(x_c) - C(0)\} = \pi^o$, (say). Thus note that under bilateral monopoly, **at least for this particular example, social welfare is maximized**; what is indeterminate is how this maximum surplus will be divided between the two; the higher is the agreed upon w, the higher will be the share of the seller S and conversely, the lower is the agreed upon w, the higher will be the share of the buyer B.

10.10 Efficiency and Competitive Equilibria

The interest in competitive equilibrium lies in the crucial relationship between competitive equilibria and efficient states. The relationship is thought to be of such great importance that the results providing this connection are collected under the common title **Fundamental Theorems of Welfare Economics**. We have built up all the apparatus to provide a brief account of these results. We begin with a description of the setting and some definitions.

Consider a collection of M consumers with an aggregate resource stock of n goods given by the vector W; a **feasible state** for this collection is a distribution of this stock among them: that is, an allocation $\{x^i \in \Re_+^n, i = 1, 2, \cdots, M\}$ such that $\sum_i x^i = W$. Let $\mathcal{F}$, denote the set of all possible feasible states. Let us also consider that each individual i has a continuous real valued utility indicator $U^i : \Re_+^n \to \Re$; we assume that the utility function is increasing. A feasible state $\{\overline{x}^i\} \in \mathcal{F}$ is said to be **efficient or Pareto Optimal** if there is no other feasible state $\{x^i\}$ such that $U^i(\overline{x}^i) \leq U^i(x^i)$ for all i with strict inequality for some i. In other words, to qualify as an efficient state, it must be the case that one cannot make any one better off unless it is at someone else's expense. Given a distribution of the total stock W among the individuals, say $\{w^i\}$, a feasible state $\{\hat{x}^i\} \in \mathcal{F}$ is said to be a **competitive state** at $\{w^i\}$ if there is a price vector $p \in \Re_+^n$ such that $\hat{x}^i$ solves, for each i max $U^i(x)$ subject to $p.x \leq p.w^i$.

We shall be concerned with the following two propositions.

Proposition 10.3   The First Fundamental Theorem: Any competitive state is an efficient state.

Proof Consider a competitive state $\{\overline{x}^i\}$ and if possible, assume that it is not an efficient state. Then there must be another feasible state $\{y^i\}$ such that $U^i(y^i) \geq U^i(\overline{x}^i)$ with strict inequality for some i, say $i = 1$. Also since $\{\overline{x}^i\}$ is competitive, there is $\overline{p}$, a price configuration and an endowment configuration $\{w^i\}$ such that $\overline{x}^i$ solves $\max U^i(x)$ subject to $\overline{p}.x \leq \overline{p}.w^i$. Also given the fact that $U^i(.)$ is increasing, we must have $\overline{p}.\overline{x}^i = \overline{p}.w^i$; and we must have $\overline{p}.y^i \geq \overline{p}.w^i \forall i$ with strict inequality for $i = 1$; the conclusion for $i = 1$ is easy to see: better points cannot be available; for the remaining, if y^i could be bought for less than $\overline{p}.w^i$ then by spending the rest one can actually purchase some bundle strictly better than y^i and hence better than $\overline{x}^i$: which is a contradiction. Hence, we must have $\sum_i \overline{p}.y^i > \sum_i \overline{p}w^i$ which contradicts the fact that $\sum_i y^i \leq \sum_i w^i$. Hence, no such y^i can exist. This proves the assertion. •

Remark 10.2   Note that we have not made any assumption about the convexity of preferences; the only restriction which has played a role is the assumption that the utility function is increasing (that more is always better). There is one assumption, which is not made explicit here: the existence of no interdependence between the decision-makers. Economists refer to this as the absence of externalities. In the presence of externalities, this result may no longer hold good and we have a case of **market failure.** *That is, the competitive market is unable to generate an efficient state. In the presence of externalities therefore, if the equilibrium is not efficient it means that by redistributing resources we can make someone better off and leave no one worse off.*

The Second Fundamental Theorem: For any efficient state $\{\overline{x}^i\}$, there is a price $\overline{p}$ and a distribution of resources $\{w^i\}$ such that $\{\overline{x}^i\}$ is a competitive state at $\{w^i\}$.

Remark 10.3   Stated as above, the result is not correct. We need to impose several additional restrictions before we can formally state and prove a similar result. First, we need to impose that the preferences are convex, that is, the utility function is quasi-concave.

With the above additional requirement, we have:

Claim 10.10.1   For any efficient state $\{\overline{x}^i\}$, there is a price $\overline{p}$ such that $\overline{p}.x \geq \overline{p}.\overline{x}^i \forall x$ such that $U^i(x) \geq U^i(\overline{x}^i)$.

Proof Define $C_i(x) = \{y \in \Re^n_+ : U^i(y) > U^i(x)\}$ and $C = \sum_i C_i(\overline{x}^i)$; that is, $y \in C \Rightarrow y = \sum_i y^i, y^i \in C_i(\overline{x}^i)$. Note that C is non-empty (by virtue of the increasing nature of the utility function), convex (since each U^i is quasi-concave); also note that $Y = \{y : y = \sum_i y^i, \{y^i\} \in \mathcal{F}\}$ is non-empty and convex too. Further $C \cap Y = \emptyset$ since otherwise, $\{\overline{x}^i\}$ cannot be efficient. Hence, by the standard Separation Theorem there is a plane which separates them: that is, there is $p \neq 0$ and a real number α such that $p.z \leq \alpha, \forall z \in \mathcal{F}$ and $p.y \geq \alpha \forall y \in C$. First of all note that since $\{\overline{x}^i\} \in \mathcal{F}$, $\overline{x} = \sum_i \overline{x}^i \in Y$ and hence, $p.\overline{x} \geq \alpha$; let ϵ_i be the vector with $\epsilon > 0$ in the i-th place, zeros everywhere else. Note that $\bar{x} + \epsilon_i \in C$ but $\bar{x} + \epsilon_i \notin Y$; thus we must have

$$p.\bar{x} + \epsilon p_i \geq \alpha \geq p.\bar{x}. \tag{10.16}$$

The above equation leads to two conclusions: first of all $p_i \geq 0 \forall i$ and second, $p.\bar{x} = \alpha$. Next, suppose that for some individual i say $i = 1$, we have for $y^1 \in C_1(\bar{x}^1)$, $p.y^1 < p.\bar{x}^1$. Consider $y^1 + \lambda_1 \epsilon_j$ and $\bar{x}^i + \lambda_i \epsilon_j$ for $i \neq 1$ where the λ_is are so chosen that

$$\alpha = p.\bar{x} = p.\sum_i \bar{x}^i > p.y^1 + \sum_{i \neq 1} p.\bar{x}^i + p_j \sum \lambda_i \epsilon \tag{10.17}$$

or that $p.\bar{x}^1 > p.y^1 + p_j \epsilon \sum_i \lambda_i$ so that we must have

$$\sum_i \lambda_i < \frac{p.\bar{x}^1 - p.y^1}{\epsilon p_j}$$

provided, of course, $p_j \neq 0$: note that such a j will always exist since $p \neq 0$.

However, note too that $Z = y^1 + \lambda_1 \epsilon_j + \sum_{k \neq 1}(\bar{x}^k + \lambda_k \epsilon_j) \in C$ and hence $p.Z \geq \alpha = p.\sum_i \bar{x}^i$; which is a contradiction to (10.17). Hence, no such y^1 can exist and this proves the claim. •

Note that what we have is slightly less than what is required at a competitive equilibrium. To close the remaining gap, we need to impose some more restrictions; before stating these note one consequence of $\{\bar{x}^i\}$ being efficient, that is,

$$\sum_i \bar{x}^i = W$$

From feasibility we know that there should be a weak inequality, that is, $\sum_i \bar{x}^i \leq W$; but if there is a strict inequality in any component, then that leftover amount can be given to someone to make that person better off and leave no one worse off: hence the earlier distribution could not have been efficient. Now for the additional assumptions:

(i) For any $x \in \Re_+^n$, the set $\{y \geq 0 : U^i(y) > U^i(x)\}$ is an open subset of $\Re_+^n$ for each i.

(ii) There is a 'cheaper' point at (p, x), that is, for $x \geq 0$ and $p \in \Re_+^n$, $p \neq 0$ there is $y \geq 0$, such that $p.y < p.x$.

Claim 10.10.2 Suppose that there is a cheaper point at $(p, \bar{x}^i)$ for each i where p is as shown to exist by virtue of Claim 10.10.1 and also suppose that the sets $C_i(x)$ defined in the proof of Claim 10.10.1 are open sets for each i and for all $x \geq 0$. Then there is a distribution of endowments w^i such that $\sum_i w^i = W, w^i \geq 0$ such that $\{\bar{x}^i\}$ is a competitive state at $\{p, \{w^i\}\}$, that is, $\bar{x}^i$ solves $\max U^i(x)$ subject to $p.x \leq p.w^i$.

Proof We already have that $p.x \geq p\bar{x}^i$ for all x such that $U^i(x) \geq U^i(\bar{x}^i)$; if possible let there be $y \geq 0$, $y \in C_i(\bar{x}^i)$ such that $p.y = p.\bar{x}$; since we have a cheaper point at $(p, \bar{x}^i)$, there is $z \geq 0$ such that $p.z < p\bar{x}^i$; consider then $v_t = ty + (1-t)z, 0 < t < 1$. Note that $v_t \geq 0 \forall t$, and $p.v_t < p\bar{x}^i$; by virtue of $C_i(\bar{x}^i)$ being an open set, $v_t \in C_i(\bar{x}^i)$

for t 'close' to 1: contradicts Claim 10.10.1; hence there can be no such y for any individual i. Thus we may conclude that $y \geq 0$, $py \leq p.\overline{x}^i \Rightarrow U^i(y) \leq U^i(\overline{x}^i)$.

Next we solve for $\{w^i\}$ by solving the equations:

$$\sum_i w^i = W, \; p.w^i = p.\overline{x}^i \forall i$$

That there are solutions to this system, we already know; since one set of solutions is given by $w^i = \overline{x}^i$ but there may be other solutions as well. This proves the claim. •

Remark 10.4 Claims 10.10.1 and 10.10.2 together constitute the Second Fundamental Theorem of Welfare Economics. If there is no cheaper point, the result may not hold. Consider the diagram below for such a situation. Whenever the Second Fundamental Theorem holds, we claim that an efficient state may be **decentralized**. *That is, if an efficient state is thought to be attained then a set of competitive markets may be used to attain this state by quoting prices and allowing the agents to maximize their own levels of satisfaction. This is the reason why the two Fundamental Theorems have attracted different schools of economists. There are those who always believe that the free market is the cure for everything: they hold dear the First Fundamental Theorem. The socialist planners, who wanted to choose the particular efficient state to be attained, laid their trust on the Second Fundamental Theorem.*

The point C in Figure 10.1 depicts a Pareto Optimum; note, however, the fact that there is no cheaper point for A at C given the price vector $(0, 1)$; since then the budget set for A coincides with the x-axis, given C to be the endowment point. Now B wants to be at C if the price is $(0, 1)$; however given the increasing nature of A's tastes, A demands an infinite amount of the good x at these prices, since it is free. Consequently, decentralization is not possible and the Second Fundamental Theorem breaks down.

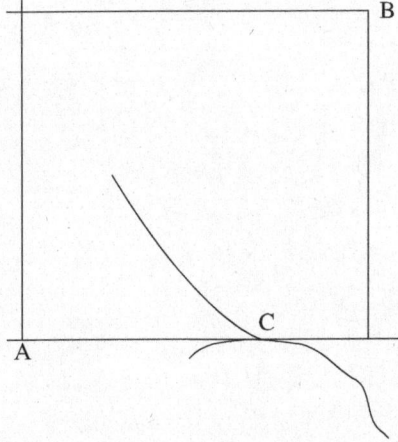

FIGURE 10.1 Pareto Optimum but not CE

KEY TERMS

Average Cost

Bilateral Monopoly

Compensated Demand Function

Cost Function

Cost Minimization Problem (Shephard's
  Lemma)

Excess

Demand Function

Existence Theorem

Expenditure Minimizing Problem

Fundamental Theorems of Welfare
  Economics

Hicks-Allen Theory

Jacobian of Excess Demand Function

Marginal Cost

Market Equilibria

Monopoly

Monopsony

Non-competitive Market Equilibria

Perfect Competition

Production Function

Profit Maximization Problem

Returns to Scale

Slutsky Equation (Fundamental Equation
  of the Theory of Value)

Social Welfare Maximization

Strict Convexity

Technology

Utility Maximization Problem

CHAPTER ELEVEN

Decision-making under Alternative Scenarios

11.1 Introduction

We have analysed decision-making when decision-makers are in complete control of what they seek to optimize: thus the objective is clear and there is nothing which is unknown nor is there anyone whose decision affects their choice. In this section, we shall examine what happens when these assumptions do not hold any longer. The first problem we shall try to analyse is the question of uncertainty: how do decision-makers choose when there is uncertainty. It should be clear that this is what the real world is like since we are always in uncertain situations and yet we need to choose or make decisions. The second set of problems that we shall examine is when there is interdependence between decision-makers, that is, one person's decisions affect the outcome of the other.

11.2 Decision-making under Uncertainty

11.2.1 Lotteries

Uncertain situations need to be properly described; consider, for example, the decision to undertake action a or b when the consequences of this choice are not known. What do we mean when we say that the consequences are not known? Is it that anything may result? If so, then such choices are beyond the scope of this text; we shall say the consequences are unknown to imply that among say, four possible outcomes we do not know which one will occur. Further, we shall also say that probability of each happening can be assigned a probability. Thus the uncertainty that we shall confront is not of the *anything may happen* variety; it is of the type where we may assign the outcomes with some probability of occurrence. Thus, it is best to describe uncertain situations or objects by the term *lottery* where we represent a lottery by the symbol $\{x, y; p, 1-p\}$ where there are two mutually exclusive and exhaustive outcomes x or y and the former occurs with a probability of p while the latter occurs with probability $1-p$. We have previously seen that the choice between objects requires that there be a ranking between them and that ranking is easily described by constructing a 'utility indicator' function. With lotteries as we have described them, can we achieve the same end? We shall see how this has been done. Let us first discuss the attitude towards such objects and a meaningful set of axioms that we shall impose on such attitudes.

11.2.2 Ranking over lotteries

First of all, we need to discuss the attitude of a decision-maker towards risky situations, where a risky situation and an uncertain situation are taken to mean one and the same thing. Consider the lottery $\{x, y; 1, 0\}$: the outcome x occurs with probability 1. We shall assume that the decision-maker, while considering such lotteries takes this to be at par with the situation where the outcome x occurs with certainty; while this may be quite a strong restriction, it seems to be the only way of treating probability one events: that is, treat them as sure outcomes and not merely as almost sure outcomes.

To proceed formally in the matter, let S be the set of all possible alternatives over which there is a ranking of the individual's preferences by the symbol $\gtrsim$, which is a binary relation defined over S and is assumed to satisfy **completeness**, **reflexivity**, and **transitivity** over the set S; we define from $\gtrsim$, the strict preference $>$ and the indifference $\approx$ in the usual manner. We assume,

Axiom 11.1 $\{x, y; 1, 0\} \approx x$

Axiom 11.2 $\{x, y; p, 1 - p\} \approx \{y, x; 1 - p, p\}$

Axiom 11.3   If $\mathcal{L} = \{x, y, ; q, 1 - q\}$ then $\{x, \mathcal{L}; p, 1 - p\} \approx \{x, y; p + q(1 - p), 1 - p - q(1 - p)\}$

The first, we have already commented on; the second merely states that the ordering of the outcomes is not important as long as the associated probabilities remain the same. Finally, if one of the outcomes is itself a lottery, then the entire lottery can be broken up as a single lottery with probabilities being assigned according to the usual rules for probabilities: this assumption is referred to as the **reduction of compound lotteries**.

We are going to demonstrate that under some conditions, we can construct a real valued continuous function $U : S \to \Re$ which satisfies the following two conditions:

(i) $s_1, s_2 \in S$ and $s_1 \gtrsim s_2 \Leftrightarrow U(s_1) \geq U(s_2)$
(ii) Let $s = \{x, y; p, 1 - p\} \in S$; then $U(s) = pU(x) + (1 - p)U(y)$.

Such a function U is said to be an **expected utility function** for the preference structure $\gtrsim$. The first requirement is to state that the function U serves as an indicator of preferences, that is, it is **order preserving**; while the second requirement is often referred to as the expected **utility property**.

Note that once we can ensure the existence of such a function, we may state the rule of decision-making in such contexts: *maximize expected utility*.

11.2.3 The expected utility function

The preference relation $\succsim$ over S apart from being reflexive, complete, and transitive is also required to satisfy two other assumptions:

Axiom 11.4 For any $a, b, c \in S$, the sets $\{p \in [0, 1] : \{a, b; p, 1 - p\} \succsim c\}\}$ and $\{p \in [0, 1] : c \succsim \{a, b; p, 1 - p\}\}$ are closed sets in $[0, 1]$.

Axiom 11.5 For any $a, b, c \in S$, $a \approx c \Rightarrow \{a, b; p, 1 - p\} \approx \{c, b; p, 1 - p\}$.

The first of the above axioms may be recalled from a similar exercise carried out while constructing an utility indicator function: the continuity assumption. The second is actually a good deal stronger than what is required. In fact, in the original Herstien and Milnor contribution, the restriction was required to hold only for $p = 1/2$ or the following was assumed:

$$\text{For any } a, b, c \in S, a \approx c, \Rightarrow \{a, b; 1/2, 1/2\} \approx \{c, b; 1/2, 1/2\}$$

It may be deduced that if this condition holds for $p = 1/2$ it holds for any $p \in [0, 1]$ (the interested reader is invited to consult the Herstein and Milnor Contribution for details). However, given our set-up, we can claim:

Claim 11.2.1 If $a, b \in S$ and $a > b$ then for every $p \in (0, 1)$, $a > \{a, b; p, 1 - p\} > b$

Proof We shall write $L(p) = \{a, b; p, 1 - p\}$. Suppose, to the contrary, that for some $\bar{p}, 0 < \bar{p} < 1$, we have $L(\bar{p}) = \{a, b; \bar{p}, 1 - \bar{p}\} \succsim a > b$; define $\mathcal{R}_a = \{c \in S : c \succsim a\}$; then we have $L(\bar{p}) \in \mathcal{R}_a$. We know that $L(p) \in \mathcal{R}_a \Rightarrow p > 0$; consider $\hat{p} = \inf_{p \in [0,1]} L(p) \in \mathcal{R}_a$: this infimum exists and $\hat{p} > 0$. Note that by virtue of Axiom 11.4, $L(\hat{p}) \in \mathcal{R}_a$; also since for $p < \hat{p} \Rightarrow L(p) \notin \mathcal{R}_a \Rightarrow a \succsim L(p)$ it follows that, once again using Axiom 11.5, that $a \succsim L(\hat{p})$ and hence $a \approx L(\hat{p})$.

Next using Axiom 11.5, we have that $L(\bar{p}) \approx \{L(\hat{p}), b; \bar{p}, 1 - \bar{p}\} \approx \{a, b; \hat{p}\bar{p}, 1 - \hat{p}\bar{p}\}$, by virtue of Axiom 11.3. Repeatedly applying Axioms 11.5 and 11.3, we have $L(\bar{p}) \approx L(\hat{p}^n \bar{p}) = \{a, b; \hat{p}^n \bar{p}, 1 - \hat{p}^n \bar{p}\}, n = 1, 2, \dots$. It, therefore, follows that $L(\hat{p}^n . \bar{p}) \succsim a$ for all n; hence as $\hat{p}^n . \bar{p} \to 0$ as $n \to \infty$, it follows using Axiom 11.4 that $L(0) \succsim a$: a contradiction since $L(0) \approx b$ by Axiom 11.1.

Hence no such $\bar{p}$ can exist and we must have $a > L(p)$ for all $p \in (0, 1)$. A similar argument will establish $L(p) > b$ as claimed. ●

We have next, the following:

Claim 11.2.2 If $a > b$ then $\{a, b; p, 1 - p\} > \{a, b; q, 1 - q\} \Leftrightarrow 1 > p > q > 0$.

Proof Suppose $a > b$ and $p > q$; then, by the previous claim we have $a > L(p) = \{a, b; p, 1 - p\} > b$ and since $q/p < 1$, we have again by the previous result $L(p) > \{L(p), b; q/p, 1 - q/p\} > b$ or using Axiom 11.3, $\{a, b; p, 1 - p\} > \{a, b; q, 1 - q\}$. In case $a > b$ and $\{a, b; p, 1 - p\} > \{a, b; q, 1 - q\}$, we may conclude from the above that $p > q$. ●

Consequently, in the set S, either all alternatives are indifferent (if there is no $a, b \in S, a > b$) or there are an infinite number of indifferent classes (since then there would be a pair $a, b \in S$ and $a > b$ with an indifferent class corresponding to each $\{a, b; p, 1 - p\}$ for each p). In particular, we have:

Claim 11.2.3 If $a > c$ and $a \gtrsim b \gtrsim c$ then there is a unique $p \in [0, 1]$ such that $b \approx \{a, c; p, 1 - p\}$.

Proof If there is a p satisfying the above, then that will be unique: this follows from the last result. Consider the sets $A = \{p \in [0, 1] : \{a, c; p, 1 - p\} \gtrsim b\}$ and $B = \{p \in [0, 1] : b \gtrsim \{a, c; p, 1 - p\}\}$; they are closed sets by Axiom 11.4. Also by the previous result $p' \in A, \bar{p} \in B \Rightarrow p' \geq \bar{p}$. Thus, consider the $\inf_{p \in A} p = \hat{p}$ and $\sup_{p \in B} p = \bar{p}$; then $\hat{p} \geq \bar{p}$; if the strict inequality holds then there is p such that $\hat{p} > p > \bar{p}$ and hence $p \notin A \cup B$. This violates completeness of the ranking $\gtrsim$ over the set S; hence we must have $\hat{p} = \bar{p}$ and further $\hat{p} \in A \cap B$. The last follows from the fact that both A, B are closed sets; hence we must have $b \approx \{a, c; \hat{p}, 1 - \hat{p}\}$. •

For any $a > b$ consider $S_{ab} = \{c \in S : a \gtrsim c \gtrsim b\}$; by virtue of what we have established above, for each $c \in S_{ab}$ there is a unique $p(c)$ such that $c \approx \{a, b; p(c), 1 - p(c)\}$. Thus on the subset S_{ab}, at least, $p(c)$ satisfies the twin properties of order preserving and expected utility; for instance, $p(c) \geq p(c') \Leftrightarrow c \gtrsim c'$; and if $c = \{x, y; q, 1 - q\}, x, y, c \in S_{ab}$; note that $x \approx \{a, b; p(x), 1 - p(x)\} = L_x$ say, while $y \approx \{a, b; p(y), 1 - p(y)\} = L_y$ and hence $c \approx \{L_x, L_y; q, 1 - q\} \approx \{a, b; qp(x) + (1 - q)p(y), 1 - qp(x) - (1 - q)p(y)\}$ so that $p(c) = qp(x) + (1 - q)p(y)$ which is the expected utility property. The final step consists of extending this construction to the whole of the set S.

The extension is achieved by first choosing $r_o, r_1 \in S$ such that $r_1 > r_o$; next choose $a, b \in S$ such that $r_o, r_1 \in S_{ab}$; for any $x \in S_{ab}$ define

$$U_{ab}(x) = \frac{p_{ab}(x) - p_{ab}(r_o)}{p_{ab}(r_1) - p_{ab}(r_o)}$$

where for any $x \in S_{ab}$, we define $p_{ab}(x)$ by:

$$x \approx \{a, b; p_{ab}(x), 1 - p_{ab}(x)\}$$

It is easy to check that the function $U_{ab}(x)$ satisfies both the required properties, that is, it is order preserving and satisfies the expected utility property. Note that $U_{ab}(r_1) = 1, U_{ab}(r_o) = 0$ for any choice of a, b. Further, one may show:

Claim 11.2.4 If $a, b, c, d \in S$ are such that $a > b, c > d$ and $r_o, r_1 \in S_{ab} \cap S_{cd}$, then for any $x \in S_{ab} \cap S_{cd}$, $U_{ab}(x) = U_{cd}(x)$.

Proof Consider $x \in S_{ab} \cap S_{cd}$; let $r_1 \gtrsim x \gtrsim r_o$; then there is a unique p_{1o} such that $x \approx \{r_1, r_o; p_{1o}, 1 - p_{1o}\}$; hence $U_{ab}(x) = p_{1o}U_{ab}(r_1) + (1 - p_{1o})U_{ab}(r_o) = p_{1o}U_{cd}(r_1) + (1 - p_{1o})U_{cd}(r_o) = U_{cd}(x)$; in case we have $x \gtrsim r_1 > r_o$; there is a unique p_1 such that $r_1 \approx \{x, r_o; p_1, 1 - p_1\}$ and as above, we have $U_{ab}(r_1) = p_1U_{ab}(x) + (1 - p_1)U_{ab}(r_o)$ and $U_{cd}(r_1)$

$= p_1 U_{cd}(x) + (1 - p_1)U_{cd}(r_o)$ so that subtracting one from the other, we obtain $p_1.[U_{ab}(x) - U_{cd}(x)] = 0$ and since $p_1 \neq 0$ (recall $r_1 > r_o$), we conclude that $U_{ab}(x) = U_{cd}(x)$. The last case when $r_1 > r_o \gtrless x$ may be considered similarly. •

Based on the above considerations, we are now ready to define the utility function $\mathcal{U} : S \to \Re$ which is **order preserving** and satisfies the **expected utility property**. Consider $r_1, r_o \in S, r_1 > r_o$ and any $x \in S$; choose $a, b \in S$ such that $x, r_1, r_o \in S_{ab}$ and define $\mathcal{U}(x) = U_{ab}(x)$.

The above provides a set of axioms under which decision-making takes place in uncertain situations. Somewhat loosely speaking, we extend the notion of objective functions pursued earlier (utilities or profits) to their expected counterparts (utilities or profits) and show that under some reasonable looking axioms, maximization of these expected utilities will be the relevant targets for decision-makers. We close this section with a discussion of how reasonable these axioms are. Alternatively, under what situations will expected utility maximization be violated? Clearly some of the axioms that we have taken will have to be violated.

Example 11.1 *Consider a decision-maker confronted with the following four alternative lotteries:*

A. get Rs 100,000 with probability 1;
B. get Rs 500,000 with probability 0.1, and Rs 100,000 with probability 0.89 and 0 with probability 0.01;
C. get Rs 100,000 with probability 0.11 and 0 with probability 0.89;
D. get Rs 500,000 with probability 0.1 and 0 with with a probability of 0.9

*It may be expected that a rational decision-maker may order the above lotteries as follows: A > B and D > C. Such a person would, of course, violate the dictates of the maximizing expected utility. This follows since the first comparison A > B would imply, if the person is an expected utility maximizer that $U(100,000) > 0.1U(500,000) + 0.89U(100,000) \Rightarrow 0.11U(100,000) > 0.1U(500,000)$ which would violate the ranking D > C, if the person was comparing expected utilities. This example is known as **Allais' Paradox** and shows that expected utility maximizing may be violated.*

Example 11.2 *A somewhat more complicated situation is attributed to **Daniel Ellsberg**. There is an urn containing 100 red balls and 200 balls which are either blue or green. The decision-maker is asked to pick a ball and is offered the following alternatives:*

A. get Rs 1000 if the ball is red;
B. get Rs 1000 if the ball is blue;
C. get Rs 1000 if the ball is not red and
D. get Rs 1000 if the ball is not blue.

> *A preference pattern $A > B$ and $C > D$ would imply a contradiction if the person happens to be an expected utility maximizer. For such a person, we have from the first, $p_r.U(1000) > p_b.U(1000)$ or that $p_r > p_b$ so that the decision-maker is assessing that p_r, the probability of picking red to be more than p_b, the probability of picking blue. But then it follows that $(1 - p_r) < (1 - p_b)$ and consequently, expected utility maximization will lead to $D > C$.*

Thus there may be situations where expected utility maximization is not the criterion for decision-making; however, the set of axioms provide a set-up where decision-makers follow expected utility in deciding the best alternative.

11.3 Risk Aversion

11.3.1 Preliminaries

We have seen how a person confronted with a risky situation, chooses among the uncertain outcomes. It may be recalled that the most common approach is to invoke the Expected Utility Theorem when the probabilities of the different outcomes are known. We have also seen that sometimes behaviour in risky situations may not conform to expected utility maximization. However, in what follows, we adopt maximization of expected utility as the behaviour rule. We shall be discussing a decision-maker's attitude towards risk and we shall also investigate whether different persons' attitudes may be compared so far as risk-taking is concerned.

11.3.2 Measures of risk aversion

Recall from our discussion of the Expected Utility Theorem that on the basis of the axioms that we mentioned, the individual decision-maker's evaluation of a lottery where outcomes are different sums of money (money gambles) is known once the particular decision-maker's utility function for money, $U(W)$, is known. We shall take such a function as our point of departure in this section. Consider any lottery $\{x, y; p, 1-p\}$ with the expected pay-off $p.x + (1 - p)y = z$. Individuals are said to be **risk averse** if

$$U(W_o + z) \geq U(\{W_o + x, W_o + y; p, 1 - p\})$$

where W_o is the current wealth and $U(.)$ satisfies the Expected Utility Theorem. In other words, the decision-maker is risk averse if he prefers a certain income equal to the expected outcome to the lottery. Note that $W_o + z = p.(W_o + x) + (1 - p)(W_o + y)$ and hence the above inequality may be written as:

$$U(p.(W_o + x) + (1 - p)(W_o + y)) \geq p.U(W_o + x) + (1 - p)U(W_o + y)$$

$$\text{for any } p \in [0, 1] \text{ and any } x, y.$$

This is, of course, equivalent to requiring the utility function of wealth $U(W)$ to be **concave** which is why risk aversion is taken to be equivalent to concavity of the utility function.

Hereafter we assume that the utility function $U(W)$ is twice continuously differentiable. The Arrow-Pratt Measure of Absolute Risk Aversion is given by

$$\rho(W) = -\frac{U''(W)}{U'(W)}$$

To appreciate why this could be thought of as a measure of risk aversion, we consider the set of all $\{(x, y)\}$ constituting lotteries $\{x, y; p, 1 - p\}$ that a decision-maker is willing to accept at the current wealth W_o given p. That is, we consider the set:

$$\{(x, y) : U(\{W_o + x, W_o + y; p, 1 - p\}) \geq U(W_o)\}$$

The **acceptance frontier** is the set of (x, y) where the above inequality is an equality:

$$A_U(W_o) = \{(x, y) : U(\{W_o + x, W_o + y; p, 1 - p\}) = U(W_o)\}$$

Along the acceptance frontier, we have thus:

$$p.U'(W_o + x) + (1 - p)U'(W_o + y).\frac{dy}{dx} = 0 \qquad (11.1)$$

Thus, evaluating at $(0, 0)$, we have

$$-\frac{dy}{dx}|_{(0,0)} = \frac{p}{1 - p}$$

Differentiating the expression (11.1) once again we have

$$p.U''(W_o + x) + (1 - p)U''(W_o + y).(\frac{dy}{dx})^2 + (1 - p)U'(W_o + y)\frac{d^2y}{dx^2} = 0 \quad (11.2)$$

Evaluating at the origin once again, we have

$$\frac{d^2y}{dx^2}|_{(0,0)} = -\frac{U''(W_o)}{U'(W_o)}.\frac{p}{(1 - p)^2} \qquad (11.3)$$

To explain, the diagram (see Figure 11.1) contains two acceptance frontier lines U_1 and U_2 relevant for two individuals: all acceptance frontiers must be tangential to the $\frac{p}{1-p}$ line at the origin. Note that a point such as L in the diagram while acceptable to U_1 is unacceptable to U_2: the second derivative in (11.3) must be larger for U_2; thus $\rho_1 < \rho_2$ where ρ_i is the Arrow-Pratt measure corresponding to the utility U_i. In this sense the measure of risk aversion is locally at least a correct measure for risk aversion.

11.3.3 Risk aversion and choice of risky assets

Consider the decision-making problem of allocating funds to a risk-free asset and a risky asset; the former has a sure return r_f per period whereas the return to the latter is a random variable $\tilde{r}$ which can be say either r_h with probability p or r_ℓ with probability $(1 - p)$. With current wealth W_o, allocating a to the risky asset means that after one period the wealth will be:

$$\tilde{W} = (W_o - a)(1 + r_f) + a.(1 - \tilde{r}) = W_o(1 + r_f) + a(\tilde{r} - r_f)$$

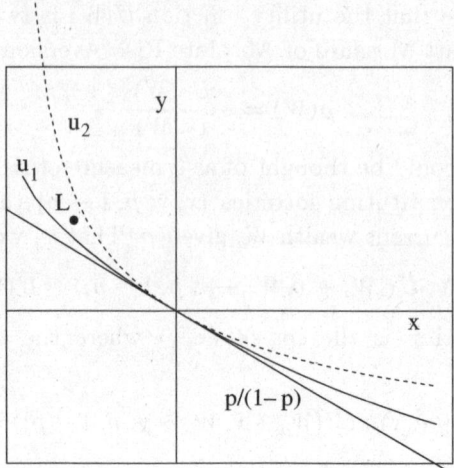

FIGURE 11.1 Acceptance Frontier

which is a random variable. The task of the decision-maker is to choose a so as to maximize

$$U(\tilde{W}) = p.U(W_o(1 + r_f) + a(r_h - r_f)) + (1 - p)\, U(W_o(1 + r_f) + a(r_\ell - r_f))$$

$$(11.4)$$

We shall write the rhs of the above expression as $E(U(\tilde{W}))$; this is due to the Expected Utility Theorem and due to the nature of the random variable specified above. If the distribution function of the random variable $\tilde{r}$ was some general $F(r)$ then, of course, we shall write $EU(\tilde{W}) = \int_A U(\tilde{W})dF$ as the Expected Utility where A is the set over which the random variable is distributed. We shall use the mathematical expectation operator E in deriving the necessary conditions for a maximum. While using the operator symbol E, it should be remembered that a derivative of the expectation is the expectation of the derivative; (11.4) may be differentiated to check the validity of this claim.

Thus, the first order condition for the maximum problem is to ensure that:

$$\frac{dEU(\tilde{W})}{da} = 0$$

or that

$$E[U'(\tilde{W}).(\tilde{r} - r_f)] = 0 \qquad (11.5)$$

To check our understanding, it may be seen that the above (11.5) reduces to $p.U'(W_h)(r_h - r_f) + (1 - p)U'(W_\ell)(r_\ell - r_f) = 0$ when (11.4) holds.

Thus at an interior maximum, where $a^\star > 0$ solves (11.5), since $U'(\tilde{W}) > 0$ it follows that the probability that $\tilde{r} - r_f > 0$ must lie in $(0, 1)$ since otherwise there is no way that the equation will hold. We shall write the optimum investment in the risky asset as $a^{\star}(W)$ when W is the current wealth.

More importantly for any positive investment in the risky asset to occur, the lhs of equation (11.5), evaluated for $a = 0$ cannot be negative:

$$U'(W_o(1 + r_f)).E(\tilde{r} - r_f) \geq 0$$

The above follows since at $a = 0 \Rightarrow \tilde{W} = W_o(1+r_f)$; the expression $E(\tilde{r}-r_f)$ is referred to as the risk premium. If the risk premium is negative $(p.r_h + (1 - p)r_\ell - r_f < 0$ in the two state case) no positive investment in the risky asset will take place. What must the risk premium be so that the entire investment is in the risky asset? As an immediate requirement it must be the case that the lhs of the equation (11.5) evaluated at $a = W_o$ cannot be negative or that

$$E[U'(W_o(1 + \tilde{r}))(\tilde{r} - r_f)] \geq 0; \tag{11.6}$$

this follows since $\tilde{W} = W_o(1 + \tilde{r})$ when $a = W_o$. Now note that, using a Taylor's series expansion we have:

$$U'(W_o(1 + \tilde{r})) = U'(W_o(1 + r_f)) + U''(W_o(1 + r_f)).(\tilde{r} - r_f) + \cdots;$$

hence

$$U'(W_o(1 + \tilde{r}))(\tilde{r} - r_f) = U'(W_o(1 + r_f))(\tilde{r} - r_f) + U''(W_o(1 + r_f)).(\tilde{r} - r_f)^2 + \cdots;$$

if $\tilde{r} - r_f$ is small then neglecting orders $(\tilde{r} - r_f)^3$ and above, we have from (11.6):

$$U'(W_o(1 + r_f)).E(\tilde{r} - r_f) \geq -U''(W_o(1 + r_f)).E(\tilde{r} - r_f)^2;$$

so that we have

$$E[(\tilde{r} - r_f)] \geq \frac{-U''(W_o(1 + r_f))}{U'(W_o(1 + r_f))}.E[(\tilde{r} - r_f)]^2 \tag{11.7}$$

Thus, note that the minimum risk premium required for full investment in the risky asset increases with ρ; the more risk averse an individual is, the higher the risk premium needs to be to elicit full investment in the risky asset. This is another way of validating the notion of ρ as a measure of risk aversion.

We say that increasing or decreasing risk aversion occurs according to:

$$\frac{d\rho}{dW} > (\text{ respectively } <) \ 0.$$

We shall show that:

Claim 11.3.1 $\frac{d\rho}{dW} < 0 \Rightarrow \frac{da^{\star}(W)}{dW} > 0.$

Proof Recall the first order condition, (11.5):

$$E[U'(W(1 + r_f) + a^{\star}(W)(\tilde{r} - r_f))] = 0;$$

since this will hold for all W, differentiating with respect to W we have:

$$E[U''(.)((1+r_f) + \frac{da^\star(W)}{dW}(\tilde{r} - r_f))(\tilde{r} - r_f)] = 0$$

or collecting terms we have:

$$\frac{da^\star(W)}{dW}.E[U''(.)(\tilde{r} - r_f)^2] = -E[U''(.)(\tilde{r} - r_f)](1 + r_f)$$

or:

$$\frac{da^\star(W)}{dW} = -\frac{E[U''(.)(\tilde{r} - r_f)](1 + r_f)}{E[U''(.)(\tilde{r} - r_f)^2]} \tag{11.8}$$

Note that on the rhs of the equation (11.8), the denominator is negative so that, the sign of the entire expression depends on the sign of the numerator. We shall try to sign this expression next.

Note that there are two possibilities:

$$\text{Either } \tilde{r} - r_f > 0 \ \ \text{OR} \ \ \tilde{r} - r_f \le 0$$

In the first case we must have

$$\rho(\tilde{W}) < \rho(W(1+r_f)) \text{ where } \tilde{W} = W(1+r_f) + a^\star(W)(\tilde{r} - r_f) \tag{11.9}$$

whereas in the second case, we must have

$$\rho(\tilde{W}) \ge \rho(W(1+r_f)). \tag{11.10}$$

We shall consider (11.9) first; note that we may rewrite it as follows:

$$-\frac{U''(\tilde{W})}{U'(\tilde{W})} < \rho(W(1+r_f))$$

or

$$U''(\tilde{W}) > -\rho(W(1+r_f)).U'(\tilde{W})$$

and hence

$$U''(\tilde{W})(\tilde{r} - r_f) > -\rho(W(1+r_f)).U'(\tilde{W})(\tilde{r} - r_f)$$

since $\tilde{r} - r_f > 0$ for the first case. Consider (11.10) next. This may be rewritten as

$$-\frac{U''(\tilde{W})}{U'(\tilde{W})} \ge \rho(W(1+r_f))$$

and hence as

$$U''(\tilde{W}) \le -\rho(W(1+r_f)).U'(\tilde{W})$$

and finally, recalling that $\tilde{r} - r_f \le 0$, we have

$$U''(\tilde{W})(\tilde{r} - r_f) \ge -\rho(W(1+r_f)).U'(\tilde{W})(\tilde{r} - r_f)$$

in this case as well. Hence we have:

$$E[U''(\tilde{W})(\tilde{r} - r_f)] > -\rho(W(1 + r_f)).E[U'(\tilde{W})(\tilde{r} - r_f)]$$

Using condition (11.5), it follows that the right side is 0 and hence the lhs is positive; this immediately signs the numerator in the rhs of (11.8) as being positive; with the minus sign, the lhs is positive as claimed. Consequently, the claim follows.•

A counterpart of the above proposition follows when the $\rho(W)$ happens to be **increasing** in W; now, of course, as is to be expected the amount of investment in the risky asset will **decrease** with W. The demonstration of this assertion follows the above closely and hence is omitted. We proceed next to another measure of risk aversion—the **measure of relative risk aversion**:

$$\rho^R(W) = -\frac{U''(W)}{U'(W)}.W$$

One may claim:

Claim 11.3.2 $\dfrac{d\rho^R(W)}{dW} > 0 \Rightarrow \eta = \dfrac{da^\star(W)}{dW}.\dfrac{W}{a} < 1.$

Proof Recall equation (11.8):

$$\frac{da^\star(W)}{dW} = -\frac{E[U''(.)(\tilde{r} - r_f)](1 + r_f)}{E[U''(.)(\tilde{r} - r_f)^2]};$$

hence, we have

$$\eta = -\frac{E[U''(.)(\tilde{r} - r_f)]W(1 + r_f)}{E[U''(.)a(W).(\tilde{r} - r_f)^2]} \tag{11.11}$$

Note that we may rewrite η as follows:

$$\eta = 1 + \frac{\dfrac{da}{dW}W - a}{a}.$$

Hence we have, using the expression (11.11):

$$\eta = 1 + \frac{W(1 + r_f)E[U''(\tilde{W})(\tilde{r} - r_f)] + aE[U''(\tilde{W})(\tilde{r} - r_f)^2]}{-aE[U''(\tilde{W})(\tilde{r} - r_f)^2]}$$

Note that the numerator on the rhs may be written as:

$$E[U''(\tilde{W})\{W(1 + r_f) + a(\tilde{r} - r_f)\}(\tilde{r} - r_f)] = E[U''(\tilde{W})\tilde{W}(\tilde{r} - r_f)]$$

while the denominator is positive; hence the sign of $\eta - 1$ is the same as the sign of $E[U''(\tilde{W})\tilde{W}(\tilde{r} - r_f)]$; we shall use the fact that

$$\frac{d\rho^R(W)}{dW} > 0$$

to sign this term. The argument is similar to the one used for the previous claim and begins by noting that

$$\tilde{r} - r_f \gtrless 0 \Leftrightarrow \tilde{W} \gtrless W(1 + r_f) \Leftrightarrow \rho^R(\tilde{W}) \gtrless \rho^R(W(1 + r_f))$$

Exactly as before, we may show that, considering separately, the cases $\tilde{r} - r_f > 0$ and $\tilde{r} - r_f \leq 0$, we must have:

$$E[U''(\tilde{W})\tilde{W}(\tilde{r} - r_f) < -\rho^R(W(1 + r_f)).E[U'(\tilde{W})(\tilde{r} - r_f)] = 0$$

so that the expression we had wanted to sign is negative and hence $\eta - 1 < 0$ which proves the claim.

Some standard forms of the utility functions $U(W)$ are as follows:

1. $U(W) = W - \frac{b}{2}W^2, b > 0$
2. $U(W) = -e^{bW}, b > 0$
3. $U(W) = \frac{b}{b-1}W^{1-\frac{1}{b}}$, for $W > 0, b > 0$

The measures of absolute and relative risk aversion should be computed for each of the above forms.

11.3.4 Global measures of risk aversion

We have seen that the measure of absolute risk aversion $\rho(W)$ could indeed be taken to be a measure of local risk aversion in the following sense: if one individual had a higher value for $\rho(W)$ compared to another at the same W then, given $p \in [0, 1]$ there would be some lotteries $\{x, y; p, 1 - p\}$ which the former would reject but the latter might accept by considering the Acceptance frontier and for small values of x, y. Alternatively, in situations of portfolio choice when there was a risk free asset and a risky asset, the person with a higher value of ρ would require a higher risk premium to allocate all into the risky asset provided the risk was small. Both of these approaches tend to characterize the measure $\rho(W)$ as a **local** measure of risk aversion. Our task in this section is to investigate whether this measure may be seen to be a measure of risk aversion regardless of the size of the risk.

We need, first of all, to show that:

Claim 11.3.3 For two risk averse individuals, A and B, with utility functions $U_A(W)$ and $U_B(W)$, $\rho_A(W) > \rho_B(W) \Leftrightarrow \exists G$ strictly increasing and concave function such that $U_A(W) = G(U_B(W))$ $(G'(.) > 0, G''(.) < 0)$.

Proof For necessity, note that since $y = U_B(W)$ is monotonic, $W = U_B^{-1}(y)$ is well-defined; hence define $U_A(W) = U_A(U_B^{-1}(y)) = G(y) = G(U_B(W))$. It follows that $U_A'(W) = G'(U_B(W)).U_B'(W) \Rightarrow G'(U_B(W)) > 0$. Differentiating once again, we have ,

$$U_A''(W) = G''(U_B(W)).(U_B'(W))^2 + G'(U_B(W)).U_B''(W)$$

and hence

$$\rho_A(W) = \rho_B(W) - \frac{G''(U_B(W)).U_B'(W)}{G'(U_B(W))} \tag{11.12}$$

so that given that $\rho_A(W) > \rho_B(W)$, $G''(.) < 0$ follows. Hence necessity follows.

The sufficiency follows by reversing the steps since if there is such a function G then, using the expression (11.12), $\rho_A(W) > \rho_B(W)$ follows. This completes the demonstration. •

We shall now show the following:

Claim 11.3.4 If there is a strictly increasing, concave function G such that $U_A(W) = G(U_B(W)) \forall W$ and the risk premium is the minimum required for B to invest fully in the single risky asset for some given level of income W, then A with the same level of income, will always invest less in the risky asset.

Proof We have already seen that when $U_A(W) = G(U_B(W)) \forall W$ and $G(.)$ is increasing and concave then $\rho_A(W) > \rho_B(W) \forall W$. Consider then a portfolio choice problem where there is one risk free asset with return r_f and a risky asset with return $\tilde{r}$, a random variable. Let B have wealth W and allocate all of it to invest in the risky asset. We have seen (11.6), for individual B, that is, $E[U'_B(W(1 + \tilde{r}))(\tilde{r} - r_f)] \geq 0$ is a necessary condition. Consider next the smallest risk premium $E(\tilde{r} - r_f)$ required for this to happen. It is easy to see that (11.6) must be met with an equality, that is, $E[U'_B(W(1 + \tilde{r}))(\tilde{r} - r_f)] = 0$, since otherwise, with a strict inequality, even a slightly smaller risk premium would satisfy (11.6) and hence would induce a full investment in the risky asset. Conversely, suppose that (11.6) is met; now note that the expression

$$\phi_i(a) = E[U_i(W(1 + r_f) + a(\tilde{r} - r_f))] \text{ for } 0 \leq a \leq W \text{ and } i = A, B$$

is strictly concave since $\phi''_i(a) = E[U''_i(W(1 + r_f) + a(\tilde{r} - r_f)).(\tilde{r} - r_f)^2] < 0$; hence (11.6) for B amounts to $\phi'_B(W) = 0$ and hence from the strict concavity, it follows that $\phi'_B(a) > 0$ for all $a < W$ and consequently (11.6) for B implies that B makes full investment in the risky asset. What does A do in this situation?

Consider then $\phi'_A(W) = E[U'_A(W(1 + \tilde{r})(\tilde{r} - r_f)]$; notice that the rhs may be split up thus:

$$E[U'_A(W(1 + \tilde{r})(\tilde{r} - r_f)|(\tilde{r} - r_f \geq 0].P[\tilde{r} - r_f \geq 0]$$
$$+ E[U'_A(W(1 + \tilde{r})(\tilde{r} - r_f)|\tilde{r} - r_f < 0].P[\tilde{r} - r_f < 0] \quad (11.13)$$

where $P[X]$ represents the probability of the event X. The first term in (11.13) may be further decomposed thus:

$$E[G'(U_B(W(1 + \tilde{r}))U'_B(.)(\tilde{r} - r_f))].P[\tilde{r} - r_f \geq 0] \quad (11.14)$$

Note that in case $\tilde{r} - r_f \geq 0$, we must have $W(1 + \tilde{r}) \geq W(1 + r_f)$ and consequently $U_B(W(1 + \tilde{r})) \geq U_B(W(1 + r_f))$ and using the concavity of the function $G(.)$ we have the expression in (11.14) to be

$$\leq G'(U_B(W(1 + r_f)))E[U'_B(W(1 + \tilde{r}))(\tilde{r} - r_f)].P[\tilde{r} - r_f \geq 0]$$

Proceeding exactly as above in the case $\tilde{r} - r_f < 0$, the second term in (11.13) may be shown to be

$$< G'(U_B(W(1 + r_f)))E[U'_B(W(1 + \tilde{r}))(\tilde{r} - r_f)].P[\tilde{r} - r_f < 0]$$

The strict inequality follows from the fact that in this situation (that is, $\tilde{r} - r_f < 0$) we have $W(1 + \tilde{r}) < W(1 + r_f)$ and given the strictly increasing property of $G(.)$ and the fact that $P[\tilde{r} < r_f] > 0$ since otherwise the problem being analysed ceases to be of interest. Combining these two expressions and using (11.13) we have

$$\phi'_A(W) < G'(U_B(W(1 + r_f))).E[U'_B(W(1 + \tilde{r})).(\tilde{r} - r_f)] = 0 \text{ (by virtue of 11.6)}$$

and consequently A will invest less than the full amount in the risky asset, as claimed. ●

It is in this sense that the measure of risk aversion $\rho(W)$ may be considered to be a global measure of risk aversion as well.

11.3.5 Portfolio choice with more than one risky asset

Next, we move to a consideration of choice over risky assets when there are two or more risky assets in addition to a risk-free asset. Given that there are many risky assets, an increase in wealth may be accompanied by more investment in some risky asset while decreasing investment in other risky assets. Or an individual may hold the same portfolio of risky assets and with increase in wealth change the mix between this portfolio and the risk-free assets. In other words, for all practical purposes, the investor acts as if there are only two assets: one risk-free and the other risky but which itself is a fixed composite of several other risky assets (**a mutual fund**). If this is exhibited we shall say that **two-fund monetary separation** takes place. To see when this may happen, consider an investor who devotes a proportion α of his initial wealth W to the risk-free asset and let β_j be the proportion of the remainder $((1 - \alpha)W)$ that is devoted to the j-th risky asset. Thus, after one period the investor's wealth becomes $\tilde{W} = \alpha W(1 + r_f) + (1 - \alpha)W \sum_j \beta_j (1 + \tilde{r}_j) = W(1 + \alpha r_f + (1 - \alpha) \sum_j \beta_j \tilde{r}_j)$ and the investor wishes to solve the following problem by choosing α, β_j:

$$\max E[U(\tilde{W}] \text{ subject to } \sum_j \beta_j = 1, \beta_j \geq 0, 1 \geq \alpha \geq 0$$

The Lagrangean for the above problem is $\mathcal{L} = E[U(\tilde{W})] + \lambda(1 - \sum_j \beta_j) + \mu(1 - \alpha)$. The first order conditions for this problem are given by:

1. $\mathcal{L}_\alpha = E[U'(\tilde{W})(r_f - \sum_j \beta_j \tilde{r}_j)] - \mu \leq 0, \ \alpha.\mathcal{L}_\alpha = 0, \ \alpha \geq 0$
2. $\mathcal{L}_{\beta_j} = E[U'(\tilde{W})(1 - \alpha)\tilde{r}_j] - \lambda \leq 0, \ \beta_j.\mathcal{L}_{\beta_j} = 0, \ \beta_j \geq 0 \ \forall j$
3. $\mathcal{L}_\lambda = (1 - \sum_j \beta_j) \geq 0, \ \lambda.\mathcal{L}_\lambda = 0, \ \lambda \geq 0$
4. $\mathcal{L}_\mu = (1 - \alpha) \geq 0, \ \mu.\mathcal{L}_\mu = 0, \ \mu \geq 0$

Note that at an interior optimum $0 < \alpha < 1$, $\beta_j > 0$; hence $\mu = 0$; also note that $\sum_j \beta_j = 1$ and hence the first and second condition reduce to

$$E[U'(W(1 + r_f + (1 - \alpha) \sum_j (\beta_j(\tilde{r}_j - r_f))))(\tilde{r}_j - r_f)] = 0 \ \forall j \qquad (11.15)$$

Two-fund monetary separation occurs if it so happens that β_j is independent of W so that it is α which keeps changing as W changes and we have a mutual fund which is made up of the risky assets in accordance with the proportion β_j being allocated to the j-th risky asset.

It may be shown that such a situation emerges only under very special conditions and a necessary and sufficient condition for two-fund monetary separation is that by Cass and Stiglitz (1970):

$$U'(W) = (A + BW)^C$$

and EITHER $B > 0, C < 0,$ with $W \geq \max[0, -A/B]$

OR $A > 0, B < 0, C > 0,$ with $0 \leq W < -A/B$

Note that the restrictions on the parameters serve to make $U''(W) < 0$ and $U'(W) > 0$ on the range of feasible values of W. It may also be shown that another form of the marginal utility of wealth, that is, $U'(W) = Ae^{BW}$ also serves the same purpose provided $A > 0$, and $B < 0$.

11.4 Interactive Decision-making: Game Theory

11.4.1 Introduction

Consider a situation where outcomes depend on the actions taken by all the decision-makers; that is the objective function (utility, say) depends not only on the actions each chooses but also on the actions chosen by others. The simplest such structure, where two decision-makers have two actions each and have to choose their actions simultaneously is as follows:

$$\begin{array}{cc} & L \qquad\quad R \\ \begin{array}{c} U \\ D \end{array} & \left(\begin{array}{cc} (1,1) & (1,2) \\ (-2,3) & (3,4) \end{array} \right) \end{array}$$

Thus, one has actions U or D while the other has actions L and R; the outcome (say, in Rupees) is specified only when both the choices are known. Let us refer to the decision-makers as Row and Column so that Row can choose either U or D and Column can choose between L and R and when Row chooses U and Column chooses L, Row and Column each get Re 1; the first entry in the table is the return to Row while the second shows the return to Column. In this setting, what should each decision-maker do? Although it may seem to be baffling at first, thinking about the problem provides us with the following analysis.

First of all note that Column will never choose L: the reason for that is that by choosing R, it can assure higher returns (than when choosing L) regardless of the choice of Row. Given this note further that the better choice by Row is to choose D (it gets Rs 3 which is higher that Re 1 Row can get by choosing U). So Row chooses D and Column chooses R.

What we have described above is a **non-cooperative game** with two **players** (decision-makers), Row and Column, where each has two strategies or actions to choose from. The table represents the outcomes or **pay-offs**. We showed how Column has a **dominant strategy**, that is, a strategy which will be the best response to any choice by Row and consequently being rational, column will always choose to **play** that. Given this, it is equally clear what Row will play, that is, D. The choices D, R constitute an **equilibrium**, properly called a **Nash equilibrium** after John Nash. We described a sensible way in which Row and Column may play the game described: it helped that one of the players had a clearly superior way of playing, that is, choosing what we called a dominant strategy. But such strategies may not exist, what would be a sensible way of playing the game? And do all games have an equilibrium of the type mentioned? We turn to an examination of these issues next. As we shall see, just as we found in the previous section, we do not need to learn new tricks; we just need to apply the tools developed earlier and see how they apply in these changed scenarios.

11.4.2 Games in normal form

To specify a game, we need to specify three sets: the set of players N, the set of strategies S_i for each player $i \in N$ and the pay-offs of each player $\pi_i : \prod_{i \in N} S_i \to \Re$.

Note that it is the domain of definition of the pay-offs that is important and it is because of this aspect that such problems are defined to be problems of '**interactive decision-making**'. The game described in the last section by the matrix of pay-offs is reproduced below (Game 1):

$$
\begin{array}{c c}
 & \begin{array}{c c} L & \quad R \end{array} \\
\begin{array}{c} U \\ D \end{array} & \begin{pmatrix} (1,1) & (1,2) \\ (-2,3) & (3,4) \end{pmatrix}
\end{array}
$$

This is a game in **normal form** specifying the players, their strategies and pay-offs. The most well-known two-person game where both players have dominant strategies is the **prisoners' dilemma** and may be represented as below (Game 2):

$$
\begin{array}{c c}
 & \begin{array}{c c} C & \quad NC \end{array} \\
\begin{array}{c} C \\ NC \end{array} & \begin{pmatrix} (-1,-1) & (4,-2) \\ (-2,4) & (3,3) \end{pmatrix}
\end{array}
$$

It has the following story attached to it. Two persons (Row and Column) are arrested on the suspicion of committing a petty crime; they have two options to either confess (C) or not confess (NC); if both confess, they are treated leniently and a penalty of 1 unit is imposed on each; if they both do not confess they are compensated for their loss of time by being given a compensation of 3 units each; however, if only one of them confesses, that person is rewarded with 4 units while the other is penalized by 2 units. This celebrated game has a unique point to make; note that C or confessing is a dominant strategy for both Row and Column; since by confessing each can get a pay-off of either −1 (if the other also plays C) or 4 (if the other plays NC) but by choosing NC one gets either −2 or 3, respectively; and consequently each chooses to play C. However, if they had both chosen NC, they would have been better off; the reason why that does not happen is because if we know that our opponent is contemplating NC, we would be better off playing C. This is how the game will be played and C,C is the **dominant strategy equilibrium**.

If there are dominant strategies, as in either in Game 1 or in Game 2, we can reason how to play the game. When both have dominant strategies then it is clear that both should use them. However, if only one has a dominant strategy, as in game 1, we know that the player with a dominant strategy will always play that: so that in the pay-off matrix only one column is relevant. Then, given this, we can figure out what is the best strategy for the other. We proceeded by eliminating dominated strategies: so the equilibrium in game 1, R,D is called an **iterated dominant strategy** equilibrium.

Consider next, Game 3:

$$
\begin{array}{c}
 & \begin{array}{cc} B & \quad T \end{array} \\
\begin{array}{c} B \\ T \end{array} & \left(\begin{array}{cc} (1,2) & (0,0) \\ (0,0) & (2,1) \end{array} \right),
\end{array}
$$

which has the following story attached to it: Column and Row have to decide how to spend an evening; they have two options: either attend a Boxing Match (B) or to attend a theater (T). They cannot discuss the matter but while Row prefers the theater and Column prefers the boxing match, they would rather spend time together, which explains the pay-offs. This game is well known as the **Battle of the Sexes**.

Note that there are **NO** dominant strategies; however, given that Row chooses T, the best response for C would be to choose T as well; while given that Column chooses B, the best response of Row would be to choose B as well. Thus while B is a best response to B, T is a best response to T as well. So (B,B) and (T,T) are two pairs of best response strategies, each best response to one another, and such a set of strategies, one for each player is said to constitute a **Nash equilibria**. Thus the Battle of the Sexes has two Nash equilibria. Therefore, the obvious way to play the game or to decide which strategy to choose becomes somewhat difficult.

Formally, the strategies S, T for two players constitute a Nash equilibrium, if S is the best response to T and T is the best response to S. More generally, the array $\{\hat{s}_i\}, \hat{s}_i \in S_i$, constitutes a Nash equilibrium if $\pi(\hat{s}_1, \cdots, \hat{s}_i, \cdots, \hat{s}_n) \geq \pi(\hat{s}_1, \cdots, s_i, \cdots, \hat{s}_n)$ $\forall s_i \in S_i, \forall i$. Do games always have Nash equilibria? We shall investigate this matter next.

We introduce the notion of 'mixed strategies' first. Consider any of the games discussed above, say game 3; there are two strategies for each player: choose either B or T; but consider the following way of choosing a strategy: toss a coin and then decide that if heads appears the player, say C, chooses B; otherwise C chooses T. Thus, we may consider C to have chosen to play a **mixed strategy** (0.5, 0.5) (assuming that the coin is a fair one) where the first refers to the probability of playing B and the other refers to the probability of playing T. One could think of spinning a roulette wheel and then deciding and hence a mixed strategy of $(p, 1 - p)$ for any $p, 0 \leq p \leq 1$ can be considered. The returns, now become expected returns, of course; and for example, if C chooses to play B with probability p and T with probability $(1 - p)$, his expected returns in case R chooses B is $2p + (1 - p).1$. We shall now refer to B, T as '**pure strategies**'; note that when a player has a dominant strategy, mixing a dominant strategy with another one is never a good idea. But in Game 3, this may work. In fact, it should be computed that in Game 3 apart from the two pure strategy Nash equilibria, there is a mixed strategy Nash equilibrium.

Consider next, Game 4: **Matching Pennies**.

$$
\begin{array}{cc}
 & \begin{array}{cc} H & \qquad T \end{array} \\
\begin{array}{c} H \\ T \end{array} & \begin{pmatrix} (1, -1) & (-1, 1) \\ (-1, 1) & (1, -1) \end{pmatrix}
\end{array}
$$

Note that there is no pure strategy Nash equilibrium; however, if Row chooses H with probability p (and T with probability 1 – p) and Column chooses H with probability q and T with probability 1 – q, we may ask, for a given q what is what is the best that Row can do?

The expected returns that Row can ensure for herself are as follows:

$$p[q + (1 - q)(-1)] + (1 - p)[q(-1) + (1 - q).1] = 2p(2q - 1) + 1 - 2q$$

Thus, if $q < 1/2$, the best response of Row is to choose $p = 1$ since that would give the maximum expected returns; on the other hand, if $q > 1/2$ the best response is to choose $p = 0$ since that provides the highest expected returns. Finally, if $q = 1/2$, any choice of $p \in [0, 1]$ is the same for Row. A similar calculation shows Column's expected returns are given by:

$$2q(1 - 2p) + 2p - 1$$

and consequently, for any value of $p < 1/2$, expected returns are maximized by choosing $q = 1$; while for $p > 1/2$, Column's expected returns are maximized by choosing $q = 0$ and for $p = 1/2$, any choice of $q \in [0, 1]$ will be the same for Column. We represent (see Figure 11.2) these responses in the $p - q$ plane: the line marked R (C) stands for the best response by R (respectively, C) to various choices by C (R, respectively). The lines cross only at $1/2, 1/2$; thus $p = 1/2$ is the best response to $q = 1/2$ and *vice versa*. Hence $(1/2, 1/2)$ constitutes a Nash equilibrium: in fact, it is the only Nash equilibrium.

A similar diagram may be constructed for Game 3 (The Battle of the Sexes) and it may be seen that the R and C lines drawn will intersect at three points: two of them will correspond to the pure strategies we have noted above; the third will be a mixed strategy. As we shall see, if there are a finite number of pure strategies, then there will always be a Nash equilibrium. The existence of a Nash equilibrium may be

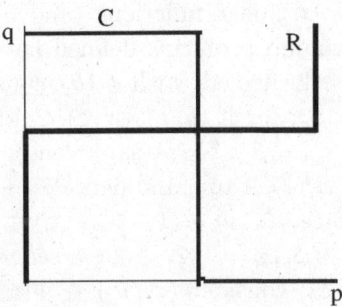

FIGURE 11.2   Mixed Strategy Nash Equilibrium

demonstrated under a somewhat more general set-up but the argument will essentially
be the same. We present the proof of the situation when there are 2 players R, C each
with a finite number of pure strategies m, n respectively; further let the strategies be
denoted by $s_1^R, \cdots, s_m^R$, and $s_1^C, \cdots, s_n^C$, respectively. Let the pay-offs be denoted by
π_{jk}^R, π_{jk}^C when R chooses s_j^R and C chooses s_k^C.

The set of strategies S^R, S^C are, of course, defined as follows: $S^R = \{y : y =
\sum_i \alpha_i s_i^R, \alpha_1 \geq 0, \sum_i \alpha_i = 1\}$; alternatively, we may consider the set of strategies for
each to be a probability distribution over the finite pure strategies. Thus, a strategy
for R will be a point $\alpha \in S_m = \{\alpha = (\alpha_i) : \alpha_i \geq 0, \sum_{i=1}^{m} \alpha_i = 1\}$ and a strategy of C
will be a point $\beta \in S_n = \{\beta = (\beta_i) : \beta_i \geq 0, \sum_{i=1}^{n} \beta_i = 1\}$. It should be understood that
choosing α means the mixed strategy of playing s_i^R with probability α_i; consequently,
a choice α, β implies the expected return

$$\sum_{i=1}^{m} \sum_{j=1}^{n} \alpha_i \beta_j \pi_{ij}^R \text{ to } R$$

and

$$\sum_{i=1}^{m} \sum_{j=1}^{n} \alpha_i \beta_j \pi_{ij}^C \text{ to } C.$$

Define $\Phi : S_m \times S_n \to 2^{S_m \times S_n}$ thus: $\Phi(\alpha, \beta) = \{(u, v)\}$ where $u \in \phi_1(\beta), v \in \phi_2(\alpha))$ where
$\phi_1(\beta)$ solves the problem:

$$\max_{\alpha \in S_m} \sum_{i=1}^{m} \sum_{j=1}^{n} \alpha_i \beta_j \pi_{ij}^R \tag{11.16}$$

and $\phi_2(\alpha)$ solves the problem:

$$\max_{\beta \in S_n} \sum_{i=1}^{m} \sum_{j=1}^{n} \alpha_i \beta_j \pi_{ij}^C \tag{11.17}$$

The map Φ has some properties which are useful to note. First, note that the set
$S_m \times S_n$ is a **convex and compact** subset, being the Cartesian product of such sets.

Next, since the maximum is of linear functions (and hence, continuous functions) over a compact set, the maximum problems defined have solutions for any $(\alpha, \beta) \in S_m \times S_n$; the map Φ is thus well defined, with $\phi_1(\beta), \phi_2(\alpha) \neq \emptyset$. In addition, $\Phi(\alpha, \beta)$ is a **convex subset** of $S_m \times S_n$; for if $(u, v), (w, z) \in \Phi(\alpha, \beta)$ then $u, w \in \phi_1(\beta) \Rightarrow \sum_{i=1}^{m} \sum_{j=1}^{n} u_i \beta_j \pi_{ij}^R = \sum_{i=1}^{m} \sum_{j=1}^{n} w_i \beta_j \pi_{ij}^R = A(\text{say})$; consequently, for any $\theta, 0 \leq \theta \leq 1$, $\sum_{i=1}^{m} \sum_{j=1}^{n} (\theta u_i + (1 - \theta) w_i) \beta_j \pi_{ij}^R = A$ too, and hence $\theta u + (1 - \theta) w \in \phi_1(\beta)$. Similarly, $\theta v + (1 - \theta) z \in \phi_2(\alpha)$ and hence $\theta(u, v) + (1 - \theta)(w, z) \in \Phi(\alpha, \beta)$, as claimed.

Finally, let $(\alpha^r, \beta^r) \in S_m \times S_n, r = 1, 2, \cdots$ be a sequence with (α^o, β^o) as a limit point; then $(\alpha^o, \beta^o) \in S_m \times S_n$; consider $(u^r, v^r) \in \Phi(\alpha^r, \beta^r)$; let u^o, v^o be a limit point of the sequence $(u^r, v^r) \in S_m \times S_n$: a compact set, hence $(u^o, v^o) \in S_m \times S_n$. We shall show that $(u^o, v^o) \in \Phi(\alpha^o, \beta^o)$. For if this is not so, then either $u^o \notin \phi(\beta^o)$ or $v^o \notin \phi(\alpha^o)$. If the former holds, then there must be $\bar{u} \in S_m$ such that

$$\sum_{i=1}^{m} \sum_{j=1}^{n} \bar{u}_i \beta_j \pi_{ij}^R > \sum_{i=1}^{m} \sum_{j=1}^{n} u_i^o \beta_j \pi_{ij}^R$$

since $u^o \in S_m$; but then we must have

$$\sum_{i=1}^{m} \sum_{j=1}^{n} u_i^r \beta_j \pi_{ij}^R < \sum_{i=1}^{m} \sum_{j=1}^{n} \bar{u}_i \beta_j \pi_{ij}^R$$

for all r large enough; which is a contradiction to the definition of the sequence (u^r, v^r); so no such $\bar{u} \in S_m$ can exist. We prove the remaining case similarly and hence the claim follows. The map Φ is thus **closed**.

The map $\Phi : S_m \times S_n \to 2^{S_m \times S_n}$ thus satisfies all the conditions required by the **Kakutani Fixed Point Theorem** and hence must allow for a fixed point $(\bar{\alpha}, \bar{\beta}) \in \Phi(\bar{\alpha}, \bar{\beta})$, that is, $\bar{\alpha} \in \phi_1(\bar{\beta})$ and $\bar{\beta} \in \phi_2(\bar{\alpha})$: which proves that $(\bar{\alpha}, \bar{\beta})$ constitutes a Nash equilibrium.

Thus we have the following proposition:

*Proposition 11.1 Consider a game with a **finite** number of players (say M); further each player i has a **finite number of pure strategies**, say i_k, denoted by s^i; let $\pi^i(s_{1_j}^1, \cdots, s_{i_r}^i, \cdots, s_{i_s}^M)$ denote the returns from playing their pure strategies. Then there is a Nash equilibrium (possibly in mixed strategies) for this game.*

We have provided a proof for the case when $M = 2$; note that extending the argument beyond 2 does not really involve anything substantial: only the notation becomes difficult! However, when the number of pure strategies is not known to be finite, we need to impose some further restrictions on the set of strategies: we need them to be **non-empty, convex, and compact**. We need to be able to define a map such as Φ: the best response map to ensure that the pay-offs are continuous **functions** in their own strategies, so that a 'best' response is possible (maximizing a continuous function on a compact domain ensures a solution). We also need to ensure that the best response map has the desired properties and we need to ensure that pay-offs vary continuously with all strategy choices. These assumptions are enough to ensure the existence of a Nash equilibrium; to enable the best response to be a unique response, of course, we need to ensure some other restrictions on the pay-off functions.

11.4.3 Refinements of Nash equilibria

Thus Nash equilibria exist under fairly reasonable conditions. When they are unique, there is some reason to expect that rational players with adequate information would be able to work out that playing Nash equilibrium strategies is a good way of playing the game: in the sense that if everyone plays their Nash equilibrium strategies, there would be **no ex-poste regret**. In fact, this property is sometimes used to define a Nash equilibrium. When multiple Nash equilibria exist (as in the Battle of the Sexes) notice that it is difficult to predict what the players might play on the basis of the normal form alone. There have been attempts at eliminating some Nash equilibria as being inadmissible and thus reducing the number of Nash equilibria. These attempts are collected together under the title of 'refinements'.

We shall begin by looking at an alternative form of representation of a game: **extensive** form. The extensive form representation shows how the game is played and is usually employed when moves are made sequentially. But as we shall see, it is an alternative way of representing a non-cooperative game and there are those who begin by looking at extensive form games first. An extensive form representation is made up of the following: the players, a game tree and an assignment of points or nodes of the tree to each player for their moves, lists of actions available to the decision-makers at their decision-making nodes, information sets, pay-offs at the terminal nodes, and if initially, nature has to move (that is, the game may begin with a particular state of the world chosen by nature), then a probability assessment over the possible moves by nature. Formally, there must be:

1. a list of players and may be Nature.
2. a game tree consisting of a finite set T; every element of T is denoted by t and is called a **node**; there is also a binary relation $\prec$ which is **asymmetric, transitive**, and satisfies the following in addition: if $t, t', t'' \in T$ and $t \prec t''$ and $t' \prec t''$ then either $t \prec t'$ or $t' \prec t$.

   $P(t') = \{t \in T : t \prec t'\}$ the **predecessors** of t'; also $t \in P(t') \Rightarrow t'$ is a **successor** of t. Thus $S(t) = \{t' \in T : t \prec t'\}$ is the set of successors of t.

   If t has no predecessor, t is an initial node; while if t has no successor, then t is a terminal node. If t is not a terminal node, then we call t a **decision** node. Given the above property of the relation $\prec$, it should be checked that every t which is not an initial node, has a unique node $p(t)$ such that $p(t) \prec t$ and if $t' \prec t$ and $t' \neq p(t)$ then $t' \prec p(t)$. $p(t)$ is the immediate predecessor of t; similarly $s(t)$ is the immediate successor of t.
3. Each decision node is assigned to one and only one decision-maker (player or nature) (at each node x, a decision-maker $d(x)$ chooses what happens

next). Further writing X as the set of all decision nodes, for each $t \in X$ there is a finite set of actions $A(t)$ and a function $\alpha_t : s(t) \to A(t)$.

4. The nodes t which are not terminal can be partitioned into **information sets** with the following properties: if t, t' are in the same information set then $t \notin P(t')$ and $t' \notin P(t)$, the decision-maker attached to t and t' is the same and $A(t) = A(t')$.

5. An assignment of pay-offs for each player is given at each terminal node.

6. Whenever nature is assigned a node, either initial or decision, there is an associated probability distribution over the actions $A(t)$.

Let us try to understand the formalism of the above. The extensive form is usually represented by a diagram like a tree which has several distinct points called **nodes**; at each node (except terminal) some designated player makes a move. Branching out from the nodes are the various alternative moves which are possible. A node is a successor to another node, if it appears later in the sequence of play. Initial nodes are those for which there is no predecessor; and terminal nodes are those which have no successor. The pay-offs are presented at the end of each terminal node. The ranking among nodes ($a \prec b$ is to be understood as a is a predecessor of b or b is a successor of a) is asymmetric since a is a predecessor of b implies that b cannot be a predecessor of a; the relation is transitive; and finally if two distinct nodes are predecessors of a third, then one of the two nodes must be a predecessor of the other.

The extensive form also allows us to state what information each player has at the time he makes a move; we do this through the use of **information sets**; an information set is usually a collection of nodes at which **the same** player makes a move; suppose there are two nodes belonging to the same information set and at these nodes player 2 moves; then none of these nodes is a successor to the other nor is there any distinctive feature about these nodes and player 2 cannot distinguish between these two nodes; she knows only that it is her turn to move; she also does not know which particular node she is at and thus does not know the exact history leading up to her move. If all information sets are singletons: then each player knows the exact history prior to making the move. The nature of the information sets for a particular game also provides a classification of what information is available to each player. Games of **perfect information** are games where all information sets are singletons; whereas games of **imperfect information** are those where at least one information set is not a singleton.

Consider Game 5 (see Figure 11.3):

There are two players R and C; the initial node consists of R making a move; she can move T or D; it is then C's turn and he can move either L or B and the game ends; of course, the pay-offs are as shown above and depend not only on what move C makes but then also upon what R chooses. The information sets are singletons and at each node, the player making the move knows what the previous history has been: thus C knows whether R has moved T or D when it is his turn to move so that C is aware exactly where he is. In contrast, consider Game 6.

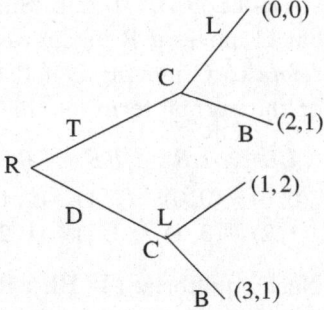

FIGURE 11.3 Extensive Form with Singleton Information Sets

This is almost the same game except that when it is C's turn to move, C does not know which node he is at: this is designated by joining the two nodes as shown (in Figure 11.4). So C does not know whether R has moved T or D and has to decide what is the best move independently of whatever move R has made. It should be noted that in this situation it is as if R and C were made to choose simultaneously: then one has to choose exactly as above. What are the Nash equilibria for these games?

One way to proceed to answer this question is to construct the normal form for these games. Consider Game 6 first. The normal form for this game is as below:

$$
\begin{array}{cc}
 & L \qquad B \\
\begin{array}{c} T \\ D \end{array} & \left(\begin{array}{cc} (0,0) & (2,1) \\ (1,2) & (3,1) \end{array} \right)
\end{array}
$$

R has two strategies T and D; since C moves next but does not know the move adopted by R, C too has two strategies L and B. Now it is clear that D is a dominant strategy for R and hence the unique Nash equilibrium for Game 6 is for R to play D and C to play L.

Next, consider Game 5; in this situation, note that C playing L after R has played T differs from C playing L after R has played D. Accordingly, we need to consider these two as distinct. Thus given that R has two moves T and D; C has four strategies

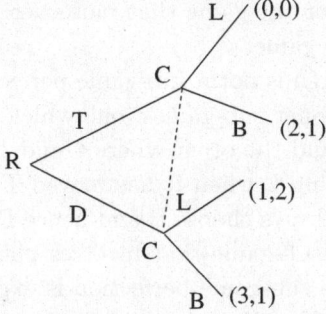

FIGURE 11.4 Extensive Form with an Information Set Containing Two Nodes

to choose from and we list them as LL, LB, BB, BL. Since there are two L or B, the first of the couple refers to what C moves if R has moved T and then the move of C if R has moved D; thus LB, refers to C moving L if R moves T and moving B if R moves D. Now we are ready for the normal form for this game:

$$
\begin{array}{c}
\quad\quad LL \quad\quad LB \quad\quad BB \quad\quad BL \\
\begin{array}{c} T \\ D \end{array}
\left(
\begin{array}{cccc}
(0,0) & (0,0) & (2,1) & (2,1) \\
(1,2) & (3,1) & (3,1) & (1,2)
\end{array}
\right)
\end{array}
$$

Note that this game has two Nash equilibria: (T, BL): R choosing T and C choosing BL; or (D, LL): R choosing D and C choosing LL. Thus, the difference between the information structure at the time C moves has changed the equilibrium structure of the game: this is not surprising since these are two different games. We may have analysed game 5 differently. This is possible because of the simple information structure: at each stage, the mover knows exactly what has happened before and what will happen afterwards. To take this aspect into account, consider for example what happens at the nodes just prior to the terminal nodes, when C moves: if C is moving after R has moved T, C will surely move B; thus the effect of R moving T is to ensure the pay-off (2,1) as C will surely move B. Similarly, if R moves D, C will surely move L and hence the pay-off from R moving D is (1,2). Clearly how, we can move step backwards to conclude that R will move U and the game will end with the pay-off of (2,1). This method, known as the method of **backward induction** works because of the information sets being singletons; thus we can, by assuming rational behaviour, figure out what each will move from the last step backwards and hence arrive at what the first mover will decide. The shrewd reader would have noticed that we seem to have disposed off one of the Nash equilibria that we obtained by looking at the normal form: (D, LL). What we have shown is that of the two Nash equilibria only one is **sub-game perfect**, that is, (T, BL).

Formally, a **proper sub-game** of an extensive form game is a node t and all its successors $S(t)$ such that t is not an initial node, the information set containing t contains t alone and finally, all the successors $t' \in S(t)$, the information sets containing t' are members of $S(t)$. A **sub-game perfect** Nash equilibrium for an extensive form game is a Nash equilibrium for the game that moreover yields a Nash equilibrium in every proper sub-game of the game.

To understand why (D, LL) is not a sub-game perfect Nash equilibrium in Game 5, note that there are two proper sub-games: one which begins at the node where C moves after R has moved T and the other where C moves after R has moved D. The strategy LL involves C choosing L when R has moved T: is patently irrational; since once R has chosen T, C will always choose B and hence D, LL fails to qualify the sub-game perfect test: there is a sub-game that involves playing something which is not the best response. Sometimes sub-game perfection is explained in terms of including only those strategies which are 'credible' but the check is whether the strategy involves best responses in all sub-games.

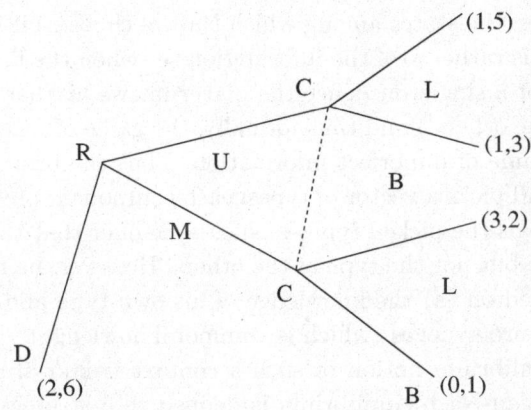

FIGURE 11.5 An Exercise

Sub-game perfection is the most common 'refinement' of the notion of Nash equilibrium. There are many other refinements as well. Interested readers will be directed to more detailed works on the subject. Note, however, that in games such as Game 6, there are no proper sub-games; since at the nodes where C moves both belong to a single information set. Hence, the Nash equilibrium is also trivially sub-game perfect.

We end this section with Game 7 (see Figure 11.5):

For the above note that player R has three pure strategies, D, M, and U; how many strategies does C have? (Hint: C has 2 pure strategies.) Note that there is a single information set containing both the moves of C. What is the normal form for this game? What are the Nash equilibria and the sub-game perfect Nash equilibria?

11.4.4 Bayesian-Nash equilibria

A particular extension of the Nash equilibrium which is used in many contexts is the Bayesian-Nash equilibrium. This extension is useful in discussing situations when agents or players do not know each other's 'types'. Up to this point, the basic assumption has been that the players knew their rivals' pay-offs and their characteristics: formally this information was taken to be 'common knowledge', that is, each knew that the other knew that the others knew ... Such games are referred to as games of **complete information**. But in many situations this is not the case and we then have a game of **incomplete information**. Consider the situation when the players do not know the others' pay-offs; in this situation we need to consider the belief that each player may have about the nature of the rival's pay-off. A widely used approach is due to Harsanyi (1967–8) and it is as follows. One imagines that each player's characteristics are determined by the realization of some random variable: usually this is formalized by introducing Nature into the game, Nature moves first and picks the characteristics of a particular player; this is observed only by the player concerned.

The prior possibilities (the states among which Nature chooses) is common knowledge among all players. It is rather as if the information set when the first actual player has to make a move is not a singleton (since the player knows his/her own characteristics but does not know the others') and thus formally, the game of incomplete information is converted into a game of imperfect information. Thus the basic assumption is that Nature moves first and picks a vector of types: each component of which is the type for a player. Nature reveals the picked types in such a manner that the i-th player comes to know his own type but not the type of the others. However, he forms a belief about the others' types based on (a) the knowledge of his own type and (b) the probability distribution over Nature's moves (which is common knowledge).

The relevant equilibrium notion in such a context is known as a Bayesian-Nash equilibrium. A Bayesian-Nash equilibrium is defined as a strategy profile and beliefs specified for each player about the types of the other players that maximizes the expected pay-off for each player given their beliefs about the other players' types and given the strategies played by the other players. We shall conclude by providing a simple example.

Consider the following simultaneous move game:

$$
\begin{array}{c}
\quad\quad L \quad\quad\quad R \\
\begin{array}{c} U \\ D \end{array}
\left(
\begin{array}{cc}
(3, \gamma) & (2, 1-\gamma) \\
(0, \gamma) & (4, 1-\gamma)
\end{array}
\right);
\end{array}
$$

where $\gamma \in \{1, 0\}$. Thus, there are two types of players choosing columns (say, player 2): for one type (say t_1, $\gamma = 1$ and for the other, say t_0, $\gamma = 0$). The player choosing rows (player 1) is of a single type and that is common knowledge. Player 2 knows whether he belongs to type t_i, $i = 1, 0$ but player 1 does not.

As is clear, if player 2 is of type t_1, he chooses to play $s_{21}^\star = L$ while if he is of type t_0, he plays $s_{20}^\star = R$: these are the dominant strategies for the respective types of player 2. Let player 1 believe that player 2 is of type t_1 with probability p (and hence of type t_0 with probability $1 - p$); then player 1's expected pay-off from choosing U is $3.p + (1 - p).2 = p + 2$ whereas the expected pay-off from choosing D is $4.(1 - p) = 4 - 4p$. Note that $p + 2 \gtrless 4 - 4.p \Leftrightarrow p \gtrless 0.4$; thus, if player 1 believes $p > 0.4$, player 1 plays U otherwise he plays D.

If, for example, $p = 0.5$, that is player 1 believes that both types of player 2 are equally likely (the fact that Nature is unbiased is common knowledge), then $s_1^\star = U, s_{21}^\star = L, s_{20}^\star = R, p = 0.5$ constitute a Bayes-Nash equilibrium.

11.4.5 Repeated games: Folk theorem

Let us return to the prisoner's dilemma game (Game 2); what happens if two play this game repeatedly? What would be the possible Nash equilibria then? If the game is played repeatedly and the pay-offs in each period are as in Game 2, note that playing (C,C) in each period is a Nash equilibrium.

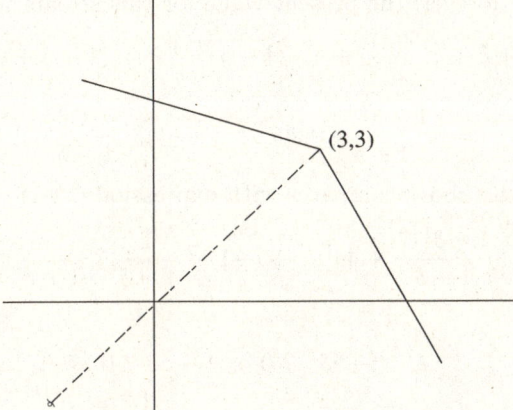

FIGURE 11.6   Pay-off Frontier for Prisoners' Dilemma

If the game is repeated for a finite number of times, then playing (C,C) in each period is the only Nash equilibrium. Consider what happens in the last period: it is exactly like the game played once and hence both play their dominant strategies (C,C); thus in the previous period too (C,C) is the only option and hence the same argument may be repeated for each preceding period.

If we plot the pay-offs to each player in Figure 11.6 we have the following situation:

Note that the point (3,3) lies on the frontier of the pay-offs but cannot be attained in equilibrium: what is attained is an interior point (1,1) when both use the strategy C. One of the properties investigated when games are repeated infinitely is often whether frontier pay-offs are attainable. Or to put it differently, are there strategies which when used are the best response against one another and at the same time help in attaining pay-offs out on the frontier. The **Folk theorem** of Game Theory asserts that it is possible to attain such pay-offs by using some appropriate strategies.

Consider the following strategy (S^*): Play NC initially and continue to play NC till the opponent plays C once; thereafter play C forever.

Let us examine whether this strategy yields the desired outcome or whether both players adopting this strategy constitute a Nash equilibrium. First of all, we need a discount factor to convert future pay-offs to their present value; let this be $\delta < 1$ for both the players. Then playing NC forever against the strategy S^* yields a stream of pay-offs $3, 3, \cdots$ and hence has present value

$$\frac{3}{1 - \delta} \tag{11.18}$$

Consider a player contemplating a deviation in say the T-th period by playing C: then this player earns 3 in every period up to period $T - 1$, earns a return 4 in period T

and thereafter gets -1 forever: the present value for this stream works out to be:

$$3[1 + \delta + \delta^2 + \cdots + \delta^{T-2}] + 4\delta^{T-1} + \delta^T[1 + \delta + \cdots]$$

$$= 3\frac{1 - \delta^{T-1}}{1 - \delta} + 4\delta^{T-1} + (-1)\delta^T\frac{1}{1 - \delta}$$

The question is how the above compares with expression (11.18). That is:

$$3\frac{1 - \delta^{T-1}}{1 - \delta} + 4\delta^{T-1} + (-1)\delta^T\frac{1}{1 - \delta} \ ? \ \frac{3}{1 - \delta}$$

or

$$\delta^{T-1} - 5\delta^T \ ? \ 0 \text{ or } 1 - 5\delta \ ? \ 0$$

Note then that whenever $\delta > 1/5$ it never pays to deviate and the best response against S^* is to play NC always. Consequently, adopting this strategy would lead to players earning right on the pay-off frontier and we have induced that from adopting Nash equilibrium strategies. However, note that it does involve an assumption that the game is played infinitely often.

We shall end our preliminary excursions into this very interesting area by making the following remarks.

Remark 11.1   Instead of assuming that the game is played infinitely often, what if we were to assume that there is always a probability $\delta < 1$ that the game will be played once more? Note that the same calculations will do given the discussion on how decision-makers view such situations. Things would be somewhat complicated by the inclusion of a discount factor, of course. But the general nature of the argument should go through. Thus repeating a game may enable the players to attain pay-offs on the frontier: thus cooperative behaviour may result from repeating non-cooperative games often if the discount rate is not too small. This is the Folk theorem.

Remark 11.2   However, there is an embarrassment of riches so to speak: once can devise strategies such that any point on the frontier $(a, b)a > -1, b > -1$ is attained in Nash equilibria by adopting strategies similar to S^*. The reader is encouraged to construct such strategies.

Further Readings for Section II

David Gale's (1960) classic *Theory of Linear Economic Models*, chapter 2, constitutes the basic reference for Chapter 7; Gale did not consider the role of characteristic roots worthy of discussion and dismissed the topic entirely from his treatment; yet in economics, these roots figure prominently. To include them we adopt a method used in Mukherji (1985) to pass from the Gale structures to discussing characteristic roots.

For the theory of several variables, a reference to the relevant chapters of Apostol (1974) may be considered for any further details. For static optimization, the fundamental paper by Kuhn and Tucker (1950) may be be consulted. Some other books containing detailed examination of Static Optimization theory are Takayama (1985) and Mangasarian (1994). The papers by Arrow et al. (1961) and Mukherji (1989) contain treatments of quasi-concave programming.

The expected utility formulation is due to a simplification of the Herstein and Milnor (1953) paper. For more readings on decision-making under uncertainty, the relevant chapter in Mas-Colell et al. (1995) and references may be considered. For the Theory of Risk Aversion, Arrow (1970), chapter 3 and Pratt (1964) are basic. Also for more details on Portfolio Theory and finance, Huang and Litzenberger (1988) is a standard reference. We considered the Expected Utility Theorem as the foundation for decision-making uncertainty. For a treatment of Non-expected Utility Theory, see Starmer (2000).

For Game theory, there are many texts now: Gibbons (1992) and Fudenberg and Tirole (1991) are two examples, the former being a more elementary text. A very useful reference for material on Game Theory is Aliprantis and Chakrabarti (1999).

Much of the applications in this section constitute the bulk of classical microeconomics; for details and further readings, we can recommend Koopmans (1957), Essay 1, McKenzie (2002), Debreu (1959), and the relevant chapters in the encyclopedic Mas-Colell, Whinston, and Green (1995).

KEY TERMS

Acceptance Frontier	Information Sets
Allais' Paradox	Matching Pennies
Arrow-Pratt Measure of Absolute Risk Aversion	Measure of Relative Risk Aversion
	Mutual Fund
Backward Induction	Nash Equilibrium
Battle of the Sexes	Nodes
Bayesian-Nash Equilibria	Non-cooperative Game
Decision-making under Uncertainty	Pay-offs
Expected Utility Function	Prisoners' Dilemma
Extensive Form Representation	Risk Aversion
Game Theory	Risky Asset
Global Measure of Risk Aversion	Strategies
Incomplete Information	Sub-game Perfect
Folk Theorem	Two-fund Monetary Separation

SECTION THREE

- **Introduction**
- **Dynamical Systems**
- **Dynamic Optimization**
- **Economic Applications III:**
  Economic Dynamics

CHAPTER TWELVE

Introduction: Objective and Tools

We shall mainly be concerned in the last section with problems where the notion of time plays an important role. Many problems of decision-making involve making decisions at different points of time. Such problems are intrinsically more difficult than the class of problems considered in the earlier sections. Being an introduction to the class of such problems, we provide an elementary treatment of these topics. As done earlier in the text, before moving on to applications, we develop the mathematics associated with these problems: the tools of Section III. First, we need to study dynamical systems and then consider optimization over time.

At the heart of any dynamic problem is the rule which governs the laws of change over time: price adjustment occurs according to the level of excess demand in the market; or in another context, the rules governing investment (changes in the capital stock) may be specified in some manner. The question of interest is whether these rules generate solutions which have determinate long-term behaviour. To specify any rule we need to specify the variables of interest, be their prices or capital stock and then define how the change is determined. Loosely speaking, this defines a dynamical system. We shall consider this in the chapters that follow. We need to then analyse how the solution to this system of equations behaves and it may be of interest to find out whether the initial conditions matter.

The next problem is that of decision-making over time: maximization of some objective over some length of time, at each instance of which we need to make a decision of some sort. This opens up the study of a separate class of problems and we choose to analyse such problems through control theory and introduce students to Pontryagin's celebrated Maximum Principle. In fact, but for the study of discrete dynamical systems, this entire section is heavily dependent on the contributions of two Mathematicians from Russia: Liapunov and Pontryagin.

KEY TERMS

Dynamical Systems
Notion of Time
Optimization over Time

CHAPTER THIRTEEN

Dynamical Systems

13.1 Continuous Time Processes

13.1.1 Introduction

Before proceeding formally, let us attempt to understand what we shall analyse in this section. Consider a rule specifying the change of position over time of some object. Can we then predict where the object will be at some instant of time, if we know where it started from? What is given is the derivative at each point of time; from that we need to uncover the position at each point of time. Technically speaking, we have been given a differential equation and we need to find out the solution or the trajectory. This process of going from the differential equation or the dynamical system to the trajectory is through a process of integration, Riemann integration, that we have discussed earlier. What makes matters somewhat different in our consideration is that we are not interested in actually obtaining the trajectory. We are interested in some properties of the trajectory or solution and we shall exhibit what these properties are. The particular trajectory may not be available because we may not know the exact form of the function which defines the rule; we may have only some properties of the rule.

Let us make matters more specific. Consider a rule of change of position, also called an autonomous differential equation on $\Re$ given by

$$\dot{x} = \alpha x(t), x(t) \in \Re, \alpha \text{ is some constant} \tag{13.1}$$

Let $\phi(t) = C.e^{\alpha.t}$ where C is some constant; it is easy to check that the function $\phi(t)$ is differentiable and $\in \Re \ \forall \ t$; further

$$\dot{\phi}(t) = \alpha\phi(t)$$

so that $\phi(t)$ is a solution or trajectory of (13.1) for any $(t_1, t_2) \in I$.

Consider any $r \in \Re$. Note that by specifying that

$$\phi(0) = r \Rightarrow C = r;$$

consequently, the solution to (13.1) with r as initial point is given by $r.e^{\alpha t}$. Note that fixing the initial point, or the point from where the trajectory starts, fixes the trajectory uniquely.

In fact, a strong property is the **uniqueness of the solution** with respect to the **initial point**. Suppose that $s \in \Re$; then the solution to (13.1) with s as initial point is $\psi(t) = s.e^{\alpha t}$. Suppose further that s is such that for some $\bar{t} > 0$ we have $r.e^{\alpha \bar{t}} = s$. Then $\psi(t) = r.e^{\alpha \bar{t}}.e^{\alpha t} = r.e^{\alpha(t+\bar{t})} = \phi(t + \bar{t})$. Thus, $\psi(t)$ coincides with $\phi(t + \bar{t})$.

In fact, there can be only one trajectory through each point. Two trajectories cannot cross: the reason for that is easy to see for if they did cross then at the point of crossing, derivatives would be different but that cannot be since the rule itself defining the motion is presumably well defined. So if two trajectories meet at a point then they must be the same trajectory with a difference in initial points at most.

The rule specified by (13.1) is an example where the rule is exactly specified and we could easily solve for the trajectory. If instead, we had a rule such as

$$\dot{x} = F(x(t)), x(t) \in \Re, \tag{13.2}$$

where we know only some properties of the function F, we may not be able to obtain the trajectory explicitly. Before anything else then we need to be able to say that there is a trajectory, in principle: that is, the properties we have on the function F allow us to conclude that a solution or trajectory exists.

For rule (13.2), $\bar{x} \in \Re$ which is such that $F(\bar{x}) = 0$: then $\bar{x}$ is said to be an equilibrium for rule (13.2). One of the things which we may try to answer regarding (13.2) is whether the trajectory, if it exists, approaches $\bar{x}$ as t becomes large. In some cases we can answer the question definitively, just on the basis of some properties of the function F. Finally, given the introductory nature of this account, we shall mainly confine attention to rules of motion on the plane, that is, where the domain of discourse is $\Re^2$; we shall indicate the results when higher dimensions are involved wherever this is possible. A word of caution though: motion on the plane is amenable to a detailed analysis in very general terms and these results, in most cases, do not hold in higher dimensions. But first, we shall look at the solutions to some particular forms.

13.1.2 Solutions to some standard forms

Classically, particular forms of differential equations were subjected to great scrutiny by mathematicians. As we have argued earlier, this approach has not been very useful in economics. However, some forms may appear from time to time. We shall discuss some of them and the only point of concern here is whether the given differential equation has a solution.

Consider, for example, the equation we started out with:

$$\frac{dy}{dx} = \alpha y \tag{13.3}$$

where α is a constant and $y, x \in \mathfrak{R}$. Notice that, this is a case when **separation of variables** is possible; we can now write:

$$\frac{dy}{y} = \alpha dx$$

Next, considering the integrals (Riemann) on both sides we obtain

$$\ln y = c + x \text{ where } c \text{ is a constant}$$

Or, we have $y = Ce^{\alpha x}$, writing $C = e^c$, as the **general solution** to (13.3). We have already commented on how the constant C will be determined. This pattern is very important and we shall come across it frequently.

Consider next, a very general form of differential equation of the first order (does not involve derivatives of orders higher than unity):

$$\frac{dy}{dx} + \phi y = \psi \qquad (13.4)$$

where ϕ, ψ are functions of x alone. When the rhs of (13.4) is not zero, the equation is said to be of **non-homogenous** form; in contrast when the rhs is zero, we have the **homogenous** form:

$$\frac{dy}{dx} + \phi y = 0$$

We first consider this homogeneous part and note that the variables are separable so that we can write:

$$\frac{dy}{y} = -\phi dx$$

and hence, exactly as above:

$$\ln y = c + e^{-\int \phi dx} \text{ or } y = C.e^{-\int \phi dx}$$

We next substitute the expression

$$y = ve^{-\int \phi dx}$$

into (13.4), where v is a function of x. Now (13.4) becomes:

$$\frac{dv}{dx}e^{-\phi dx} = \psi$$

and as before

$$v = C + \int \psi e^{\int \phi dx}$$

Thus the general solution to (13.4) is given by:

$$y = Ce^{-\int \phi dx} + e^{-\int \phi dx} \int \psi e^{\phi dx} dx$$

A final form that we shall look at is a general form of (13.3); it is general in the sense that $y \in \Re^n$, $n > 1$, $x \in \Re$; we have

$$\frac{dy}{dx} = Ay \qquad (13.5)$$

where A is a square matrix of order n; in other words, we have n simultaneous equations of the form:

$$\frac{dy_i}{dx} = \sum_{j=1}^{n} a_{ij} y_j \quad i = 1, 2, \cdots, n$$

where we have written the matrix $A = (a_{ij})$; noting the solution to (13.3), we expect that the solution to the current n-dimensional form should be similar:

$$y(x) = e^{Ax} . C$$

where $e^{Ax} = I + A.x + \frac{1}{2}A^2.x^2 + \cdots + \frac{1}{n!}A^n.x^n + \cdots$, $C \in \Re^n$, a constant. Note also that

$$\frac{de^{Ax}}{dx} = A.e^{Ax}$$

so that

$$\frac{dy(x)}{dx} = Ay(x)$$

as one may expect it to be. To see how to actually compute the solutions, (how does one compute e^{Ax}, for example) it may be easier to consider the case when $n = 2$, that is,

$$\frac{dy_i}{dx} = a_{i1} y_1 + a_{i2} y_2 \quad i = 1, 2$$

Consider a trial solution to the above given by

$$y = \begin{pmatrix} k_1 \\ k_2 \end{pmatrix} e^{\lambda x} = K.e^{\lambda x}$$

For this to be a solution, we must have $\lambda K e^{\lambda x} = A K e^{\lambda x}$ or $\lambda K = AK$; that is, we must have λ to be a characteristic root of the matrix A and K to be the corresponding characteristic vector. For the case $n = 2$, there are the following cases:

1. λ_i, $i = 1, 2$ are distinct and real;
2. λ_i, $i = 1, 2$ are equal (and hence, real); and
3. λ_i, $i = 1, 2$ are complex conjugate, that is, $\lambda_1 = \alpha + \iota\beta$, $\lambda_2 = \alpha - \iota\beta$

Accordingly, the characteristic vectors will need to be worked out. In the first case, note that the equation

$$\det A(\lambda) = \det \begin{pmatrix} \lambda - a_{11} & -a_{12} \\ -a_{21} & \lambda - a_{22} \end{pmatrix} = 0$$

has two distinct roots λ_i; consider then a non-trivial solution to $A(\lambda_i).K = 0$, say K^i; note that we may choose $k_1^i = 1$, $k_2^i = \frac{\lambda_i - a_{11}}{a_{12}}$, provided of course, $a_{12} \neq 0$. In case

$a_{12} = 0$, $\lambda_i = a_{ii}$ and we may choose $k_1^1 = 1$, $k_2^1 = 0$; $k_1^2 = 0$, $k_2^2 = 1$. Now the solution to the system (13.4) for $n = 2$ is given in this case by:

$$y(x) = \alpha K^1 e^{\lambda_1 x} + \beta K^2 e^{\lambda_2 x}$$

In the second case, let the common value of the characteristic root be λ; we shall show that the solution is of the form

$$y_1(x) = (A_1 + A_2 x)e^{\lambda x}; \; y_2(x) = (B_1 + B_2 x)e^{\lambda x}$$

where the constant coefficients A_i, B_i are related to one another in a particular manner which needs to be determined as follows. Note that if these are to be solutions, then they must satisfy the equation (13.4). Hence we must have, after factoring out $e^{\lambda x}$:

$$\begin{pmatrix} a_{11}(A_1 + A_2 x) + a_{12}(B_1 + B_2 x) \\ a_{21}(A_1 + A_2 x) + a_{22}(B_1 + B_2 x) \end{pmatrix} = \begin{pmatrix} \lambda(A_1 + A_2 x) + A_2 \\ \lambda(B_1 + B_2 x) + B_2 \end{pmatrix}$$

This, in turn, means that we must have the equations:

$$x\{A_2(\lambda - a_{11}) - a_{12}B_2\} + \{(\lambda - a_{11})A_1 - a_{12}B_1 + A_2\} = 0 \qquad (13.6)$$

and

$$x\{(\lambda - a_{22})B_2 - a_{21}A_2\} + \{(\lambda - a_{22})B_1 - a_{21}A_1 + B_2\} = 0 \qquad (13.7)$$

and these conditions must hold for every x; this means that the terms within braces must be zero; that is, we must have:

- $\{A_2(\lambda - a_{11}) - a_{12}B_2\} = 0$,
- $\{(\lambda - a_{11})A_1 - a_{12}B_1 + A_2\} = 0$,
- $\{(\lambda - a_{22})B_2 - a_{21}A_2\} = 0$, and
- $\{(\lambda - a_{22})B_1 - a_{21}A_1 + B_2\} = 0$.

Hence, we have the following:

- $B_2 = \dfrac{\lambda - a_{11}}{a_{12}} A_2$,
- $B_1 = \dfrac{(\lambda - a_{11})A_1 + A_2}{a_{12}}$, and
- $B_2 = \dfrac{a_{21}}{\lambda - a_{22}} A_2$,
- $B_1 = \dfrac{a_{21}}{\lambda - a_{22}} A_1 - \dfrac{1}{\lambda - a_{22}} B_2$.

Assuming that $a_{12} \neq 0$, $\lambda \neq a_{22}$; if these conditions are not met, we need to proceed in an obvious manner directly from the equations (13.6) and (13.7). Now since λ is a characteristic root of the matrix A, we must have

$$\frac{\lambda - a_{11}}{a_{12}} = \frac{a_{21}}{\lambda - a_{22}}$$

so (fortunately!) the two equations for B_2 coincide; using these, the last requirement can be written as

$$B_1 = \frac{\lambda - a_{11}}{a_{12}} A_1 - \frac{\lambda - a_{11}}{a_{12}(\lambda - a_{22})} A_2.$$

Now this coincides with the earlier requirement on B_1 iff $-(\lambda - a_{11}) = \lambda - a_{22}$, that is, iff

$$\lambda = \frac{a_{11} + a_{22}}{2}$$

which is true iff λ is a repeated characteristic root of the matrix A. Thus we have indeed worked out the general solution to the equation (13.4) when $n = 2$ and there is a repeated characteristic root.

We consider next, what happens when the roots λ_i are complex $\lambda_i = \alpha \pm \iota\beta$. The situation is much like the case of distinct characteristic roots; the characteristic vectors V^i satisfying $AV^i = \lambda_i V^i$ are complex too; in fact V^i are complex conjugate. The solution will then be written as $y(x) = B_1 V^1 e^{(\alpha + \iota\beta)x} + B_2 V^2 e^{(\alpha - \iota\beta)x} = e^{\alpha x}[B_1 V^1 e^{(\iota\beta)x} + B_2 V^2 e^{-(\iota\beta)x}]$ where B_i are complex conjugate numbers. We note one feature of this solution for future reference: it is the exponential of the real part of the solution which determines what happens when the variable x becomes large; this is so because the term in the square brackets contains $e^{\pm \iota x} = \text{Cos } x \pm \iota \text{ Sin } x$ which does not contribute to unbounded behaviour as x becomes large.

13.1.3 Definitions and propositions

Next, we consider conditions under which there will always be a solution with desirable properties. Such an investigation is of paramount importance when the functional forms are not known. Let $x \in \Re^n$ where $\Re^n$ denotes the n-dimensional Euclidean space, $F : \Re^n \to \Re^n$. Consider the system

$$\dot{x}(t) = F(x(t)) \tag{13.8}$$

where $t \in [0, \infty) = I$, $\dot{x}(t) = (\dot{x}_i(t), i = 1, 2 \cdots n)$ and

$$\dot{x}_i(t) = \frac{dx_i}{dt}$$

The equation (13.8) is referred to as an **autonomous system** of first order differential equations.

We shall say that $x^* \in \Re^n$ is an **equilibrium point** of (13.8) if

$$F(x^*) = 0$$

Let $(t_1, t_2) \in I$; $\phi(t_1, t_2) \to \Re^n$ is said to be a **solution** of (13.8) if

      (i) $\phi(t)$ is differentiable on (t_1, t_2); and

      (ii) $\dot{\phi}(t) = F(\phi(t))$ on (t_1, t_2)

Let $x^o \in \Re^n$; we shall be interested in a solution to (13.8) beginning from x^o (the initial **point**), that is, a solution $\phi(t)$ such that $\phi(0) = x^o$.

The first problem, then, is to determine conditions under which the system of equations (13.8) has a solution with some x^o as initial point. This is done through the following proposition:

Proposition 13.1   Let $F(.)$ be continuous on $\Re^n$ with continuous first order partial derivatives $(\dfrac{\partial F_i}{\partial x_j})$ on $\Re^n$ and $x^o \in \Re^n$. Then there exists a function $\phi : [0, t) \to \Re^n$ for some t such that $\phi(.)$ is continuous on $[0, t)$ with $\phi(0) = x^o$ such that $\phi(t)$ is a solution of the system (13.8) which is continuous and unique with respect to the initial point x^o.

Remark 13.1   Instead of the existence and continuity of partial derivatives $\dfrac{\partial F_i}{\partial x_j}$, one may require instead that the function $F(.)$ satisfy a **Lipschitz condition**, that is, there exists a constant $k > 0$ such that $|F(x^1) - F(x^2)| < k.|x^1 - x^2| \; \forall x^1, x^2 \in \Re^n$.

Remark 13.2   It should be pointed out that Proposition 13.1 guarantees the existence of solution on $[0, t)$; ideally we would like the solution to exist on $[0, \infty)$; in other words, we would like to be able to **continue** the solution beyond $[0, t)$; this is possible if the function $F(.)$ satisfies some additional restrictions. A condition which permits this type of continuation is that the function $F(.)$ be **bounded** on $\Re^n$.

We shall write the solution to (13.8) with x^o as initial point, as $\phi(t, x^o)$ or alternatively as $x(t, x^o)$. Let x^* be an equilibrium point of (13.8). We shall say that the **solution $x(t, x^*)$ (or the equilibrium x^*) is stable** if for every $\epsilon > 0$, there is $\delta > 0$ such that for all solutions $x(t, x^o)$ for which $|x^o - x^*| \leq \delta$, $|x(t, x^o) - x(t, x^*)| \leq \epsilon \; \forall t \geq 0$.

The solution $x(t, x^*)$ (or the equilibrium x^*) is said to be **locally asymptotically stable** if it is stable and there exists $\alpha > 0$ such that $|x^o - x^*| \leq \alpha \to \lim_{t \to \infty} |x(t, x^o) - x(t, x^*)| = 0$.

The solution $x(t, x^*)$ (or the equilibrium x^*) is said to be **globally asymptotically stable** if it is stable and if $\lim_{t \to \infty} |x(t, x^o) - x(t, x^*)| = 0$ for any $x^o \in R^n$.

To consider the **stability of the system** (13.8) consider the following:

Claim 13.1.1   Let $x(t, x^o)$ be a solution to (13.8). If $\lim_{t \to \infty} x(t, x^o) = \bar{x}$ for some finite $\bar{x} \in R^n$, then $\bar{x}$ is an equilibrium of (13.8).

Proof Suppose this is not so; that is, $F_k(\bar{x}) \neq 0$ for some k. By hypothesis, note that $F_k(x(t, x^o)) \to F_k(\bar{x})$. Consequently, given any $\epsilon > 0$, there is T such that

$$F_k(\bar{x}) - \epsilon < F_k(x(t, x^o)) < F_x(\bar{x}) + \epsilon, \text{ for all } t > T$$

Let $|F_k(\bar{x})| = 2\eta > 0$; then ϵ can be so chosen that $|F_k(x(t, x^o))| > \eta$ for all $t > T$. Thus either $\dot{x}_k(t) > \eta$ or $\dot{x}_k(t) < -\eta$ for all $t > T$.

Integrating from t to $t + h$, for $t > T$ and any $h > 0$, we have

$|x_k(t+h) - x_k(t)| > \eta.h$ for all $t > T$, and any $h > 0$:
which contradicts the convergence of $x_k(t)$. This establishes the claim. •

The claim allows us to define the **system (13.8) as globally stable**, if for all x^o in the domain of definition of the function $F(.)$, $x(t, x^o)$ converges.

Next, consider solutions $x(t, x^o)$ which are contained in some compact set $C \subset R^n$ for all $t > 0$. Then $\exists$ a sub-sequence of the solution, $x(t_s, x^o)$ such that $\lim_{s \to \infty} x(t_s, x^o) = \bar{x}$ for some $\bar{x} \in C$. $\bar{x}$ is called a **limit point of the solution** $x(t, x^o)$. Further, $x(t, \bar{x})$ is called a **limit path** of the system (13.8). For limit paths, we have the following:[1]

Remark 13.3 $x(t, \bar{x})$
$= x(t, \lim_{s \to \infty} x(t_s, x^o))$, *(from the definition of $\bar{x}$)*
$= \lim_{s \to \infty} x(t, x(t_s, x^o))$, *(from the continuity of the solution with respect to the initial point)*
$= \lim_{s \to \infty} x(t + t_s, x^o)$ *(from the property of uniqueness of the solution with respect to the initial point)*.

Thus, **every point of the limit path** is a **limit point of the solution**.

We can now define the process (13.8) as quasi-globally stable if for all x^o in the domain of definition of the function $F(.)$, the limit points of $x(t, x^o)$ are equilibrium points of (13.8). Consequently, we have:

Claim 13.1.2 *If the equilibrium points of (13.8) are **isolated**, then quasi-global stability of (13.8)* $\Rightarrow$ *global stability of (13.8).*

Proof If (13.8) is quasi-globally stable, then every limit point of the solution $x(t, x^o)$ for any x^o in the domain of definition of $F(.)$ is an equilibrium point for (13.8); for some x^o, let $x(t, x^o)$ have two distinct limit points x^1, x^2. Then these are equilibrium points for (13.8), and are consequently, isolated points. Thus there exist open neighbourhoods $N(x^1)$ and $N(x^2)$ in R^n such that $N^c(x^1) \cap N^c(x^2) = \emptyset$ where $N^c(.)$ denotes the closure of $N(.)$. Further, given that the equilibrium points of (13.8) are isolated, the neighbourhoods $N(.)$ can be so chosen that $N^c(x^i) - x^i$ contain no other equilibria, $i = 1, 2$. Now let the subsequence $x(t_s, x^o)$ converge to x^1 as $s \to \infty$; since there is another distinct limit point, x^2, there must be another subsequence $x(t_s + h_s, x^o)$ such that $x(t, x^o) \in N(x^1)$ for all t satisfying $t_s \leq t < t_s + h_s$ with $x(t_s + h_s, x^o) \notin N(x^1)$, and $x(t_s + h_s, x^0) \in N_B^c(x^2)$ where $N_B^c(.)$ denotes the boundary of $N^c(.)$. Since $N_B^c(.)$ is a closed and bounded subset, $x(t_s + h_s, x^o)$ must have a limit point in $N_B^c(.)$ and this must be an equilibrium point of (13.8) which is distinct from x^2, given quasi-global stability. But this contradicts the definition of the sets $N(x^i)$. Hence, there cannot be two distinct limit points for the solution to (13.8): and the claim follows by virtue of Claim 13.1.1 •

The basic method for establishing quasi-global stability or global stability for a process such as (13.8) is by way of Liapunov's second method and we describe this

---

[1] See Uzawa (1961).

next. A function $V : R^n \to R$ is said to be a **Liapunov function** for (13.8), if the following conditions hold:

(i) $V(x)$ is a continuous function on R^n;
(ii) $V(x(t, x^o))$ converges as $t \to \infty$ for all x^o in the domain of definition of $F(.)$;
(iii) $\dot{V}(x(t, x^o))$ exists and is zero iff $x(t, x^o)$ is an equilibrium for (13.8).

Claim 13.1.3 *If a Liapunov function exists for (13.8), and if given any initial point, the solution to (13.8) is bounded, then (13.8) is quasi-globally stable.*

Proof Suppose, there is a Liapunov function for the system (13.8). Then by (ii) mentioned above, $\lim_{t \to \infty} V(x(t, x^o)) = V^*$ for some finite V^*, given some x^o. Further, since $x(t, x^o)$ is bounded, $\exists$ a sub-sequence $x(t_s, x^o)$ such that $x(t_s, x^o) \to x^*$ as $s \to \infty$ for some x^*. Consequently, by the property (i) of Liapunov functions, $V(x(t_s, x^o)) \to V(x^*)$. Again by (ii), $V(x^*) = V^*$. Next, consider $x(t, x^*)$: a limit path and use Remark 10.1, to conclude that $V(x(t, x^*)) = V^*$ for all t. By property (iii), every point of $x(t, x^*)$ must be an equilibrium for (13.8), including its initial point, that is, x^*. Since this is true for any arbitrary limit point of $x(t, x^o)$, the claim follows. ●

13.1.4 The linear case

We turn next to some special cases. Consider first of all, the case when $F(x) = A.x$ where A is some constant $n \times n$ matrix. Our system (13.8), then reduces to the linear system:

$$\dot{x} = A.x$$

It should be noted that this system is the exact counterpart of the equation (13.1) and, therefore, amenable to a direct solution. Recall too, that we have discussed solutions to this equation earlier and noted the role played by the characteristic roots of the matrix A. Because of this, it is not surprising that the classical result in this connection is as follows:

Proposition 13.2 *The solution $x(t, 0)$ is globally asymptotically stable iff the real parts of all characteristic roots of A are negative.*

Consequently, the results for stable matrices noted in Section 7.6.2 of this text apply. The use of the linear system arises in a natural way when one tries to analyse solutions to (13.8) with the initial point x^o close to an equilibrium x^*. For then, using a Taylor's series expansion of the functions $F_i(.)$ around the equilibrium x^*, we have the following:

$$\dot{x}_i \approx \sum_j F_{ij}(x^*)(x_j - x_j^*)$$

Next writing the matrix of partial derivatives $F_{ij}(x^*)$ as A and a change of variables, $x_j - x_j^* = \theta_j$ we have the system $\dot{\theta} = A.\theta$ which is of course, the linear system. Consequently, an immediate application of the Proposition 13.2 leads to the following:[2]

Proposition 13.3 For the system (13.8), a sufficient condition for local asymptotic stability of x^* is that A (as defined above) should have all characteristic roots with real parts negative.

How does one check whether a given matrix A has all characteristic roots with real parts negative? Recall from Section 7.6.2 that there are two results used in this connection: Propositions 7.1 and 7.2. We re-state them for ease of reference:

Proposition 7.1 For a square matrix A, a necessary and sufficient condition for all characteristic roots to have real parts negative is that there be a positive definite matrix B such that $B.A + A^T.B$ is negative definite.

Recall the following notation:

$$s_i = (-1)^i[\text{ sum of all principal minors of order i of } A]$$

Using the above we may state the other result thus:

Proposition 7.2 The following conditions are necessary and sufficient for all the characteristic roots of A to have their real parts negative $s_i > 0 \forall i$ and:

$$\det \begin{pmatrix} s_1 & s_3 \\ 1 & s_2 \end{pmatrix} > 0,$$

$$\det \begin{pmatrix} s_1 & s_3 & s_5 \\ 1 & s_2 & s_4 \\ 0 & s_1 & s_3 \end{pmatrix} > 0, \cdots,$$

and

$$\det \begin{pmatrix} s_1 & s_3 & \cdots \\ 1 & s_2 & \cdots \\ \cdots & \cdots & \cdots \\ \cdots & \cdots & s_n \end{pmatrix} > 0$$

In the above statement, we take $s_i = 0 \forall i > n$. Thus for $n = 2$ the necessary and sufficient conditions are $s_1, s_2 > 0$ and $\det \begin{pmatrix} s_1 & 0 \\ 1 & s_2 \end{pmatrix} > 0$ which is not a separate restriction ($s_3 = 0$ since there is no principal minor of order 3); thus the necessary and sufficient conditions are $s_1 > 0, s_2 > 0$: that is, the trace $(-s_1)$ should be negative

---

[2] Note that this is a **sufficient** condition for local stability; this is not necessary since even with characteristic root zero, it may be possible to have stability. Consider for instance, $\dot{x} = -x^3$; note that $x = 0$ is the unique equilibrium and is globally stable; however, when one linearizes around equilibrium, the linear system is $\dot{x} = 0$. It is for this reason that some authors have described this condition as being necessary and sufficient for **linear approximation stability**.

and the determinant s_2 should be positive. Similarly, for $n = 3$, the necessary and sufficient conditions are $s_1, s_2, s_3 > 0$ and $s_1.s_2 > s_3$; note that now $s_4 = s_5 = 0$ so that the second determinant mentioned in the statement of the Routh-Hurwitz Theorem does not add a separate restriction.

13.1.5 Motion on the plane

The next special case is when instead of $\Re^n$ for arbitrary n, we confine attention to the case when $n = 2$: that is, we shall consider motion on the plane. Such situations are of great importance and it would be of some interest to collect together the results available in this connection. Apart from being of interest independently, there is another very important reason why we must consider motion on the plane with great care. This is because for $n > 2$, general results are not possible.

Recall the definition of a solution $x(t, x^o)$ to (13.8) and of a limit point to the solution. The term ω-limit set is used to describe the set of limit points of $x(t, x^o)$. Formally, the ω-**limit set** of $x(t, x^o)$ is the set $L(x^o) = \{y : \exists$ a subsequence $x(t_s, x^o)$ such that $x(t_s, x^o) \to y$ as $s \to \infty\}$. Note that if x^* is an equilibrium of (13.8), then $L(x^*) = x^*$. On the other hand, global stability would be equivalent to requiring $L(x^o) = x^*$ for any x^o. It should also be pointed out that if $\bar{x} \in L(x^o)$, then by virtue of Remark 13.3, $x(t, \bar{x}) \in L(x^o) \forall t$. A point x^o is said to be in a **closed orbit** if $x(t, x^o) = x^o$ for some $t > 0$. Closed orbits are also called **cycles**. We shall try to characterize the nature of the limit sets for systems such as (13.8) when $n = 2$. The classical *Poincaré-Bendixson Theorem*[3] achieves this:

Proposition 13.4   In case of (13.8) being a system in $\Re^2$ with the functions $F(.)$ continuously differentiable, then any non-empty compact limit set which contains no equilibrium point is a closed orbit.

Thus, whenever the solution is bounded, it either has an equilibrium as a limit point or it converges to a closed orbit, which is called **a limit cycle** in these circumstances. This is perhaps the most clear-cut result that can be obtained for the case $n = 2$.

The treatise by Andronov et al. (1966) contains an exhaustive account of the dynamics on the plane. To be more specific about the nature of ω-limit sets we have the following:

*Proposition 13.5   On the plane, the structure of **non-empty** ω-limit sets is known to be one of the following:*[4]

(i) Consists of a single equilibrium, or

---

[3] See, for example, Hirsch and Smale (1974), p. 248.

[4] See, for instance, Andronov et al. (1966), p. 362.

(ii) Consists of one closed orbit, or

(iii) A union of equilibria and paths tending to them.

Note that for convergence to equilibrium, we need to eliminate not only limit cycles but case (iii) as well which are the most complicated types. To enable us to understand the implications of the above, it would be of some help to note that first, **for the linear case**, Proposition 13.2, allows us to obtain a particularly simple necessary and sufficient condition for convergence. Recall the following from the last section:

Remark 13.4   For the linear case, when $n = 2$, a necessary and sufficient condition for convergence is that both the following conditions should hold: (i) trace of the matrix should be negative and (ii) the determinant of the matrix should be positive.

Remark 13.5 · For local stability analysis, where the functions $F(.)$ are linearized around x^*, an equilibrium, the conditions indicated above, reduce to the following:

$$(i)\, \frac{\partial F_1(x^*)}{\partial x_1} + \frac{\partial F_2(x^*)}{\partial x_2} < 0$$

and

$$(ii)\; det \begin{pmatrix} \dfrac{\partial F_1(x^*)}{\partial x_2} & \dfrac{\partial F_1(x^*)}{\partial x_2} \\[2ex] \dfrac{\partial F_2(x^*)}{\partial x_1} & \dfrac{\partial F_2(x^*)}{\partial x_2} \end{pmatrix} > 0$$

The above conditions, if satisfied, guarantee that if the solution $x(t, x^o)$ has an initial point x^o 'close' to the equilibrium x^*, then the solution to the original system (13.8) will converge. In case this nearness to equilibrium initially cannot be guaranteed then these results are inapplicable, since the original system is, in general, non-linear. But the nature of the characteristic roots of the matrix in (ii), the Jacobian of $F(x_1, x_2)$ evaluated at equilibrium x^* (which we shall denote by $\nabla F(x^*)$) would play some role in determining the nature of limit sets in the non-linear case as well.

Consider the characteristic roots of $\nabla F(x^*)$. A **focus** is an equilibrium or fixed point where the characteristic roots are complex conjugates; the equilibrium is a stable focus when the real parts of these roots are negative; it is an unstable focus when the real parts are both positive; the equilibrium is called a **node** when these characteristic roots are both real and of the same sign; again it is a stable node if the real roots are both negative and an unstable node if the real roots are positive. Sometimes stable focii and nodes are called **sinks**; unstable nodes and focii are called **sources**. A **saddle point** is an equilibrium when the characteristic roots are both real but of opposite sign. A **centre** is an equilibrium when the characteristic roots are pure complex conjugates with real parts zero.

We provide next some diagrams to exhibit the behaviour of the solutions in some cases; consider the case when the characteristic roots at the equilibrium are real and unequal and for the sake of definiteness, say, negative: all orbits will approach the

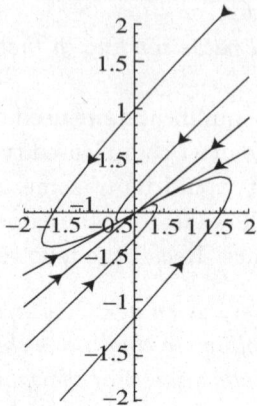

FIGURE 13.1   Stable Node (Unequal Negative Real Roots)

equilibrium, as is seen from Figure 13.1, tangentially to two sets of straight lines; this is a **stable node or a sink**; the equilibrium in the diagram is taken to be the origin below.

In case the roots are real, unequal, and positive, the situation will be as above, with the arrows pointing the other way: an unstable node. Consider, next the case when the roots are real but of opposite sign: we call this a **saddle point** (Figure 13.2).

Note that there is a single orbit which approaches the equilibrium (once again taken to be the origin): this orbit is the y-axis; in other cases, the orbit will approach another straight line through the origin. This straight line is an orbit along which the motion is away from the equilibrium.

If the roots are not equal but are repeated: this is the case of a **degenerate node**. One can divide the cases depending on whether the rank of the 2×2 characteristic matrix of the Jacobian is 0 or 1. In the former case, all orbits are straight lines; in the

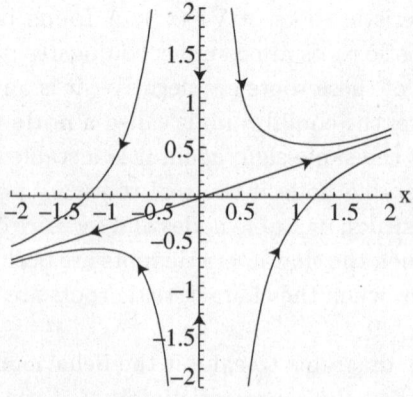

FIGURE 13.2   A Saddle Point (Real Roots of Opposite Sign)

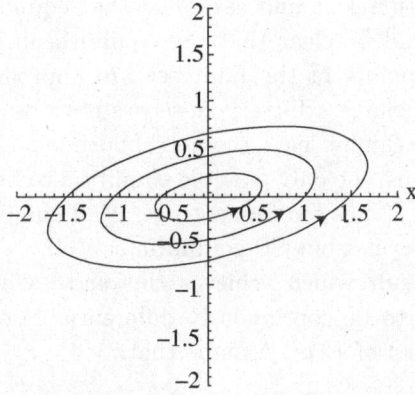

FIGURE 13.3   Centre (Pure Imaginary Roots)

second case, there is one orbit which is a straight line and other orbits are tangential and approach the origin (in case the repeated root is negative) tangentially to this orbit.

Consider next, the cases when the characteristic roots are complex. Consider first, the case of purely imaginary characteristic roots, that is, real parts are zero. We usually refer to such a situation as **centre or vortex points** (see Figure 13.3). No orbit approaches the equilibrium.

When the real parts are non-zero, the equilibria are called, **spiral or focal points**; we have drawn the diagram for unstable focal point (see Figure 13.4), a source, that is, with characteristic roots real parts positive; if the real parts of these roots had been negative, the orbits would have spiralled inwards towards the equilibrium instead of away from it.

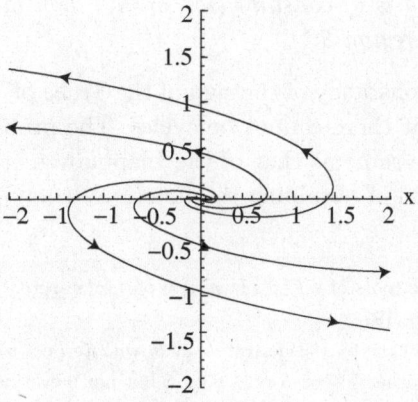

FIGURE 13.4   Focal Point (Complex Conjugate Roots)

Returning to Proposition 13.5 and assuming that equilibria are non-degenerate, consider case (iii): it should be clear that the equilibria in case (iii) can neither be sinks nor sources. Since points in the limit set are approached by the solution or trajectory, approaching a source arbitrary closely is not possible. Approaching a sink means that the trajectory cannot leave the neighbourhood as well and we are in case (i) of Proposition 13.5. Thus, the only possible equilibria to be in limit sets of the type case (iii) are saddle points. Thus for convergence, if all equilibria are hyperbolic,* we need to rule out not only cycles but SP equilibria as well.

We first present a result which achieves this end:[5] Consider (13.8) where the functions F_i are assumed to be continuously differentiable on the plane $\Re^2$ and the matrix $\nabla F(x)$, the Jacobian of $F(x)$. Assume that:[6]

O1: There is a unique equilibrium $(\bar{x}_1, \bar{x}_2)$ to (13.8).
O2: Trace of $\nabla F(x) < 0$ for all $x \in \Re^2$.
O3: Determinant of $\nabla F(x) > 0$ for all $x \in \Re^2$.
O4: Either $F_{11}.F_{22} \neq 0$ for all $x \in \Re^2$ or $F_{12}.F_{21} \neq 0$ for all $x \in \Re^2$.

Proposition 13.6   Under the conditions O1–O4, the unique equilibrium $(\bar{x}_1, \bar{x}_2)$ is globally asymptotically stable.

It is clear that we still need to satisfy a lot of conditions, even for the relatively simple case of motion on the plane, to achieve convergence. It may be of some interest to break down the implications of the various conditions employed in Olech's Theorem to get at what is exactly required. For this purpose, we first consider the elimination of cyclical orbits.

Remark 13.6   Dulac's criterion: If we can find a function $\theta(x_1, x_2)$ which is continuously differentiable on the region S and for which

$$\frac{\partial \theta(x_1, x_2) F_1(x_1, x_2)}{\partial x_1} + \frac{\partial \theta(x_1, x_2) F_2(x_1, x_2)}{\partial x_2}$$

is not identically zero and is of constant sign on S,[7] then there is no closed orbit for the system (13.8) on the region S.[8]

Note, therefore, the constancy of the sign of the Trace of $\nabla F(x)$, which is basically condition O2, implies that there cannot be cycles. The problem with the application of Dulac's criterion is the same as that of the Liapunov method we described above: there is no general method of obtaining the function $\theta(.)$.

---

* That is, all characteristic roots of $\nabla F(x^*)$ have real parts non-zero.

[5] See Olech (1963) and Ito (1978).

[6] We shall denote by $F_{ij}(x)$ the partial derivative of F_i with respect to x_j.

[7] The sign could thus be either ≥ 0 on S or ≤ 0 on S but not 0 everywhere in the region.

[8] See, for instance, Andronov et al. (1966), p. 305; another condition, the Bendixson's criterion, which serves the same purpose, is a special case of Dulac's criterion when $\theta(x_1, x_2) = 1$.

To proceed, it may be worthwhile to take into account Poincaré's Indices.[9] Consider a simple closed curve S which does not pass through equilibrium and N, any point on it; consider the tangent to the trajectory of (13.8) through N (F_1, F_2) and consider the rotation of the point N on the closed curve S: the vector (F_1, F_2) will rotate continuously and when we return to the point N (since S is a closed curve), the vector would have rotated through an angle $2\pi j$ where j is an integer. It is shown that the integer j is independent of the shape of the closed curve and is called the **index of the closed curve** S with respect to the vector filed (F_1, F_2). If the closed curve encircles an equilibrium then the index is determined by the nature of the equilibrium and is hence referred to as the **Poincaré Index** of the equilibrium. In case the equilibrium x^* is such that the determinant of the matrix $(\nabla F(x^*)$ does not vanish, it may be shown that the Poincaré Indices for a node, focus or centre are all $+1$ while for a saddle point it is -1.[10] Several other interesting conclusions may be drawn; we include a sample:

(i) The index of a closed curve not enclosing an equilibrium is zero.
(ii) The index of a closed curve surrounding a number of equilibrium points is equal to the sum of the indices of these points.
(iii) The index of a closed orbit is $+1$.

Some important corollaries: first of all, if there is a single equilibrium inside a closed orbit, it must be a node or a focus; if inside a closed orbit there are many equilibria, then they must be odd in number with the number of SPs being one less than the number of nodes and focii.

13.1.6 Lotka-Volterra system of equations

An important application of the theory of the last section is the Lotka-Volterra system of equations or alternatively, the Predator–Prey Models. We shall also find this to be a useful way of introducing another topic that we shall be concerned with later. This has to do with the theory of bifurcations. This is a study of how the dynamics of a system undergoes changes when some parameters change.

Consider an environment made up of two species of life-forms, one of which preys on the other: the prey and the predator. Let the population of the prey be designated by x while that of the predator by y. The basic assumption is that in the absence of the predator, the population of the prey **grows** at a constant proportional rate a; and on the other hand, in the absence of the prey, the population of the predator **decays** at a constant proportional rate b (here both a and b are assumed positive). In the presence of both the prey and predator, adjustments to this basic story have to be

---

[9] See, Andronov et al. (1966), p. 300 for details.

[10] For these conclusions, the non-vanishing of the determinant of the Jacobian at equilibrium is crucial; if this is not satisfied, it is not necessary that indices be $+$ or $-$ 1; see Andronov et al. (1966), p. 300–5.

made and we have

$$\dot{x} = x(a - \alpha y) \text{ and } \dot{y} = y(\beta x - b) \tag{13.9}$$

where α, β are also assumed to be positive and are to be interpreted as the effect of the presence of one population on the other.

There are two equilibria for the above system of equations:

$$(x = 0, y = 0) \text{ Trivial Equilibrium or (TE)}$$

and

$$(x = b/\beta, y = a/\alpha) \text{ Non-trivial Equilibrium or (NTE)}$$

We are interested in what happens to the solution, $z(t) = (x(t), y(t))$ to the system (13.9) beginning from an initial configuration $z^o = (x^o, y^o)$; we shall represent this solution by $z(t, z^o)$.

Local and global stability

We note first of all, the following local stability properties of the equilibria mentioned above:

Claim 13.1.4 For the system (13.9), TE is a SP while NTE is a centre.

Proof The Jacobian of the rhs of the system (13.9) is given by:

$$\begin{pmatrix} a - \alpha y & -\alpha x \\ y\beta & \beta x - b \end{pmatrix}$$

It is then straightforward to check that at TE, the characteristic roots are:

$$(a, -b);$$

while at NTE, the characteristic roots are purely imaginary:

$$(\iota \sqrt{a.b}, -\iota \sqrt{a.b}).$$

The claim follows. ●

Next, we note that:

Claim 13.1.5 With any $z^o = (x^o, y^o) > (0,0)$ as initial point, the solution to the system (13.9), $\phi_t(z^o)$, is a closed orbit around NTE $(b/\beta, a/\alpha)$.

Proof:[11] Consider

$$V(t) = \{\beta x(t) - b \log x(t)\} + \{\alpha y(t) - a \log y(t)\}$$

---

[11] Hirsch and Smale (1974), p. 262, Theorem 1.

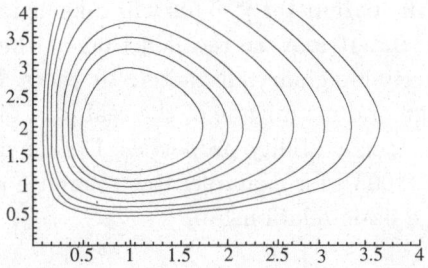

FIGURE 13.5   Closed Orbits in the Lotka-Volterra Model

and consider the derivative $\dot{V}$ along the solution $z(t) = (x(t), y(t))$ to the system(13.9) and note that

$$\dot{V} = (\beta x - b)\frac{\dot{x}}{x} + (\alpha y - a)\frac{\dot{y}}{y} = 0;$$

Thus the function V remains constant along the solution to the system (13.9) the value of this constant is defined by the initial point. Thus $V(t) = V^o = (\beta x^o - b \log x^o) + (\alpha y^o - a \log y^o)$ for all t; also note that $V(z) = V(x, y) = (\beta x - b \log x) + (\alpha y - a \log y)$ is strictly convex and attains a global minimum value, say V^*, at the NTE $(b/\beta, a/\alpha)$; thus if the initial point is not the NTE, we have $V^o > V^*$; consequently, the trajectory cannot approach NTE; nor can it approach any other equilibrium, since along the solution $x(t), y(t)$ must remain positive as otherwise $V(t)$ would become unbounded. In addition, $x(t), y(t)$ must also remain bounded, since otherwise, $V(t)$ would become unbounded. Thus the ω-limit set is non-empty and compact, and cannot contain any equilibrium; by the Poincaré-Bendixson Theorem, the ω-limit set must be a closed orbit. Thus there are two possibilities: either the closed orbit is approached in the limit (a limit cycle) or the trajectory itself is a closed orbit.

In case of a limit cycle $\mathcal{L}$, it must be the case that for every $z \in \mathcal{L}$, $V(z) = V^o$; in addition for some neighbourhood $\mathcal{N}$ of z^o, $z \in \mathcal{N} \to \phi_t(z) \to \mathcal{L}$.[12] Consequently, we must have $V(z) = V^o$ for all $z \in \mathcal{N}$: this cannot be since the function $V(z)$ is not constant over open sets. So there cannot be a limit cycle. The only possibility then is that $\phi_t(z^o)$ is a closed orbit. ●

The above claim may be seen from Figure 13.5. The diagram has been constructed using the following values of the parameters $a = 2, \alpha, \beta, b = 1$. This model has been used to explain why the population of some species like the above constantly keep chasing one another and never settles down to any fixed values. It should be pointed out that this is not really the end of the story: the conclusions really depend very crucially on the formulation; if instead of the equation used, we had $\dot{x} = x(a - \gamma x - \alpha y)$ for the equation governing the change in the variable x, the results will change considerably,

---

[12] For this property of limit cycles, see Hirsch and Smale (1974), p. 251.

depending on the sign of the parameter γ. This will take us into details which we may wish to avoid for the present. It may be recalled that at the NTE, the Jacobian for the system (13.9), has purely imaginary characteristic roots. Introduction of the term γ is likely to affect the sign and magnitude of the real part of the characteristic roots and, therefore, change the local stability properties. Details of what happens globally are available in Mukherji (2005), for example; this reference also contains an analysis of the general forms of the basic relationship.

13.2 Discrete Processes

13.2.1 Preliminary definitions

In the previous sections we considered the rule of change or the equation of motion to be in a continuous time framework. We shall now provide a consideration of the situation when the equations are in a discrete time framework. The problem that we shall encounter is that general results are not obtainable; we can at best exhibit some of the well-known tendencies that such processes may generate. To be able to do that, we need to proceed somewhat formally and thus we begin with some definitions.

Definition 13.1 Let $f : X \to X$ be continuous; $X = [a, b]$; then (X, f) defines a **dynamical system** *with $x_{n+1} = f(x_n)$.*

For any $x \in X$, $f^0(x) = x$, $f^n(x) = f(f^{n-1}(x))$, $n = 1, 2, ...$; $\gamma(x) = \{(f^n(x)\}_{n \geq 0}$: **trajectory through** x. If $f^k(x) = x$ and $f^r(x) \neq x \forall r < k, r > 0$ then x is **periodic with period** k; and $\{x, f(x), ...f^k(x)\}$ is the **cycle through** x with period k. Let $z \in (a, b)$ be a **fixed point (or equilibrium point)**, that is, $f(z) = z$. Thus, a fixed point of the map f is periodic with period unity. We are interested in characterizing the asymptotic behaviour of the trajectory through $x \in X : \gamma(x)$; this is described by the ω-limit set $\omega(x) = \{y \in X : \lim_{r \to \infty} f^{n_r}(x) = y$ for some sub-sequence $\{f^{n_r}(x)\} \in \gamma(x)\}$. An **attractor** for the map $f(.)$ is a closed set $F \subset X$ such that $\omega(x) = F$ for some x in a non-degenerate interval in X. Thus attractors depict the long-term behaviour of trajectories which begin from some non-negligible subsets of X in a sense to be made precise soon.

We first note that if the trajectory converges, then it must converge to a fixed point of the system:

Claim 13.2.1 If $f^n(x) \to \bar{x}$ as $n \to \infty$ then $\bar{x} = z$: a fixed point.

Suppose, to the contrary, that the claim is false and $f^n(x) \to \bar{x} \neq z$. Suppose that $|f(\bar{x}) - \bar{x}| = \delta > 0$. Then from the convergence of $f^n(x)$ we have for any $\epsilon > 0$ a $n(\epsilon)$ such that for all $n > n(\epsilon)$

$$\left| f^{n+1}(x) - \bar{x} \right| < \epsilon \forall n > n(\epsilon)$$

or, $\forall n > n(\epsilon)$, $|f(f^n(x)) - f(\bar{x}) + f(\bar{x}) - \bar{x}| < \epsilon$.

Now note that ϵ is arbitrary and that in the last line, the first two terms move arbitrarily close due to the continuity of the function f and we have arrived at a contradiction. Hence the Claim 13.2.1.

13.3 Stability of Periodic Points

We saw earlier that if the trajectory from some point $x \in X$ converges, then it can only do so to a fixed point of the dynamical system. To check whether any fixed point may be an attractor, it would be convenient to impose an additional restriction on the dynamical system (X, f):

The map $f : X \to X$ is differentiable on X.

Unless stated to the contrary, we shall assume that the above is satisfied. We now have the following:

Proposition 13.7   Let $x^\star$ be an equilibrium (fixed point) for the dynamical system (X, f). Then $x^\star$ is locally asymptotically stable $\Rightarrow f'(x^\star) \leq 1$; and $f'(x^\star) < 1 \Rightarrow x^\star$ is locally asymptotically stable.

To see this, note that $|x_{n+1} - x^\star| = |f(x_n) - x^\star| = |f'(\overline{x_n})| \, |x_n - x^\star|$ where $\overline{x_n} = \lambda_n x_n + (1 - \lambda_n)x^\star$ for some λ_n, $0 \leq \lambda_n \leq 1$. So if $|f'(x^\star)| < 1$, and x_n is sufficiently close to $x^\star$, there exists $k < 1$ such that $|x_{n+1} - x^\star| = k^{n+1} |x_0 - x^\star|$ and the rhs tends to zero as n becomes large. Note that the derivative being equal to 1 makes matters inconclusive.

As defined above, the attractor could be some closed subset F of the domain space X such that $\omega(x) = F$ for x belonging to some non-negligible subset in X. Specifically, the set F could be a cycle C through some point $y \in X$; the cycle C or the trajectory through $y, \gamma(y)$ or the periodic point y is asymptotically stable if there is some non-degenerate interval V with $y \in V$ such that $\omega(x) = \gamma(y) \forall x \in V$. In this connection, we also note that if $\overline{x}$ is a periodic point of the system (X, f) with period r, then $\overline{x}$ is a fixed point or equilibrium point of the system (X, f^r). That is, we have distinct points $C = \{x_1, ...x_r\}$, forming a cycle, where $\{\overline{x} = x_1, x_2 = f(x_1), ..., x_{r-1} = f(x_{r-2}), x_r = f(x_{r-1}), x_1 = f(x_r)\}$. It is clear that $f^r(x_1) = x_1$; so applying Proposition 13.7 to the system (X, f^r), we have that if $|f^{r\prime}(x_1)| < 1$ then x_1 is locally asymptotically stable for the system (X, f^r); and hence for the system (X, f), the cycle C is locally asymptotically stable. It should be clear that $|f^{r\prime}(x_1)| = |f'(x_1).f'(x_2)...f'(x_r)|$ and hence we have the following:

Claim 13.3.2   If $C = \{x_1, ...x_r\}$ is a cycle of period r for the dynamical system (X, f) then

$$|f'(x_1).f'(x_2)...f'(x_r)| < 1 \to C$$

is locally asymptotically stable for the system (X, f).

To sum up, note that for the trajectory with x_o as initial point, there are the following possibilities:

(i) x_o is a fixed point;

(ii) x_o is a periodic point (that is, $f^r(x_o) = x_o$ for some integer r);

(iii) x_o is eventually periodic, that is, $f^m(x_o)$ is periodic;

(iv) x_o is asymptotically periodic, that is, the sequence $\{f^n(x_o)\}$ contains a sub-sequence converging to some periodic point;

(v) x_o is aperiodic, that is, none of the above types. The orbit or trajectory from x_o is then called aperiodic.

Consider next, maps $f(x, \mu) : X \to X$ where μ is a parameter; it will be convenient to capture how the behaviour of the maps changes when the parameter changes; consider a particular fixed point $\xi(\mu)$ (that is, $\xi(\mu)$ solves $f(x, \mu) = x$); as will be clear from the discussion above, the thing to check is the derivative $f'(\xi(\mu))$. Several types of changes may occur: a fixed point may lose stability or it may gain stability, when parameters change; or new fixed (or periodic) points may appear or disappear when parameters pass through some value. The values of the parameters when such changes occur are known as points of **bifurcation**. We provide examples of such bifurcations.

Transcritical bifurcations

Let $f(x, \mu) = (1 + \mu)x + x^2 = x(1 + \mu + x)$; note that $\xi(\mu) = 0$ or $-\mu$. Note also that $f'(0, \mu) = 1 + \mu$ while $f'(-\mu, \mu) = 1 - \mu$. Thus we may consider μ to have great impact when it passes through 0: for example, 0 is stable for $\mu < 0$ and small while it is unstable for $\mu > 0$ and small; while the fixed point $-\mu$ is stable for $\mu > 0$ and small and unstable for $\mu < 0$ and small; thus as μ passes through the value zero from negative to positive, the stable fixed point 0 loses stability while the the fixed point $-\mu$ gains stability.

Flip bifurcations

Let $f(x, \mu) = -(1+\mu)x+x^3 = x(-1-\mu+x^2)$; $\xi(\mu) = 0, \pm\sqrt{(1+\mu)}$; $f'(0, \mu) = -(1+\mu)$; when μ is small and negative, 0 is stable; when μ changes sign 0 loses stability. The other fixed points are not really affected when μ passes through 0 and hence are not of any interest at the moment. But note that once $\mu > 0$ and small, there is a 2-period cycle $\pm\sqrt{(\mu)}$ which exists only for $\mu > 0$; further $f'(+\sqrt{(\mu)}).f'(-\sqrt{(\mu)}) = (2\mu-1)^2 < 1$ if μ is small.

Fold bifurcations

Let $f(x, \mu) = \mu + x - x^2$; since at the fixed point, we must have $x^2 = \mu$, it follows that there are no fixed points when $\mu < 0$ but there are two fixed points when $\mu > 0$. When $\mu > 0$, the values of the derivative at the fixed points are given by $1 \pm \sqrt{(\mu)}$: thus for small μ, one of the fixed points is stable and the other unstable.

13.3.1 The logistic map

General results being not possible, we shall analyse in some detail, a well-known example to illustrate the various possibilities. Consider $f(x) = K.x(1-x)$ For $1 \le K \le 4$, $f : X \to X$ where $X = [0, 1]$. For equilibrium or fixed points, we need to locate the solution to the equation:

$$f(x) = x$$

or

$$Kx(1-x) = x$$

Thus the equilibria are:

$$x_1^\star = 0 \; ; \; x_2^\star = \frac{K-1}{K}$$

Further, for local stability considerations, by virtue of Proposition 13.7, we need to note that:

$$f'(x^\star) = K(1 - 2x^\star).$$

Note then that the critical point of the logistic map $\overline{x}$ is given by $\overline{x} = 1/2$ and is independent of the value of the parameter K; further we note that the equilibrium $x_1^\star$ is locally unstable if $K > 1$ and is locally stable when $K < 1$; on the other hand, the equilibrium $x_2^\star$ is locally stable when $1 < K < 3$. Thus, different values of K correspond to quite different dynamics.

Next, we ask, the question whether there are some 2-period cycles: that is, do there exist distinct points p, q such that they solve $p = f(q)$ and $q = f(p)$; since $f(.) = Kx(1-x)$, we need to solve the following equations:

$$p = K(q - q^2) \text{ and } q = K(p - p^2)$$

These imply that

$$p + q = \frac{K+1}{K} \text{ and } p.q = \frac{1+K}{K^2}$$

These equations imply that p, q must be the roots of the quadratic

$$x^2.K^2 - (K+1)Kx + (K+1) = 0$$

In other words, p, q are the roots

$$\frac{K(K+1) \pm \sqrt{K^2(K+1)(K-3)}}{2K^2}$$

Thus p, q exist iff $K > 3$.

Now when are these 2-period cycles stable? Again using the results of the last section, we need to compute $|f'(p).f'(q)|$, that is, $|K^2(1-2p)(1-2q)| = |-K^2 + 2K + 4| < 1$ for $3 < K < 3.5$. Thus, just as K crosses the value 3, the fixed point $x_2^\star$ loses stability and a stable 2-cycle appears.

Note that as we keep changing the values of the parameter K, we pass through a series of flip bifurcations, and period doubling (a stable 2-period cycle giving way to a stable 4-period, in turn giving way to a stable 8-period cycle and so on). Let us denote by a_k the value of the parameter K when a stable 2^k-period cycle first appears; we have then:

$$a_1 = 3, a_2 = 3.449499, a_3 = 3.544090, a_4 = 3.564407\ldots$$

The sequence a_n has a limit $a^\star$ and it is known that

$$a_k \sim a^\star - c\mathcal{F}^{-k}$$

where

$$a^\star = 3.569946\ldots c = 2.6327\ldots\mathcal{F} = 4.669202$$

and that this behaviour is common in such situations; further $\mathcal{F}$ is known as Feigenbaum's constant and for a large class of one parameter family maps, has the same value. Thus, one may get an approximation of $a^\star$ as soon as we know some of the first values of the a_k's by using

$$\frac{\mathcal{F}a_3 - a_2}{\mathcal{F} - 1}$$

When $K \in (a^\star, 4)$ contains infinitely many values for which there are stable cycles: first we get to observe even period cycles and then odd period cycles. The first to appear is a 3-period cycle which is stable which leads to a period doubling and a stable 6-period cycle; outside this range, there are no stable periodic paths: this is the chaotic stage. To exhibit the various stages, we present the bifurcation diagram (see Figure 13.6).

The horizontal axis represents various values of the parameter K from 3 to 4; on the vertical axis we represent the limiting configuration: note that from $K = 3$, the limiting configuration is a stable 2-cycle, then a 4-cycle and so on; thereafter, the limiting configuration is diffused, filling up space; beyond, this range, the three cycles appear, only to become diffused once again. This is the chaotic regime. In fact, the

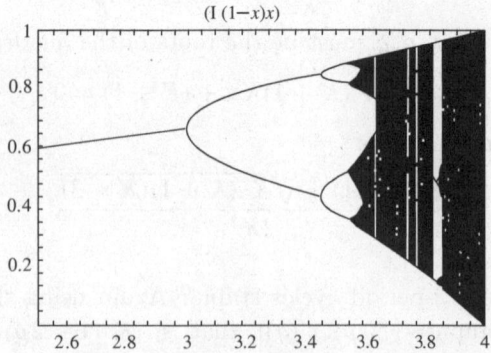

FIGURE 13.6 Bifurcation Diagram for the Logistic Map, $3 < K < 4$

existence of a 3-cycle is taken to signify a chaotic regime. Thus, simple examples are capable of exhibiting complicated behaviour.

One of the implications of the above is that in a discrete framework, we seem to be facing great difficulties: our lack of knowledge of functional forms and parameter values make prediction impossible. Not only do we need to know these things, the worrying thought seems to be that we need pretty exact values for them. For the logistic curve, for example, we need to know what the value of K is quite exactly before we can say what the dynamics will lead to. And finally, we chose the logistic equation because it has been well researched; but these conclusions are not specific to that formulation.

EXERCISES

Consider first the solution to some linear equations of the first order. Solve: (y' denotes $\frac{dy}{dx}$)

1. $y' - \alpha y = e^{mx}$ $\alpha \neq m$, constants.
2. $y' - \alpha y = e^{\alpha x}$ α constant.
3. $y' - \dfrac{2x}{x^2 + 1} y = 2x(x^2 + 1)$.

Consider next, some equations on the plane; in each case, sketch a phase diagram and discuss the nature of the equilibria and obtain the solution.

1. $\dot{x} = x - y$ and $\dot{y} = 3x - y$.
2. $\dot{x} = -y$ and $\dot{y} = x$; $x(0) = 1$, $y(0) = 1$.
3. $\dot{x} = 2xy - 2y^2$ and $\dot{y} = x - y^2$

Next some more questions on motion on the plane:

1. Consider $\dot{x} = \mu x - 1$ and $\dot{y} = 3x - y$; examine the nature of the equilibrium as μ changes values through 1; refer to Hirsch and Smale (1974) and read about Hopf Bifurcations.
2. Consider the system $\dot{x} = x(2 - x - 3y)$ and $\dot{y} = y(1 + 2x - 4y)$; find out the equilibria and analyse their stability properties. Draw a phase diagram and discuss the long-term behaviour of the solution.

Finally some questions on discrete processes. For each case, find the equilibria and determine whether they are stable or unstable.

1. $x_{n+1} = 1.7x_n - 0.14x_n^2$.
2. $x_{n+1} = 0.2x_n - 0.2x_n^3$.
3. $x_{n+1} = 3x_n - x_n^2 + 3$
4. $x_{n+1} = 5x_n - x_n^2 - b$. Examine the behaviour of the equilibria as b changes.

KEY TERMS

Attractor

Autonomous Differential Equation

Autonomous System

Centre or Vortex Point

Closed orbit

Degenerate Node

Discrete Processes

Dulac's Criterion

Dynamical System

Equilibrium Point

Flip Bifurcation

Focus

Fold Bifurcation

Globally asymptotically stable

Homogenous form

Index of the closed curve

Initial Point	Node
Liapunov Function	Non-homogenous Form
Limit Cycle	Olech's Theorem
Limit Path	Poincaré Index
Limit Point	Poincaré-Bendixson Theorem
Limit Set	Quasi-globally Stable
Linear System	Saddle Point
Lipschitz Condition	Sinks
Locally Asymptotically Stable	Sources
Logistic Map	Spiral or Focal Points
Lotka-Volterra System of Equations	Trajectory
Predator–Prey Models	Transcritical Bifurcation
Motion on the Plane	

USEFUL WEB LINKS

http://chaos.wlu.edu/106/programs/logisticdes.html#activ (accessed on 2 August 2010): discrete maps and chaotic behaviour

http://hopf.chem.brandeis.edu/yanglingfa/pattern/rd/LN_Bifurcation.pdf (accessed on 2 August 2010);

http://www.me.rochester.edu/courses/ME406/webexamp6/bifurc3.pdf (accessed on 2 August 2010): Mathematica Notebook—transcritical bifurcation, bifurcation theory

http://math.asu.edu/~kuang/class/494MB/Lect22.pdf (accessed on 2 August 2010): Liapunov function and Dulacs criterion

http://www.aw-bc.com/ide/ (accessed on 2 August 2010);

http://wwwmaths.anu.edu.au/~briand/chaos/orbits.html (accessed on 2 August 2010): differential equations

http://www.math.byu.edu/~grant/courses/m634/f99/lec39.pdf (accessed on 4 October 2010);

http://www.math.harvard.edu/archive/118r_spring_05/handouts/poincare.pdf (accessed on 2 August 2010): Poincaré-Bendixson Theorem

http://www.scholarpedia.org/article/Predator-prey_model (accessed on 2 August 2010): Lotka-Volterra system of equations or Predator–Prey Models

CHAPTER FOURTEEN

Dynamic Optimization

14.1 Introduction to the Optimal Control Theory

Consider a decision-maker who has to choose, for various points of time, values for a particular variable. For example, a consumer, in making a consumption plan, chooses her rates of consumption for different points of time, and a firm, in making an invest-ment plan, chooses its rates of investment for different points of time. Suppose also that the decision-maker associates some rate of **net return** with each point of time for which she makes a choice. For example, the consumer might associate a discounted rate of utility with each point of time for which she chooses a rate of consumption and the firm might associate a discounted flow of profit with each point of time for which it chooses a rate of investment. The choice made by the decision-maker for any given instant may affect the values of the variable available for choice at later instants or, even if the set of values available for choice is unaffected, affect the rate of net return to the decision-maker at a later instant. For example, a consumer's current consumption decisions might determine which time-paths for future consumption are feasible and a firm's current investment decisions might, for a given time-profile of future investments, determine the future time-profile of the firm's capital stock and the discounted value of its future stream of profits.

Suppose the choice made by the decision-maker for any point of time t is repre-sented by the value $u(t)$ of the variable under the decision-maker's control, that is the **decision or control variable**. Then, the decision-maker can be thought of as choos-ing a function $u : S \rightarrow U$, where S is the set of all points of time (values of the time vari-able) for which decisions are taken and U is the possible set of values of the control vari-able. Note that the decision-maker may have to make choices for more than one control variable. If the variables are denoted as $u_1, u_2, ..., u_r$ where r is a positive integer then u in the above problem may be considered to be a vector-valued function with range U.

If S is a finite set, for example, $S = \{1, 2, ..., T\}$ where T is a positive integer, and the decision-maker wishes to choose a function u amongst a set $\mathcal{U}$ of feasible alternatives to maximize or minimize the value of an **objective functional** $g(u)$, then the decision-making problem can be written as

$$\max \ g(u) \text{ subject to } u \in \mathcal{U}$$

We call g, a functional, since it is a function whose domain is a set of functions. While analysing problems of static optimization, it may be recalled that appropriate values of variables had to be chosen; in contrast we have to now choose appropriate functions.

The number of time points in S may, however, be infinite. For example, if choices are assumed to be made for every point in an interval between two points of time t_0 and t_1 then $S = [t_0, t_1]$. In this case the theory of static optimization will no longer suffice. The theory of dynamic optimization, known as the **Theory of Optimal Control**, which we outline in this chapter, provides us with the tools to study problems in which the set S is of the form $[t_0, t_1]$ or $[t_0, \infty)$. The latter refers to the case where choices are assumed to be made for every point of time t_0 onwards.

There is another fundamental aspect of the theory which needs to be mentioned at the outset. Remember that for any given instant t, the time-profiles of choices which are feasible t onwards and the net return to the decision-maker for any given profile of choices t onwards may depend on choices made for points of time preceding t. This implies that, in principle, the choices made t onwards may depend on an infinite number of choices made in the past. In the theory of dynamic optimization, however, the decision-making problem considered is always such that, for any given point of time t, the dependence of the choice constraints and the rates of net return t onwards on the choices made before t can be expressed as a dependence of these constraints and rates on the values at t of a finite number of variables known as **state variables**. One can think of the vector of values of these variables at t as the **state** (of things) at t^1 which, determined by the history before t, determines choices made for subsequent points of time.

Suppose the consumer in our example is choosing a consumption plan for the rest of her lifetime and the discounted rate of utility she associates with any point of time is simply a function of her chosen rate of consumption for that point of time. Then, it is conceivable that the only way her history of consumption before an instant t can affect her consumption plan t onwards is through its influence on her level of wealth at t. The wealth of the consumer would then be counted as a state variable for the consumer's decision-making problem. Similarly, rates of investment chosen by a firm before an instant t may influence the firm's choice of rates of investment after t only through its effect on the firm's capital stock at t. The value of the firm's capital stock is then a state variable for the firm's decision-making problem.

Further, if $x_1, x_2, x_3, ..., x_n$ are n state variables (n being a positive integer) in a decision-making problem then the problem can be studied using the theory of optimal control only if the rate of change of each of these variables with respect to time at any given instant can be written as a function of the values of the control variables, the

---

[1] More formally, this is known as the **state of the system** at t.

state variables and the time variable itself at that instant. Thus, it must be possible to define n ordinary differential equations in normal form:

$$\dot{x}_i = f_i(x, u(t), t), \quad i = 1, 2, 3, ..., n$$

where $x = (x_1, x_2, x_3, ..., x_n)$, $u(t) = (u_1(t), u_2(t), u_3(t), ..., u_r(t))$ and $u_j(t)$ $(j = 1, 2, ..., r)$ denotes the value of u_j at time t. These equations, which are the equations of motion for the state variables, are known as the **state equations** for the decision-making problem.

14.2 A Basic Optimal Control Problem

Suppose a decision-maker is assumed to choose values of r real variables $u_1, u_2, ..., u_r$ for each instant of time over some time interval.[2] u is the (column) vector of **control variables** for the decision-maker so that $u^T = (u_1, u_2, ..., u_r)$. Any pattern of choices made by the decision-maker over any time-interval $I \subseteq R$ can then be represented as a vector-valued function $u : I \to R^r$.

We will assume that the decision-maker can choose a function $u : I \to R^r$ only if it is a **piece-wise continuous function** on I. This means that u can have at most a finite number of points of discontinuity on any finite time interval $[a, b] \subseteq I$. Moreover, at each such point of discontinuity τ, both the limits, $\lim_{t \to \tau-} u(t)$ and $\lim_{t \to \tau+} u(t)$ must exist (finitely) and the discontinuity must arise because $\lim_{t \to \tau-} u(t) \neq \lim_{t \to \tau+} u(t)$.[3] Thus, attention is restricted to time-paths of control variables which can exhibit at most a finite number of 'jumps' over a finite time interval.

Any such piece-wise continuous function $u : I \to R^r$ will be referred to as a **control** for the decision-maker.

Suppose $x_1, x_2, ..., x_n$ are n real variables, $x^T = (x_1, x_2, ..., x_n)$, t_0 and t_1 are real constants $(t_1 > t_0)$, x^0 and x^1 are n-tuples of real numbers and $U \subseteq R^r$. Suppose $F : R^n \times R^r \times R \to R$ and $f_i : R^n \times R^r \times R \to R$ $(i = 1, 2, ..., n)$ are continuous functions and have continuous partial derivatives with respect to x_j $(j = 1, 2, ..., n)$ at all $(x, u, t) \in R^n \times R^r \times R$. Let $f^T = (f_1, f_2, ..., f_n)$.

Suppose we are told that the decision-maker chooses a control which solves the following optimization problem:

Problem (A): $\max\limits_{\substack{u(t) \\ t \in [t_0, t_1]}} \int_{t_0}^{t_1} F(x(t), u(t), t)dt$

$$\text{s.t. } \dot{x}(t) = f(x(t), u(t), t)$$
$$x(t_0) = x^0, x(t_1) = x^1$$
$$u(t) \in U$$

---

[2] The time-interval for which choices are made may or may not be decided by the decision-maker.

[3] This implies that any point of discontinuity must be an interior point of I.

How do we interpret Problem (A)? What does it tell us about the nature of the problem facing the decision-maker?

The first line of Problem (A) indicates that for the maximization exercise the decision-maker has to choose values of the control variables in the fixed time interval $[t_0, t_1]$. This implies that the decision-maker chooses among only those controls which have the fixed interval $[t_0, t_1]$ as their domain. The last line of Problem (A) indicates that, given any instant t, the decision-maker has to choose the value of u at t from a given set U. U is known as the **control region** for this problem.

The vector differential equation in Problem (A) indicates that the variables x_1, x_2, ..., x_n are the **state variables** for the problem and x the vector of state variables. The underlying scalar differential equations

$$\dot{x}_i(t) = f_i(x_1(t), x_2(t), ..., x_n(t), u_1(t),$$
$$u_2(t), ..., u_r(t), t), \ i = 1, 2, ..., n$$

are known as the **state equations** for the problem.

The second and third lines of Problem (A) are taken to imply the following:

The decision-maker can choose a control $u : [t_0, t_1] \to U$ only if it is possible to define a unique trajectory or time path for the vector of state variables $x : [t_0, t_1] \to R^n$ such that:

(i) $x(t)$ is continuous on $[t_0, t_1]$,
(ii) $x = x(t)$ is a solution of the vector differential equation $\dot{x} = f(x, u(t), t)$ on all intervals $I \subseteq [t_0, t_1]$ on which u is continuous (that is except at points of discontinuity of u, $\dot{x}(t) = f(x(t), u(t), t)$ for all $t \in [t_0, t_1]$),[4] and
(iii) $x(t_0) = x^0$ and $x(t_1) = x^1$.

$x(t)$ is defined to be the **trajectory** associated with the control $u(t)$. (i) and (ii) imply that $x(t)$ is a **piece-wise continuously differentiable function** on $[t_0, t_1]$. The value of the vector-valued variable x at any time t is known as the **state at time** t. (iii) implies that the state at the **initial time** t_0 is x^0 and the state at the **terminal time** t_1 must be x^1. The point x^0 is, therefore, known as the **initial state** and the point x^1 is known as the **terminal state**.

Any control which satisfies the constraints on choice of control implied by Problem (A) is known as an **admissible control** for the given problem. If $x : [t_0, t_1] \to R^n$ is the trajectory associated with an admissible control $u : [t_0, t_1] \to U$ then (x, u) is known as an **admissible pair** for the given problem.

Let $\mathcal{U}$ denote the set of admissible controls for the given problem.

[4] At the endpoints t_0 and t_1 of the domain of definition of $x(t)$, $\dot{x}(t)$ refers to the appropriate one-sided derivatives.

Since $F(x, u, t)$ is a continuous function on $R^n \times R^r \times R$, and for any admissible control u with associated trajectory x, $x(t)$ is a continuous function on $[t_0, t_1]$ and $u(t)$ is a piece-wise continuous function on $[t_0, t_1]$, $\Phi(t) = F(x(t), u(t), t)$ is a piece-wise continuous function of t on $[t_0, t_1]$. Remember that for a piece-wise continuous function, both right-hand and left-hand limits exist at each point of discontinuity. Therefore, for any admissible control u with associated trajectory x, the definite integral $\int_{t_0}^{t_1} F(x(t), u(t), t) dt$ is defined.

In Problem (A), $F(x(t), u(t), t)$ denotes the rate of **net return** to the decision-maker at an instant t when the decision-maker chooses a control u with associated trajectory x. The objective of the decision-maker is to choose a control from $\mathcal{U}$ which maximizes the integral sum of instantaneous net returns obtained from the interval $[t_0, t_1]$.

$$\max_{u \in \mathcal{U}} \int_{t_0}^{t_1} F(x(t), u(t), t) dt$$

This integral sum of net returns is known variously as the **objective functional**, **criterion functional**, **performance criterion**, and **performance index**.

Note that the value of the objective functional for any admissible control is independent of the values of the control variable at each point of discontinuity of the control. By convention, therefore, it is assumed that for any control u, at each point of discontinuity τ, $u(\tau) = \lim_{t \to \tau^-} u(t)$.

A solution to Problem (A) will constitute an **optimal control** for the decision-maker. If $x : [t_0, t_1] \to R^n$ is the trajectory associated with an optimal control $u : [t_0, t_1] \to U$ then (x, u) is known as an **optimal pair** for the given problem. Since the above decision-making problem is a problem of finding an optimal control it is referred to as an **optimal control problem**. In the 1950s, a Russian mathematician L.S. Pontryagin and his associates considered this and a related set of optimal control problems, and were able to derive results outlining sets of conditions which an optimal control must necessarily satisfy in these problems. The set of conditions outlined for each problem involved the maximization of a function called the **Hamiltonian** for the problem and, therefore, came to be known as **Pontryagin's Maximum Principle** for the problem.

14.3 Necessary Conditions

14.3.1 Some special assumptions

The proof of Pontryagin's Maximum Principle for Problem (A) is outside the scope of this text. The purpose of this section is to develop some idea about what the necessary conditions for an optimal control might look like. In the course of the discussion, we will make various special assumptions about the set of admissible controls and the existence and nature of solutions for Problem (A) and for other related problems.

Needless to say, a *proof* of the Maximum Principle for Problem (A) requires none of the additional assumptions introduced in this section.[5]

Let τ be any real number belonging to $[t_0, t_1]$ and let y be any n-tuple of real numbers. Consider the following variant of Problem (A):

$$\max_{\substack{u(s) \\ s \in [\tau, t_1]}} \int_\tau^{t_1} F(x(s), u(s), s) ds$$

$$\text{s.t. } \dot{x}(s) = f(x(s), u(s), s)$$

$$x(\tau) = y, x(t_1) = x^1$$

$$u(s) \in U$$

Call this Problem $(A(\tau, y))$. Problem $(A(\tau, y))$ differs from Problem (A) in two respects: the initial time is τ instead of t_0 and the initial state is y instead of x^0.[6] Given any real number $\tau \in [t_0, t_1]$ and any $y \in R^n$, let $\mathcal{U}(\tau, y)$ denote the set of admissible controls for Problem $(A(\tau, y))$.

For all $\tau \in [t_0, t_1]$, define $Q(\tau) = \{y | y \in R^n \land \mathcal{U}(\tau, y) \neq \phi\}$. An initial state $y \in R^n$ belongs to $Q(\tau)$ iff there is at least one admissible control for the Problem $(A(\tau, y))$. We will assume that for all $\tau \in [t_0, t_1]$, $Q(\tau)$ is a non-empty set in R^n and $\Omega = \cup_{\tau \in [t_0, t_1]}\{(\tau, y) | y \in Q(\tau)\}$ is an open set in $[t_0, t_1] \times R^n$.[7] Note that the latter assumption implies that $Q(\tau)$ is an open set in R^n, for all $\tau \in [t_0, t_1]$.

Let us suppose that there exists a function $V(\tau, y) = \max_{u \in \mathcal{U}(\tau, y)} \int_\tau^{t_1} F(x(s), u(s), s) ds$, for all $(\tau, y) \in \Omega$, which, given any initial time $\tau \in [t_0, t_1]$ and any initial state $y \in Q(\tau)$ gives us the maximum value of the objective functional which can be achieved by any admissible control in Problem $(A(\tau, y))$. That is, $V(\tau, y)$ gives us the maximum returns attainable by the decision-maker on the interval $[\tau, t_1]$ (that is, subsequent to time τ) given that the state at time τ is y. We shall call $V : \Omega \to R$ the **value function** associated with Problem (A). Assume that all partial derivatives of V are continuous on $\Omega_I = \cup_{\tau \in (t_0, t_1)}\{(\tau, y) | y \in Q(\tau)\}$. Since Ω is an open set in $[t_0, t_1] \times R^n$ it follows that Ω_I is an open set in R^{n+1}.

14.3.2 A maximization condition

Suppose $u^* : [t_0, t_1] \to U$ is an optimal control for Problem (A) with associated trajectory $x^* : [t_0, t_1] \to R^n$. Suppose $\tau \in [t_0, t_1]$. Consider Problem $(A(\tau, x^*(\tau)))$.

---

[5] To develop a better understanding of the discussion in this section, readers might wish to assume, during their first reading, that there is a single control variable, and a single state variable, that is $n = r = 1$.

[6] There is another insignificant difference: s rather than t, as in Problem (A), is now used to denote the time variable.

[7] We will define, for any positive integer m, a set S to be an open set in X ($S \subseteq X \subseteq R^m$) iff $(\forall v)(v \in S \to (\exists \varepsilon)(\varepsilon \in R_{++} \land \{x | x \in X \land d(x, v) < \varepsilon\} \subseteq S))$.

Given any three non-empty sets X, Y and Z such that $Z \subseteq X$ and given any function $f : X \to Y$, let us denote the function $\{(x, y) \mid x \in Z \land y = f(x)\}$ (the restriction of f to Z) simply as $f : Z \to Y$. We can then claim that, given any $\tau \in [t_0, t_1]$, $u^* : [\tau, t_1] \to U$ is an optimal control for Problem $(\mathrm{A}(\tau, x^*(\tau)))$.

The reason for the above claim is the following. Note that if $u^* : [\tau, t_1] \to U$ is not an optimal control for Problem $(\mathrm{A}(\tau, x^*(\tau)))$ then there must exist some admissible pair $(\tilde{x}, \tilde{u})$ for Problem $(\mathrm{A}(\tau, x^*(\tau)))$ such that

$$\int_{\tau}^{t_1} F(\tilde{x}(s), \tilde{u}(s), s)ds > \int_{\tau}^{t_1} F(x^*(s), u^*(s), s)ds \tag{14.1}$$

We can then define the following admissible control $\hat{u} : [t_0, t_1] \to U$ for Problem (A):

$$\hat{u}(s) = u^*(s), s \in [t_0, \tau]$$
$$= \tilde{u}(s), s \in (\tau, t_1]$$

But, from (14.1) it would then follow that $u^* : [t_0, t_1] \to U$ cannot be an optimal control for Problem (A) because

$$\int_{t_0}^{t_1} F(\hat{x}(s), \hat{u}(s), s)ds$$
$$= \int_{t_0}^{\tau} F(x^*(s), u^*(s), s)ds + \int_{\tau}^{t_1} F(\tilde{x}(s), \tilde{u}(s), s)ds$$
$$> \int_{t_0}^{t_1} F(x^*(s), u^*(s), s)ds$$

Suppose $t \in (t_0, t_1)$ and consider a 'small' positive real number $\Delta t < t_1 - t$. Precisely how small Δt is required to be will be indicated later.

Suppose $v \in U$. Note that because $f(x, v, s)$ and $\partial f_i(x, v, s)/\partial x_j$ $(i = 1, 2, ..., n; j = 1, 2, ..., n)$ are continuous functions of x and s on $R^n \times R$, for sufficiently small Δt there exists a unique solution on $[t, t + \Delta t]$ to the initial value problem:

$$\dot{x} = f(x, v, s), \ x(t) = x^*(t)$$

Let $x^v : [t, t+\Delta t] \to R^n$ be the solution to the above initial value problem on $[t, t+\Delta t]$. Since $x^v(s)$ is a continuous function on $[t, t + \Delta t]$, $(t, x^v(t)) = (t, x^*(t)) \in \Omega$ and Ω is an open set in $[t_0, t_1] \times R^n$, therefore, for sufficiently small Δt, $(t + \Delta t, x^v(t + \Delta t)) \in \Omega$. Given that the value function V is defined on Ω, suppose $(\hat{x}, \hat{u})$ is an optimal pair for Problem $\mathrm{A}((t + \Delta t, x^v(t + \Delta t)))$. The following is then an admissible control for Problem $\mathrm{A}((t, x^*(t)))$:

$$\tilde{u}(s) = v, s \in [t, t + \Delta t]$$
$$= \hat{u}(s), s \in (t + \Delta t, t_1]$$

Suppose $\tilde{x} : [t, t_1] \to R^n$ is the trajectory associated with $\tilde{u}$. Since $\tilde{x}(t + \Delta t) = x^v(t + \Delta t)$, $(\hat{x}, \hat{u})$ is an optimal pair for Problem $\mathrm{A}((t + \Delta t, x^v(t + \Delta t)))$ and $(\tilde{x}(s), \tilde{u}(s)) =$

$(\hat{x}(s), \hat{u}(s))$ for all $s \in (t + \Delta t, t_1]$, therefore,

$$V(t + \Delta t, \tilde{x}(t + \Delta t)) = \int_{t+\Delta t}^{t_1} F(\tilde{x}(s), \tilde{u}(s), s)ds \qquad (14.2)$$

Since $\tilde{u}$ is an admissible control for Problem $(A(t, x^*(t)))$ and $u^* : [t, t_1] \to U$ is an optimal control for the same problem it follows that

$$\int_t^{t_1} F(x^*(s), u^*(s), s)ds \geq \int_t^{t+\Delta t} F(\tilde{x}(s), \tilde{u}(s), s)ds + \int_{t+\Delta t}^{t_1} F(\tilde{x}(s), \tilde{u}(s), s)ds \quad (14.3)$$

Since $u^* : [t + \Delta t, t_1] \to U$ is an optimal control for Problem $(A(t + \Delta t, x^*(t + \Delta t)))$, from (14.2) and (14.3) it follows that,

$$\int_t^{t+\Delta t} F(x^*(s), u^*(s), s)ds + V(t + \Delta t, x^*(t + \Delta t)) \geq$$
$$\int_t^{t+\Delta t} F(\tilde{x}(s), \tilde{u}(s), s)ds + V(t + \Delta t, \tilde{x}(t + \Delta t)) \qquad (14.4)$$

Think of the decision-maker as facing the following problem: The decision-maker has the option of choosing the values of the control u^* in the interval $[t, t + \Delta t]$ or of choosing any given point $v \in U$ for each point of the interval $[t, t + \Delta t]$. Suppose it is the case that for the interval $[t + \Delta t, t_1]$ the optimal control for Problem $(A(t + \Delta t, x(t + \Delta t)))$ will always be chosen, where $x(t + \Delta t)$ is the state at $t + \Delta t$ resulting from her choices for the interval $[t, t + \Delta t]$. If the decision-maker is interested in maximizing her net returns over $[t, t_1]$ what should be her choices for the interval $[t, t + \Delta t]$?

What (14.4) tells us is that, in this truncated decision-making problem as well, the decision-maker should choose the values of u^* for the interval $[t, t + \Delta t]$. This will fetch at least as much net returns over $[t, t_1]$ as choosing any given point $v \in U$ for each point of $[t, t + \Delta t]$.

Next, suppose that u^* is continuous on $[t_0, t_1]$. Then, it follows that the function $F(x^*(s), u^*(s), s)$ is continuous at $s = t$. Therefore, for sufficiently small Δt,

$$F(x^*(s), u^*(s), s) \cong F(x^*(t), u^*(t), t),$$
$$\text{for all } s \in [t, t + \Delta t]$$

Similarly, because $\tilde{u}(s) = v$ for all $s \in [t, t + \Delta t]$, and $\tilde{x}(t) = x^*(t)$, therefore, for sufficiently small Δt,

$$F(\tilde{x}(s), \tilde{u}(s), s) \cong F(x^*(t), v, t), \text{ for all } s \in [t, t + \Delta t]$$

Therefore, given (14.4), for sufficiently small Δt it follows that

$$F(x^*(t), u^*(t), t).\Delta t + V(t + \Delta t, x^*(t + \Delta t)) \geq F(x^*(t), v, t).\Delta t + V(t + \Delta t, \tilde{x}(t + \Delta t))$$
$$(14.5)$$

By assumption, all partial derivatives of V exist and are continuous on Ω_I and Ω_I is an open set in R^{n+1}. Therefore, for sufficiently small Δt,[8]

$$V(t + \Delta t, x^*(t + \Delta t)) \cong V(t, x^*(t)) + V_y(\tau, y)|_{(\tau, y) = (t, x^*(t))} \cdot [x^*(t + \Delta t) - x^*(t)]$$
$$+ V_\tau(\tau, y)|_{(\tau, y) = (t, x^*(t))} \cdot \Delta t \tag{14.6}$$

Since, by assumption, u^* is continuous in a neighbourhood of t (x^* is a continuously differentiable function in a neighbourhood of t), for sufficiently small Δt,

$$x^*(t + \Delta t) \cong x^*(t) + f(x^*(t), u^*(t), t) \cdot \Delta t \tag{14.7}$$

Similarly, given that $\tilde{x}(t) = x^*(t)$, for sufficiently small Δt,

$$V(t + \Delta t, \tilde{x}(t + \Delta t)) \cong V(t, x^*(t)) + V_y(\tau, y)|_{(\tau, y) = (t, x^*(t))} \cdot [\tilde{x}(t + \Delta t) - x^*(t)]$$
$$+ V_\tau(\tau, y)|_{(\tau, y) = (t, x^*(t))} \cdot \Delta t \tag{14.8}$$

and

$$\tilde{x}(t + \Delta t) \cong x^*(t) + f(x^*(t), v, t) \cdot \Delta t \tag{14.9}$$

From (14.5) to (14.9) it follows that, for sufficiently small Δt,

$$F(x^*(t), u^*(t), t) \cdot \Delta t + V_y(\tau, y)|_{(\tau, y) = (t, x^*(t))} \cdot f(x^*(t), u^*(t), t) \cdot \Delta t$$
$$\geq F(x^*(t), v, t) \cdot \Delta t + V_y(\tau, y)|_{(\tau, y) = (t, x^*(t))} \cdot f(x^*(t), v, t) \cdot \Delta t \tag{14.10}$$

Refer back to the discussion after (14.4). The net returns to the decision-maker from time t onwards consist of the net returns in $[t, t + \Delta t]$ and the subsequently maximized net returns in $[t + \Delta t, t_1]$. The decision-maker's choice of values for the control variables in $[t, t + \Delta t]$ has two effects on her net returns in $[t, t_1]$:

(i) There is a direct effect in the form of net returns in $[t, t + \Delta t]$. For sufficiently small Δt, this is equal to $F(x^*(t), u^*(t), t) \cdot \Delta t$ if the decision-maker chooses values of the control u^* and this is equal to $F(x^*(t), v, t) \cdot \Delta t$ if the decision-maker chooses $v \in U$ for every point in $[t, t + \Delta t]$.

(ii) There is an indirect effect on the net returns in $[t + \Delta t, t_1]$ through the effect on the state at $t + \Delta t$. For sufficiently small Δt, $V_y(\tau, y)|_{(\tau, y) = (t, x^*(t))}$ is the (row) vector of rates of change in the maximum attainable returns time $t + \Delta t$ onwards with respect to the state variables. For sufficiently small Δt, the vector of changes in the state variables over the interval $[t, t + \Delta t]$ is $f(x^*(t), u^*(t), t) \cdot \Delta t$ if the decision-maker chooses values of the control u^* and is $f(x^*(t), v, t) \cdot \Delta t$ if the decision-maker chooses $v \in U$ for every point in $[t, t + \Delta t]$.

Since by choosing the values of the control u^* in $[t, t + \Delta t]$ the decision-maker gets at least as much net returns over $[t, t_1]$ as by choosing any given point $v \in U$ for all

---

[8] Throughout this chapter, we will denote the row vector of partial derivatives of a function Φ with respect to the components of a (row or column) vector of variables v by Φ_v.

points in $[t, t + \Delta t]$, therefore, the sum of the two effects for the choice of u^* will be at least as much as the sum of these two effects for the choice of any given $v \in U$. This is indicated by (14.10).

Suppose we define $\lambda(s) = V_y(\tau, y)|_{(\tau, y) = (s, x^*(s))}$, for all $s \in (t_0, t_1)$. Then, from (14.10) it follows that

$$F(x^*(t), u, t) + \lambda(t).f(x^*(t), u, t) \text{ is maximized on } U \text{ at } u = u^*(t) \qquad (14.11)$$

(14.11) is then a condition which (under the special assumptions made in this section) has to be necessarily satisfied by the optimal pair for Problem (A), (x^*, u^*). We now proceed to derive another such condition.

14.3.3 A differential equation

Assume that, given any $\tau \in (t_0, t_1)$ and any $y \in Q(\tau)$, there exists a unique and continuous optimal control for Problem $(A(\tau, y))$. Let $c(s, \tau, y)$ denote the value of the optimal control for Problem $(A(\tau, y))$ at s, for all $s \in [\tau, t_1]$. Moreover, let $z(s, \tau, y)$ denote the state at s along the trajectory associated with the optimal control for Problem $(A(\tau, y))$, for all $s \in [\tau, t_1]$. By assumption, for any given $(\tau, y) \in \Omega_I$, $c(s, \tau, y)$ and $z(s, \tau, y)$ are continuous functions of s on $[\tau, t_1]$. Suppose that all partial derivatives of the components of c and z with respect to the components of y exist and are continuous functions of s, for any given $(\tau, y) \in \Omega_I$. Suppose at all $(s, x, u) \in \Omega \times U$ partial derivatives of $F(x, u, s)$ and $f_i(x, u, s)$ $(i = 1, 2, ..., n)$ with respect to the components of u exist and for any given (x, u) are continuous functions of s on $[\tau, t_1]$.

From the discussion above, it is obvious that if $u^* : [t, t_1] \to U$ is an optimal control for Problem $(A(t, x^*(t)))$, where $t \in (t_0, t_1)$ and Δt is a positive real number less than $(t_1 - t)$, then:

$$V(t, x^*(t)) = \int_t^{t+\Delta t} F(x^*(s), u^*(s), s)ds + V(t + \Delta t, x^*(t + \Delta t))$$

In an analogous manner it can be shown that for all $(\tau, y) \in \Omega_I$ and for all $\Delta t \in (0, t_1 - \tau)$,

$$V(\tau, y) = \int_\tau^{\tau+\Delta t} F(z(s, \tau, y), c(s, \tau, y), s)ds + V(\tau + \Delta t, z(\tau + \Delta t, \tau, y))$$

Therefore,[9]

$$\begin{aligned}
V_y(\tau, y) = \int_\tau^{\tau+\Delta t} &[F_z(z(s, \tau, y), c(s, \tau, y), s).z_y(s, \tau, y) \\
&+ F_c(z(s, \tau, y), c(s, \tau, y), s).c_y(s, \tau, y)]ds \\
&+ V_z(\tau + \Delta t, z(\tau + \Delta t, \tau, y)).z_y(\tau + \Delta t, \tau, y)
\end{aligned}$$

---

[9] $c_y(s, \tau, y)$ and $z_y(s, \tau, y)$ denote the matrices $\left(\frac{\partial c_i(s,\tau,y)}{\partial y_j}\right)_{r \times n}$ and $\left(\frac{\partial z_i(s,\tau,y)}{\partial y_j}\right)_{n \times n}$, respectively.

The equation tells us that the change in the maximum value attainable time τ onwards resulting per unit change in any state variable at τ is the sum of three components: the change in the net returns attained in the interval $[\tau, \tau + \Delta t]$ due to the resulting change in the values of the optimal control in $[\tau, \tau + \Delta t]$, the change in the net returns attained in the interval $[\tau, \tau + \Delta t]$ due to the resulting change in the values of the state variables in $[\tau, \tau + \Delta t]$, and the change in the maximum value attainable in $[\tau + \Delta t, t_1]$ due to the resulting change in the state at $\tau + \Delta t$.

From our assumptions regarding the continuity of the partial derivatives of the functions F, c and z, it follows that the integrand in the above equation is a continuous function of s. Therefore, for sufficiently small Δt,

$$
\begin{aligned}
V_y(\tau, y) \cong \ &[F_z(z(\tau, \tau, y), c(\tau, \tau, y), \tau).z_y(\tau, \tau, y) \\
&+ F_c(z(\tau, \tau, y), c(\tau, \tau, y), \tau).c_y(\tau, \tau, y)]\Delta t \\
&+ V_z(\tau + \Delta t, z(\tau + \Delta t, \tau, y)).z_y(\tau + \Delta t, \tau, y)
\end{aligned} \tag{14.12}
$$

By definition, $z(\tau, \tau, y) = y$, for all $(\tau, y) \in \Omega_I$. Therefore, $z_y(\tau, \tau, y) = I$, the identity matrix. Also, by definition, for all $(\tau, y) \in \Omega_I$, for all $\Delta t \in (0, t_1 - \tau)$ and for all $i \in \{1, 2, ..., n\}$,

$$
z_i(\tau + \Delta t, \tau, y) = z_i(\tau, \tau, y) + \int_\tau^{\tau + \Delta t} f_i(z(s, \tau, y), c(s, \tau, y), s)ds
$$

Therefore, for all $i \in \{1, 2, ..., n\}$ and for all $j \in \{1, 2, ..., n\}$,

$$
\begin{aligned}
\frac{\partial z_i(\tau + \Delta t, \tau, y)}{\partial y_j} = \ &\frac{\partial y_i}{\partial y_j} + \int_\tau^{\tau + \Delta t} \left[\sum_{h=1}^n \frac{\partial f_i(z(s, \tau, y), c(s, \tau, y), s)}{\partial z_h} \frac{\partial z_h(s, \tau, y)}{\partial y_j} \right. \\
&\left. + \sum_{k=1}^r \frac{\partial f_i(z(s, \tau, y), c(s, \tau, y), s)}{\partial c_k}.\frac{\partial c_k(s, \tau, y)}{\partial y_j} \right] ds
\end{aligned}
$$

Since, given $(\tau, y) \in \Omega_I$, z, c, $\partial c_k/\partial y_j$, $\partial z_h/\partial y_j$, $\partial f_i/\partial c_k$ and $\partial f_i/\partial z_h$ $(h = 1, 2, ..., n; i = 1, 2, ..., n; j = 1, 2, ..., n; k = 1, 2, ..., r)$ are continuous functions of s on $[\tau, t_1]$ and since $\partial y_i/\partial y_j = 0$ for all $i \neq j$, therefore, for sufficiently small Δt,[10]

$$
z_y(\tau + \Delta t, \tau, y) \cong I + f_y(y, c(\tau, \tau, y), \tau).\Delta t + f_c(y, c(\tau, \tau, y), \tau).c_y(\tau, \tau, y).\Delta t
$$

From (14.12) it, therefore, follows that for all $(\tau, y) \in \Omega_I$ and for all sufficiently small Δt,

$$
\begin{aligned}
V_y(\tau, y) \cong \ &[F_y(y, c(\tau, \tau, y), \tau) + V_z(\tau + \Delta t, z(\tau + \Delta t, \tau, y)).f_y(y, c(\tau, \tau, y), \tau)]\Delta t \\
&+ [F_c(y, c(\tau, \tau, y), \tau) + V_z(\tau + \Delta t, z(\tau + \Delta t, \tau, y)).f_c(y, c(\tau, \tau, y), \tau)] \\
&\times c_y(\tau, \tau, y).\Delta t + V_z(\tau + \Delta t, z(\tau + \Delta t, \tau, y))
\end{aligned} \tag{14.13}
$$

---

[10] f_y and f_c denote the $(n \times n)$-order matrix $\left(\frac{\partial f_i}{\partial y_j} \right)$ and the $(n \times r)$-order matrix $\left(\frac{\partial f_i}{\partial c_k} \right)$, respectively.

The lhs expression in (14.13) is the vector of rates of change in the maximum value attainable τ onwards with respect to the components of y, the state at τ. The rhs is the sum of three expressions. The last gives the vector of rates of change in the maximum value attainable in $[\tau + \Delta t, t_1]$ with respect to the components of y, on the assumption that the change in y is exactly reflected in the change in the state at $\tau + \Delta t$ and there is no effect on the state at $\tau + \Delta t$ of the changes in the optimal values of the control and state variables in the interval $[\tau, \tau + \Delta t]$. The second expression gives (for sufficiently small Δt) the vector of rates of change in the net return attained in $[\tau, t_1]$ with respect to the components of y when the change in the net return is entirely due to the changes in the optimal values of the control variables in the interval $[\tau, \tau + \Delta t]$ following from the changes in the components of y. The first expression, on the other hand, gives (for sufficiently small Δt) the vector of rates of change in the net return attained in $[\tau, t_1]$ with respect to the components of y when the change in the net return is entirely due to the changes in the optimal values of the state variables in the interval $[\tau, \tau + \Delta t]$ following from the changes in the components of y. This is, in turn, the sum of two terms. The first considers the change in the net return attained in the interval $[\tau, \tau + \Delta t]$ because of the change in optimal values of the state variables in this interval. The second considers the change in the maximum value attainable time $\tau + \Delta t$ onwards because of the change in the state at $\tau + \Delta t$, given that the change in the state at $\tau + \Delta t$ follows simply from the change in the optimal values of the state variables in $[\tau, \tau + \Delta t]$.

Now, let $\tau = t \in (t_0, t_1)$ and $y = x^*(t)$. Then, $z(\tau + \Delta t, \tau, y) = x^*(t + \Delta t)$ and $c(\tau, \tau, y) = u^*(t)$. Therefore, from the definition of $\lambda : (t_0, t_1) \to R^n$ it follows that:

$$\frac{\lambda(t + \Delta t) - \lambda(t)}{\Delta t} \cong -[F_x(x, u, s)|_{(x,u,s)=(x^*(t),u^*(t),t)} + \lambda(t+\Delta t).f_x(x, u, s)|_{(x,u,s)=(x^*(t),u^*(t),t)}]$$

$$-[F_u(x, u, s)|_{(x,u,s)=(x^*(t),u^*(t),t)} + \lambda(t+\Delta t).f_u(x, u, s)|_{(x,u,s)=(x^*(t),u^*(t),t)}]$$

$$\times c_y(\tau, \tau, y)|_{(\tau,\tau,y)=(t,t,x^*(t))} \tag{14.14}$$

Suppose that $u^*(t)$ is an interior point of U. Then, from (14.11) it follows that

$$F_u(x, u, s)|_{(x,u,s)=(x^*(t),u^*(t),t)} + \lambda(t).f_u(x, u, s)|_{(x,u,s)=(x^*(t),u^*(t),t)} = 0$$

Therefore, taking the limit on both sides of (14.14) as $\Delta t \to 0+$,

$$\dot\lambda^+(t) = -[F_x(x, u, s) + \lambda(s).f_x(x, u, s)]_{(x,u,s)=(x^*(t),u^*(t),t)} \tag{14.15}$$

where $\dot\lambda^+(t)$ denotes the rhs derivative of the function λ at t.

The reader is invited to verify that for all $(\tau, y) \in \Omega_I$ and for all sufficiently small $\Delta t \in (0, \tau - t_0)$,

$$V(\tau - \Delta t, y) = \int_{\tau - \Delta t}^{\tau} F(z(s, \tau - \Delta t, y), c(s, \tau - \Delta t, y), s)ds + V(\tau, z(\tau, \tau - \Delta t, y))$$

and to derive therefrom a counterpart of (14.15) in which the rhs derivative of λ is replaced with the lhs derivative.

(14.15) can, therefore, be generalized to

$$\dot{\lambda}(t) = -[F_x(x, u, s) + \lambda(s).f_x(x, u, s)]_{(x,u,s)=(x^*(t),u^*(t),t)} \qquad (14.16)$$

Note also that the above argument can easily be extended to the case where $t = t_0$ provided we interpret all derivatives at t_0 with respect to the time variable as the appropriate rhs derivatives.

(14.11) and (14.16) indicate the nature of necessary conditions for an optimal control in Problem (A). Note that by our definition, for all $i \in \{1, 2, , ..., n\}$ and all $s \in (t_0, t_1)$, $\lambda_i(s)$ is the rate of change in the maximum value attainable from time s onwards with respect to the value of the state variable x_i at s, given that the state at s is the value of the optimal trajectory at s. $\lambda_i(s)$, therefore, represents the marginal return of x_i after time s when the state variables are on their optimal trajectories at s. It can, therefore, be thought of as the maximum amount (in units in which net returns are measured) that the decision-maker is willing to pay per additional unit of the i-th state variable at s when the state variables are moving along their optimal trajectories. Consequently, $\lambda_i(s)$ is often referred to as the **shadow price** of x_i at s.

14.3.4 The backward value function

Now, suppose $\tau \in [t_0, t_1]$ and y is any n-tuple of real numbers. Consider the following variant of Problem (A):

$$\max_{\substack{u(s) \\ s \in [t_0, \tau]}} \int_{t_0}^{\tau} F(x(s), u(s), s)ds$$

$$\text{s.t. } \dot{x}(s) = f(x(s), u(s), s)$$

$$x(t_0) = x^0, x(\tau) = y$$

$$u(s) \in U$$

In the above problem the initial time and initial state are the same as in Problem (A) but the terminal time has changed from t_1 to τ and the terminal state has changed from x^0 to y. Given any $\tau \in [t_0, t_1]$ and given any $y \in R^n$, let us denote this problem as Problem $(A_B(\tau, y))$ and the set of admissible controls for this problem as $\mathcal{U}_B(\tau, y)$. Let, for all $\tau \in [t_0, t_1]$, $Q_B(\tau) = \{y | y \in R^n \wedge \mathcal{U}_B(\tau, y) \neq \phi\}$ be a non-empty set and let $\Omega_B = \cup_{\tau \in [t_0, t_1]}\{(\tau, y) | y \in Q_B(\tau)\}$ be an open set in $[t_0, t_1] \times R^n$. Finally, let $V_B(\tau, y) = \max_{u \in \mathcal{U}_B(\tau, y)} \int_{t_0}^{\tau} F(x(s), u(s), s)ds$ be defined for all $(\tau, y) \in \Omega_B$. The function $V_B : \Omega_B \rightarrow R$ gives us the maximum value of the objective functional which can be achieved by any admissible control in Problem $(A_B(\tau, y))$ for any given terminal time $\tau \in [t_0, t_1]$ and any given terminal state $y \in Q_B(\tau)$. $V_B(\tau, y)$, therefore, gives us the maximum returns attainable by the decision-maker on the interval $[t_0, \tau]$ (that is, before time τ) given that the state at time τ is y. Let us refer to V_B as the **backward value function** associated with Problem (A). All partial derivatives of V_B are assumed to be continuous on $\Omega_{BI} = \cup_{\tau \in (t_0, t_1)}\{(\tau, y) | y \in Q_B(\tau)\}$, which is an open set in R^{n+1}.

We can construct an argument involving the function V_B similar to the one involving V and conclude that for all $t \in (t_0, t_1]$,

$$F(x^*(t), u, t) - \lambda_B(t).f(x^*(t), u, t) \text{ is maximized on } U \text{ at } u = u^*(t)$$

$$\dot{\lambda}_B(t) = [F_x(x, u, s) - \lambda_B(s).f_x(x, u, s)]_{(x,u,s)=(x^*(t),u^*(t),t)}$$

where $\lambda_B(t) = V_{B_y}(\tau, y)|_{(\tau,y)=(t,x^*(t))}$.

Note that, for any $t \in (t_0, t_1]$, if $\lambda_B(t) = (\lambda_{B_1}(t), \lambda_{B_2}(t), ..., \lambda_{B_n}(t))$ then $\lambda_{B_i}(t)$ $(i = 1, 2, ..., n)$ denotes the rate of change in the maximum net return attainable before time t with respect to the value of the i-th state variable at t, given that the state variables are on their optimal trajectories at t. It can, therefore, be thought of as the marginal return before t of the i-th state variable at time t when all state variables are on their optimal trajectories.

Consider $t \in (t_0, t_1)$. What is the relation between $\lambda(t)$ and $\lambda_B(t)$?

Suppose instead of solving Problem (A) the decision-maker solves the following problem:

$$\max V_B(t, y) + V(t, y)$$

$$\text{s.t. } y \in Q(t) \cap Q_B(t)$$

Given the definitions of $Q(t)$ and $Q_B(t)$ and our assumptions about V and V_B, for any $y \in Q(t) \cap Q_B(t)$ it is true that there exists some admissible pair (x, u) in Problem (A) such that $x(t) = y$ and $\int_{t_0}^{t_1} F(x(s), u(s), s)ds = V_B(t, y) + V(t, y)$. Also, if (x^*, u^*) is an optimal pair for Problem (A) then $x^*(t) \in Q(t) \cap Q_B(t)$ and $\int_{t_0}^{t_1} F(x^*(s), u^*(s), s)ds = V_B(t, x^*(t)) + V(t, x^*(t))$. Therefore, for all $y \in Q(t) \cap Q_B(t)$, $V_B(t, x^*(t)) + V(t, x^*(t)) \geq V_B(t, y) + V(t, y)$. Since Ω and Ω_B are open sets in $[t_0, t_1] \times R^n$, $Q(t)$ and $Q_B(t)$ are open sets in R^n. Given that all partial derivatives of $V_B(\tau, y)$ and $V(\tau, y)$ are assumed to exist with respect to the components of y on Ω_t and Ω_{Bt} respectively, therefore,

$$V_{B_y}(\tau, y)|_{(\tau,y)=(t,x^*(t))} + V_y(\tau, y)|_{(\tau,y)=(t,x^*(t))} = 0$$

From the definitions of λ and λ_B it, therefore, follows that for all $t \in (t_0, t_1)$,

$$\lambda_B(t) = -\lambda(t)$$

Hence, it is now possible to argue that (14.11) and (14.16) are valid for all $t \in [t_0, t_1]$ and that

$$\lambda(t) = V_y(\tau, y)|_{(\tau,y)=(t,x^*(t))}, \, t = t_0$$

$$= V_y(\tau, y)|_{(\tau,y)=(t,x^*(t))} = -V_{B_y}(\tau, y)|_{(\tau,y)=(t,x^*(t))}, \, t \in (t_0, t_1)$$

$$= -V_{B_y}(\tau, y)|_{(\tau,y)=(t,x^*(t))}, \, t = t_1 \tag{14.17}$$

Note that the i-th component $(i = 1, 2, ..., n)$ of $-V_{B_y}(\tau, y)|_{(\tau,y)=(t,x^*(t))}$ may be interpreted as the net return that has to be sacrificed before t per unit addition to the value of the i-th state variable at t when the state at t is equal to the value of the optimal trajectory at t. It can, therefore, be thought to represent the marginal *cost* of the i-th state variable at t.

14.4 The Maximum Principle for the Basic Problem

14.4.1 The maximum principle for problem (A)

The Maximum Principle for Problem (A) states that if $u^* \in \mathcal{U}$ is an optimal control for Problem (A) with associated trajectory x^* then there exists a constant λ_0 and a continuous function $\lambda : [t_0, t_1] \to \Re^n$ such that the following conditions hold:

(i) $\lambda_0 = 0$ or $\lambda_0 = 1$.

(ii) *If $\lambda_0 = 0$ then $\lambda(t) \neq \mathbf{0}$ for all $t \in [t_0, t_1]$.*

(iii) *If for all $(x, u, s, \lambda) \in \Re^n \times \Re^r \times \Re \times \Re^n$ we define:*
$$H(x, u, s, \lambda) = \lambda_0 F(x, u, s) + \lambda . f(x, u, s)$$
then, except at points of discontinuity of u^*, for all $t \in [t_0, t_1]$,
$$\dot{\lambda}(t) = -H_x(x, u, s, \lambda)|_{(x,u,s,\lambda)=(x^*(t),u^*(t),t,\lambda(t))}.$$

(iv) *If H is defined as in (iii) then, for all $t \in [t_0, t_1]$,*
$$H(x^*(t), u, t, \lambda(t)) \text{ is maximized on } U \text{ at } u = u^*(t)$$

The Maximum Principle for Problem (A) provides necessary conditions for an optimal control. These necessary conditions resemble (14.11) and (14.16) in the previous section. Conditions (*iii*) and (*iv*) in the Maximum Principle reduce to (14.16) and (14.11), respectively, if we assume $\lambda_0 = 1$. In (14.11) and (14.16), λ could be given a particular interpretation (see 14.17) which arises out of the special assumptions made in deriving those conditions. These assumptions are, in general, not satisfied. Therefore, one must proceed with caution in interpreting λ and the conditions in the Maximum Principle in the same manner as λ and the conditions (14.11) and (14.16) in the preceding section.[11] In subsequent discussions we will, however, use this interpretation as a convenient tool with which to introduce necessary conditions for optimal controls in variants of the basic control problem.

An optimal control for problem (A) is, of course, an admissible control for the problem. Some statements of the Maximum Principle for Problem (A), therefore, also contain the following conditions:

(v) *If H is defined as in (iii) then, except at points of discontinuity of u^*, for all $t \in [t_0, t_1]$,*
$$\dot{x}^*(t) = f(x^*(t), u^*(t), t)$$

(vi) $x^*(t_0) = x^0, x^*(t_1) = x^1$.

(vii) $u^*(t) \in U$, for all $t \in [t_0, t_1]$

For all $i \in \{1, 2, ..., n\}$, let $\lambda_i : [t_0, t_1] \to \Re^n$ be such that for all $t \in [t_0, t_1]$, $\lambda(t) = (\lambda_1(t), \lambda_2(t), ..., \lambda_n(t))$. Then, for all $i \in \{1, 2, ..., n\}$, the variable λ_i whose time-path is defined by $\lambda_i : [t_0, t_1] \to \Re^n$ is known as a **co-state, adjoint, or auxiliary**

---

[11] For general results about when and to what extent such interpretations might be valid, see Seierstad and Sydsaeter (1987), pp. 210–15.

variable for the problem. The scalar differential equations

$$\dot{\lambda}_i = -\left[\frac{\partial H(x, u, s, \lambda)}{\partial x_i}\right]_{(x,u,s,\lambda)=(x^*(t),t,\lambda(t),u^*(t))}, i = 1, 2, ..., n$$

underlying the vector differential equation in (ii) are accordingly known as **co-state, adjoint, or auxiliary equations** for the problem.

The function $H(x, u, s, \lambda)$ in the statement of the Maximum Principle is known as the **Hamiltonian** (Hamiltonian function) for the problem.

If U is an open set or if we know that for all $t \in [t_0, t_1]$, $u^*(t)$ lies in the interior of U, then, provided the partial derivatives of $H(x, u, s, \lambda)$ with respect to $u_1, u_2, ..., u_r$ exist for all $(x, u, s, \lambda) \in \Re^n \times int\ U \times [t_0, t_1] \times \Re^n$, (iv) can be rewritten as:

$(iv.a)$ For all $t \in [t_0, t_1]$, $[H_u(x^*(t), u, t, \lambda(t))]_{u=u^*(t)} = 0$

Problems in which an optimal control satisfies the Maximum Principle with $\lambda_0 = 0$ are known as **abnormal problems**. Note that if we assume $\lambda_0 = 0$ in conditions (iii) and (iv) in the statement of the Maximum Principle then the net return function ceases to appear anywhere in the necessary conditions for an optimal control. The implication is that even though every optimal control maximizes the integral sum of net returns, in the case of abnormal optimal control problems, the Maximum Principle picks out a set of possible optimal controls from $\mathcal{U}$ without any reference to the net return function. For example, if there is only one admissible control then one can conclude that it is an optimal control without any reference to the net return function.

Since most optimal control problems of interest are not of the above type, in trying to characterize an optimal pair (x^*, u^*) using the Maximum Principle it is customary to begin by assuming that $\lambda_0 = 0$. Usually, one can then conclude using conditions (ii) and (iv) that, for any given $t \in [t_0, t_1]$, either the value of $u^*(t)$ satisfying (iv) cannot be the value characterizing an optimal control (because an admissible control which gives a higher value of the objective functional can be identified) or there can exist no value of $u^*(t)$ satisfying (iv). A contradiction is established, implying, given (i), that $\lambda_0 = 1$.

Given a function $\psi : \Re^n \times [t_0, t_1] \times \Re^n \to U$ suppose that condition (iv) in the Maximum Principle implies that it is possible that $u^*(t) = \psi(x^*(t), \lambda(t), t)$, for all $t \in [t_0, t_1]$. Then condition (iii) in the Maximum Principle together with the admissibility conditions (v)–(vi) imply that if $u^*(t) = \psi(x^*(t), \lambda(t), t)$ then $x = x^*(t) = (x_1^*(t), x_2^*(t), ..., x_n^*(t))$, $\lambda = \lambda(t) = (\lambda_1(t), \lambda_2(t), ..., \lambda_n(t))$ must be a solution of the $2n$-dimensional system of differential equations

$$\dot{x}_i = f(x, \psi(x, \lambda, t), t), \quad i = 1, 2, ..., n$$
$$\dot{\lambda}_i = -\frac{\partial H(x, \psi(x, \lambda, t), t, \lambda)}{\partial x_i}, \quad i = 1, 2, .., n$$

where $x = (x_1, x_2, ..., x_n)$ and $\lambda = (\lambda_1, \lambda_2, ..., \lambda_n)$, on any interval $I \subseteq [t_0, t_1]$ on which $\psi(x^*(t), \lambda(t), t)$ is continuous and must satisfy the $2n$ boundary conditions $x_i(t_0) = x_i^0$ and $x_i(t_1) = x_i^1$ $(i = 1, 2, ..., n)$ where $x^0 = (x_1^0, x_2^0, ..., x_n^0)$ and $x^1 =$

$(x_1^1, x_2^1, ..., x_n^1)$. Moreover, at any point $\tau \in [t_0, t_1]$ at which $\psi(x^*(t), \lambda(t), t)$ is discontinuous, $\lim_{t \to \tau+} x^*(t) = \lim_{t \to \tau-} x^*(t)$ and $\lim_{t \to \tau+} \lambda(t) = \lim_{t \to \tau-} \lambda(t)$.

Note that if we know that u^* is continuous then we will consider only continuous ψ and $x = x^*(t), \lambda = \lambda(t)$ will be a solution of the above system of differential equations on the entire interval $[t_0, t_1]$. It can be proved[12] that if, given any $(\bar{x}, \bar{s}, \bar{\lambda}) \in \Re^n \times [t_0, t_1] \times \Re^n$ it is true that there is a unique value of u which maximizes $H(\bar{x}, u, \bar{s}, \bar{\lambda})$ on U, then all optimal controls for problem (A) are continuous on $[t_0, t_1]$. A sufficient condition for the Hamiltonian for Problem (A) to be **regular**, that is, for the Hamiltonian for Problem (A) to have the above property, is that U is a convex set and, given any $(\bar{x}, \bar{s}, \bar{\lambda}) \in R^n \times [t_0, t_1] \times R^n$, $H(\bar{x}, u, \bar{s}, \bar{\lambda})$ is strictly concave on U.

14.5 Sufficient Conditions for an Optimal Control

Amongst all admissible controls only those which satisfy the conditions in the Maximum Principle can be optimal controls. The Maximum Principle, therefore, allows us to reduce the number of admissible controls we need to consider as possible optimal controls. Any admissible control which satisfies the conditions in the Maximum Principle and also satisfies some set of sufficient conditions for optimal controls in Problem (A) will be an optimal control for Problem (A). Two sets of sufficient conditions are in general use, one due to O.L. Mangasarian (1966) and the other due to K.J. Arrow (1968).

14.5.1 The Mangasarian sufficiency conditions for problem (A)

Let for all $(x, u, s, \lambda) \in R^n \times R^r \times R \times R^n$, $H(x, u, s, \lambda) = F(x, u, s) + \lambda.f(x, u, s)$. *Suppose that:*

(i) *U is a convex set.*
(ii) *The functions F and $f_1, f_2, ..., f_n$ have continuous partial derivatives with respect to u_i, for all $i \in \{1, 2, ..., r\}$.*
   Suppose u^* is an admissible control for Problem (A) with associated trajectory x^* and suppose there exists a continuous function $\lambda : [t_0, t_1] \to R^n$ such that:
(iii) *For all $t \in [t_0, t_1]$, $H(x, u, t, \lambda(t))$ is jointly concave in (x, u).*
(iv) *Except at points of discontinuity of u^*, for all $t \in [t_0, t_1]$, $\dot{\lambda}(t) = -H_x(x, u, s, \lambda)|_{(x,u,s,\lambda)=(x^*(t),u^*(t),t,\lambda(t))}$.*
(v) *for all $t \in [t_0, t_1]$, $H(x^*(t), u, t, \lambda(t))$ is maximized on U at $u = u^*(t)$. Then, u^* is an optimal control for Problem (A).*

Suppose conditions (i) to (iii) hold and there is an admissible pair (x^*, u^*) and a continuous function $\lambda : [t_0, t_1] \to R^n$ satisfying conditions (iv) to (v). Suppose that

[12] See Grass, et al. (2008) pp. 113–14.

(x^a, u^a) is any arbitrarily given admissible pair. Define $\Delta J = \int_{t_0}^{t_1} [F(x^*(s), u^*(s), s) - F(x^a(s), u^a(s), s)]ds$. Then, $\Delta J = \int_{t_0}^{t_1} [H(x^*(s), u^*(s), s, \lambda(s)) - H(x^a(s), u^a(s), s, \lambda(s))]ds + \int_{t_0}^{t_1} \lambda(s).[\dot{x}^a(s) - \dot{x}^*(s)]ds$.

Since conditions (i) to (iii) are satisfied and given that $F, f_1, f_2, ..., f_n$ have continuous partial derivatives with respect to x_i for all $i \in \{1, 2, ..., n\}$,

$$H(x^*(s), u^*(s), s, \lambda(s)) - H(x^a(s), u^a(s), s, \lambda(s))$$
$$\geq H_x(x, u, s, \lambda)|_{(x,u,s,\lambda)=(x^*(s),u^*(s),s,\lambda(s))}.(x^*(s) - x^a(s))$$
$$+ H_u(x, u, s, \lambda)|_{(x,u,s,\lambda)=(x^*(s),u^*(s),s,\lambda(s))}.(u^*(s) - u^a(s)),$$

for all $s \in [t_0, t_1]$. Since (x^*, u^*) satisfies (iv) therefore,[13] for all $s \in [t_0, t_1]$,

$$H_x(x, u, s, \lambda)|_{(x,u,s,\lambda)=(x^*(s),u^*(s),s,\lambda(s))}.(x^*(s) - x^a(s)) = \dot{\lambda}(s).(x^a(s) - x^*(s)).$$

Moreover, since (x^*, u^*) satisfies (v) and condition (i) holds, therefore,

$$H_u(x, u, s, \lambda)|_{(x,u,s,\lambda)=(x^*(s),u^*(s),s,\lambda(s))}.(u^*(s) - u^a(s)) \geq 0.$$

Consequently, $\Delta J \geq \int_{t_0}^{t_1} [\dot{\lambda}(s).(x^a(s) - x^*(s)) + \lambda(s).(\dot{x}^a(s) - \dot{x}^*(s))]ds$.[14] That is,[15]

$$\Delta J \geq \lambda(t_1).(x^a(t_1) - x^*(t_1)) - \lambda(t_0).(x^a(t_0) - x^*(t_0)) \qquad (14.18)$$

Since, both (x^*, u^*) and (x^a, u^a) are admissible pairs for Problem (A), it follows that the rhs expression in the above inequality is equal to 0. Since (x^a, u^a) is any arbitrarily given admissible pair, u^* must be an optimal control.

Suppose (iii) is replaced with the stronger condition: $(iii.a)$ For all $t \in [t_0, t_1]$, $H(x, u, t, \lambda(t))$ is strictly concave in (x, u).

Then, the inequality in (14.18) will be strict and u^* will be the *unique* optimal control for Problem (A).

14.5.2 The Arrow sufficiency conditions for problem (A)

Let for all $(x, u, s, \lambda) \in R^n \times R^r \times R \times R^n$, $H(x, u, s, \lambda) = F(x, u, s) + \lambda.f(x, u, s)$.

Suppose u^* is an admissible control for Problem (A) with associated trajectory x^* and suppose there exists a continuous function $\lambda : [t_0, t_1] \to R^n$ such that:

(i) *for all $(x, t) \in \Re^n \times [t_0, t_1]$, $H^0(x, t, \lambda(t)) = u \in U \max H(x, u, t, \lambda(t))$ exists and is concave in x;*

(ii) *except at points of discontinuity of u^*, for all $t \in [t_0, t_1]$,*
$\dot{\lambda}(t) = -H_x(x, u, s, \lambda)|_{(x,u,s,\lambda)=(x^*(t),u^*(t),t,\lambda(t))}$; *and*

(iii) *for all $t \in [t_0, t_1]$, $H(x^*(t), u, t, \lambda(t))$ is maximized on U at $u = u^*(t)$.*
Then, u^* is an optimal control for Problem (A).

---

[13] Except at a finite number of points where u^* is discontinuous.

[14] This definite integral of a piece-wise continuous function is properly interpreted as a sum of definite integrals over adjoining intervals on which the function is continuous.

[15] Remember that x^*, x^a, and λ are all continuous on $[t_0, t_1]$.

Seierstad and Sydsaeter (1987) provide a proof for the theorem. Following Arrow and Kurz (1970), we prove the sufficiency of conditions (i) to (iii) by making some special assumptions. Given any $(x, t) \in R^n \times [t_0, t_1]$, let there be a unique value $v^0(x, t, \lambda(t)) \in int\ U$ which maximizes $H(x, u, t, \lambda(t))$ on U. Also, let the partial derivative of v^0 with respect to x_i be defined for all $i \in \{1, 2, ..., n\}$. Note that for all $(x, t) \in R^n \times [t_0, t_1]$, $H^0(x, t, \lambda(t)) = H(x, v^0(x, t, \lambda(t)), t, \lambda(t))$. Therefore, $H^0_x(x, t, \lambda(t)) = H_x(x, u, s, \lambda)|_{(x,u,s,\lambda)=(x,v^0(x,t,\lambda(t)),t,\lambda(t))}$ because $H_u(x, u, s, \lambda)|_{(x,u,s,\lambda)=(x,v^0(x,t,\lambda(t)),t,\lambda(t))} = 0$.

Suppose (x^*, u^*) is an admissible pair and $\lambda : [t_0, t_1] \rightarrow R^n$ is a continuous function such that (i)–(iii) are satisfied. Suppose (x^a, u^a) is an arbitrarily given admissible pair. Since (i) is satisfied, therefore, for all $t \in [t_0, t_1]$, $H^0(x^a(t), t, \lambda(t)) \leq H^0(x^*(t), t, \lambda(t)) + H^0_x(x, t, \lambda(t))|_{x=x^*(t)} \cdot (x^a(t) - x^*(t))$. By definition, $H(x^a(t), u^a(t), t, \lambda(t)) \leq H^0(x^a(t), t, \lambda(t))$. Also, since ($iii$) holds, therefore, $H^0(x^*(t), t, \lambda(t)) = H(x^*(t), u^*(t), t, \lambda(t))$ and $v^0(x^*(t), t, \lambda(t)) = u^*(t)$. The latter equality implies that $H^0_x(x, t, \lambda(t))|_{x=x^*(t)} = H_x(x, u, s, \lambda)|_{(x,u,s,\lambda)=(x^*(t),u^*(t),t,\lambda(t))}$. Hence, $H(x^*(t), u^*(t), t, \lambda(t)) - H(x^a(t), u^a(t), t, \lambda(t)) \geq H_x(x, u, s, \lambda)|_{(x,u,s,\lambda)=(x^*(t),u^*(t),t,\lambda(t))} \cdot (x^*(t) - x^a(t))$. The rest of the proof follows that provided in the case of the Mangasarian sufficiency conditions.

14.6 Variants of the Basic Problem

14.6.1 Alternative conditions on the terminal state

No restrictions on $x(t_1)$

Necessary conditions: Suppose we alter Problem (A) by removing the constraint that $x(t_1) = x^1$. Note that the set of admissible controls for Problem (A) is contained in the set of admissible controls for the new problem.

Suppose (x^*, u^*) is an optimal pair for the new problem. Then, (x^*, u^*) is an optimal pair for Problem (A) if in Problem (A) we specify $x^1 = x^*(t_1)$. It follows that all the conditions in the Maximum Principle for Problem (A) must necessarily be satisfied by the optimal pair for the new problem. The Maximum Principle for the new problem, however, includes an additional condition.

Refer to our discussion in Section 14.3 in this book. $y = x^*(t_1)$ must be a solution to the problem:

$$\max_{y} \ V_B(t_1, y)$$
$$\text{s.t. } y \in Q_B(t_1)$$

Since Ω_B is an open set, therefore, $Q_B(t_1)$ is an open set. If we assume that all partial derivatives of $V_B(\tau, y)$ with respect to the components of y exist on Ω_B then $V_{B_y}(\tau, y)|_{(\tau,y)=(t_1,x^*(t_1))} = 0$. Then, from (14.17), it follows that $\lambda(t_1) = 0$.

In fact, *the statement of the Maximum Principle for the new problem is the same as that for Problem (A) except for the addition of the condition*:

($v.a$) $\lambda(t_1) = 0$

From conditions (*i*) and (*ii*) in the Maximum Principle for the new problem it immediately follows that $\lambda_0 = 1$.

In the previous section we saw that the n co-state equations in the Maximum Principle together with the n state equations form a system of $2n$ differential equations in the n state variables and n co-state variables if the optimal values of the control variables can be expressed as a function of the values of the state and co-state variables. In Problem (A), there were given $2n$ boundary conditions involving the state variables. These boundary conditions could be used to solve the system of differential equations for the trajectories of state and co-state variables. In the new problem there are only n boundary conditions involving the state variables. The Maximum Principle, however, in this case provides alternative boundary conditions, $\lambda_i(t_1) = 0$ ($i = 1, 2, ..., n$), involving the terminal values of the co-state variables which substitute for the missing n boundary conditions involving values of the state variable. Such conditions are known as **transversality conditions**.

Sufficient conditions: Consider the proof of the Mangasarian sufficiency conditions in Section 14.5. Consider (14.18). For the new problem we can no longer claim that $x^*(t_1) = x^a(t_1)$ and, therefore, (14.18) no longer implies that $\Delta J \geq 0$. However, note that if we assume that $\lambda(t_1) = 0$ then (14.18) implies that $\Delta J \geq 0$.

> Therefore, *the Mangasarian sufficiency conditions for the new problem are the same as the Mangasarian sufficiency conditions for Problem (A) except for the addition of condition (v.a). The same holds true for the Arrow sufficiency conditions as well.*

x(t₁) ≧ x¹

Necessary conditions: Suppose (x^*, u^*) is an optimal pair for the optimal control problem obtained by replacing the equality constraint on $x(t_1)$ in Problem (A) with the above inequality constraint. Then, (x^*, u^*) is an optimal pair for Problem (A) when $x^1 = x^*(t_1)$. Therefore, (x^*, u^*) satisfies all the conditions in the Maximum Principle for Problem (A).

Consider once again, the discussion in Section 14.3. $y = x^*(t_1)$ must be a solution for the following problem:

$$\max_{y} V_B(t_1, y)$$
$$\text{s.t. } y \in Q_B(t_1)$$
$$y \geq x^1$$

From the theory of static optimization we know that $V_{B_y}(\tau, y)|_{(\tau, y)=(t_1, x^*(t_1))} \leq 0$ and $V_{B_y}(\tau, y)|_{(\tau, y)=(t_1, x^*(t_1))} \cdot (x^*(t_1) - x^1) = 0$. From (14.17) it follows that

(v.b) $\lambda(t_1) \geq 0$ and $\lambda(t_1) \cdot (x^*(t_1) - x^1) = 0$.

> *The statement of the Maximum Principle for the new problem is the same as that for Problem (A) except for the addition of the transversality condition (v.b).*

Sufficient conditions: Note that for the new problem the right hand side expression in (14.18) is non-negative if condition (*v.b*) holds.

Therefore, *the Mangasarian and Arrow sufficiency conditions for the new problem are the same as those for Problem (A) except for the addition of condition (v.b).* $\mathbf{x_i(t_1)} = \mathbf{x_i^1}$, $i \in S_E$; $\mathbf{x_i(t_1)} \geq \mathbf{x_i^1}$, $i \in S_{IE}$; $S_E \cup S_{IE} \subseteq \{1, 2, ..., n\}$; $S_E \cap S_{IE} = \phi$.

In Problem (A) and in the above two variants of Problem (A) there was either no restriction or there was the same type of restriction (equality or inequality constraint) on the terminal value of each state variable. Consider instead a general problem of which Problem (A) and its above two variants are special cases. In this problem amongst the terminal values of state variables $x_i(t_1)$ ($i = 1, 2, ..., n$) some ($i \in S_E \subseteq \{1, 2, ..., n\}$) may be subject to equality constraints, some ($i \in S_{IE} \subseteq \{1, 2, ..., n\}$) may be subject to inequality constraints and the remaining, if any, would be subject to no constraints ($i \in \{1, 2, ..., n\} - (S_E \cup S_{IE})$). The discussion of necessary and sufficient conditions for an optimal control in Problem (A) and its above two variants can be extended to justify the following necessary and sufficient conditions in the general problem.

Necessary conditions:

> *The statement of the Maximum Principle for the new problem is the same as that for Problem (A) except for the addition of the transversality condition*:
>
> (*v.c*) $\lambda_i(t_1) = 0$, *for all* $i \in \{1, 2, ..., n\} - (S_E \cup S_{IE})$;
>
> $\lambda_i(t_1) \geq 0$ *and* $\lambda_i(t_1).(x_i^*(t_1) - x_i^1) = 0$, *for all* $i \in S_{IE}$

Sufficient conditions:

> *The Mangasarian and Arrow sufficiency conditions for the new problem are the same as those for Problem (A) except for the addition of condition (v.c).*

14.6.2 Addition of a salvage value function

Note that in Problem (A) and in the variants of Problem (A) discussed in the last section, it is assumed that the terminal state does not influence the decision-maker's net returns beyond the terminal time the way the state, at any time $t \in [t_0, t_1)$, influences her net returns beyond t. More precisely, any value or worth that the terminal state has for the decision-maker beyond the terminal time is only implicitly represented by the fact that terminal values of the state variables may be required to be equal to or not less than certain specified values. However, suppose that we can, in addition, associate with each value of the terminal state some explicit amount of net returns ('value') that can be extracted ('salvaged') from the use of the terminal quantities of state variables even after the terminal time. Let, for each $y \in R^n$, $S(y)$ denote this salvage value of the terminal state when the terminal state is y and let $S : R^n \to R$ be referred to as the **salvage value function**. We shall assume that all partial derivatives of S exist and are continuous on R^n.

The decision-maker now seeks to maximize not simply her net returns over the interval $[t_0, t_1]$ but the sum of these returns with the salvage value of the terminal state. Note, however, that the decision-making problem obtained from Problem (A) by replacing $\int_{t_0}^{t_1} F(x(t), u(t), t)dt$ with $\int_{t_0}^{t_1} F(x(t), u(t), t)dt + S(x(t_1))$ is actually the same as Problem (A) because the sets of admissible controls in the two problems are the same and for every admissible control $S(x(t_1)) = S(x^1)$. Suppose instead that the decision-maker faces the following problem:

$$\textbf{Problem (B):} \max_{\substack{u(t) \\ t \in [t_0, t_1]}} \int_{t_0}^{t_1} F(x(t), u(t), t)dt + S(x(t_1))$$

$$\text{s.t. } \dot{x}(t) = f(x(t), u(t), t)$$

$$x(t_0) = x^0$$

$$x_i(t_1) = x_i^1, \text{ for all } i \in S_E$$

$$x_i(t_1) \geq x_i^1, \text{ for all } i \in S_{IE}$$

$$(S_E \cup S_{IE} \subseteq \{1, 2, ..., n\}, S_E \cap S_{IE} = \phi)$$

$$u(t) \in U$$

Necessary Conditions: If (x^*, u^*) is an optimal pair for Problem (B) then it follows that (x^*, u^*) must be an optimal pair for Problem (A) with $x^1 = x^*(t_1)$. Therefore, all conditions in the Maximum Principle for Problem (A) must be satisfied by (x^*, u^*). In addition, note that $y = (y_1, y_2, ..., y_n) = (x_1^*(t_1), x_2^*(t_1), ..., x_n^*(t_1)) = x^*(t_1)$ must be a solution for the following problem:

$$\max_{y_1, y_2, ..., y_n} V_B(t_1, y_1, y_2, ..., y_n) + S(y_1, y_2, ..., y_n)$$

$$\text{s.t. } (y_1, y_2, ..., y_n) \in Q_B(t_1)$$

$$y_i = x_i^1, \text{ for all } i \in S_E$$

$$y_i \geq x_i^1, \text{ for all } i \in S_{IE}$$

$$(S_E \cup S_{IE} \subseteq \{1, 2, ..., n\}, S_E \cap S_{IE} = \phi)$$

Using (14.17) and assuming that Ω_B is an open set and all partial derivatives of $V_B(\tau, y)$ with respect to y_i ($i = 1, 2, ..., n$) exist on Ω_B, the necessity of the following transversality condition (v.d) (which holds without these assumptions) can then be proved for $\lambda_0 = 1$.

The statement of the Maximum Principle for Problem (B) is the same as that for Problem (A) except for the addition of the transversality condition:

(v.d) $\lambda_i(t_1) - \lambda_0 \partial S(x(t_1))/\partial x_i(t_1)|_{x(t_1)=x^*(t_1)} = 0$, *for all* $i \in \{1, 2, ..., n\} - (S_E \cup S_{IE})$;

$\lambda_i(t_1) - \lambda_0 \partial S(x(t_1))/\partial x_i(t_1)|_{x(t_1)=x^*(t_1)} \geq 0$ and

$(\lambda_i(t_1) - \lambda_0 \partial S(x(t_1))/\partial x_i(t_1)|_{x(t_1)=x^*(t_1)}).(x_i^*(t_1) - x_i^1) = 0$, for all $i \in S_{IE}$.

Note that (as in Problem (A)) when $\lambda_0 = 0$ net returns (including salvage values) do not matter in identifying possible optimal controls using the Maximum Principle.

Sufficient Conditions:

The Mangasarian and Arrow sufficiency conditions for Problem (B) are the same as those for Problem (A) except for the addition of condition (v.d) written with $\lambda_0 = 1$ and the following condition:

(vi) S is concave on R^n

Refer to the discussion on the Mangasarian sufficiency conditions. Let us redefine $\Delta J = \int_{t_0}^{t_1}[F(x^*(s), u^*(s), s) - F(x^a(s), u^a(s), s)]ds + S(x^*(t_1)) - S(x^a(t_1))$. If we assume that S is a concave function then (14.18) can be rewritten as

$$\Delta J \geq (\lambda(t_1) - S_y(y)|_{y=x^*(t_1)}).(x^a(t_1) - x^*(t_1)) - \lambda(t_0).(x^a(t_0) - x^*(t_0))$$

Given that (v.d) holds with $\lambda_0 = 1$, it follows that $\Delta J \geq 0$. A similar argument can be made with respect to the Arrow sufficiency conditions.

14.6.3 An important note

In Problem (A) and the variants of Problem (A) discussed till now we have assumed that the domain of the functions F and f_i $(i = 1, 2, ..., n)$ is $R^n \times R^r \times R$ and that these functions and some partial derivatives of these functions have certain properties on this domain. However, the necessary and sufficient conditions stated for optimal controls in these problems remain valid even if the domain of these functions is $R^n \times U \times [t_0, t_1]$; the functions are continuous on this domain and the relevant partial derivatives are continuous in any set D which is open in $R^n \times U \times [t_0, t_1]$ and $\{(x^*(t), u^*(t), t)|t \in [t_0, t_1]\} \subseteq D$. In fact, the results hold for even weaker restrictions on these functions.[16]

14.6.4 Variable terminal time

$t_1 \in (t_0, \infty)$

Suppose that the decision-maker can choose not only values of the control variables but also for how long she wants to exercise this choice. Then the terminal time t_1, which was fixed in Problem (A), becomes a choice variable.

Necessary conditions: Suppose $u^* : [t_0, t_1^*] \to U$ is an optimal control for the new problem with associated trajectory $x^* : [t_0, t_1^*] \to \Re^n$. It follows that (x^*, u^*) would be an optimal pair for Problem (A) if $t_1 = t_1^*$. Therefore, (x^*, u^*) must satisfy the

[16] See Seierstad and Sydsaeter (1987), note 12, chapter 2.

conditions in the Maximum Principle for Problem (A) with t_1 replaced with t_1^*. If there are alternative conditions on the terminal state, the conditions in the Maximum Principle for the corresponding variants of Problem (A) will be satisfied by (x^*, u^*) with t_1 replaced with t_1^*.

Revert back to the discussion in the section on necessary conditions. Suppose now that for all $\tau \in [t_0, \infty)$, $Q_B(\tau) = \{y | y \in R^n \wedge \mathcal{U}_B(\tau, y) \neq \phi\}$ is a non-empty set, $\Omega_B = \cup_{\tau \in [t_0, \infty)} \{(\tau, y) | y \in Q_B(\tau)\}$ is an open set in $[t_0, \infty) \times R^n$ and the function V_B is defined on Ω_B with all partial derivatives of V_B being assumed to be continuous on $\Omega_{B_I} = \cup_{\tau \in (t_0, \infty)} \{(\tau, y) | y \in Q_B(\tau)\}$. Then, $\tau = t_1^*$ must be a solution for the following problem:

$$\max_{\tau} V_B(\tau, x^1)$$

$$\text{s.t. } \tau \in (t_0, \infty)$$

Therefore,

$$V_{B_\tau}(\tau, y)|_{(\tau, y) = (t_1^*, x^*(t_1^*))} = 0$$

Let $t \in (t_0, \infty)$ and $\Delta t \in (0, t - t_0)$. Given our assumptions about V_B and Ω_B, for sufficiently small Δt,

$$V_B(t - \Delta t, x^*(t - \Delta t)) \cong V_B(t, x^*(t)) + V_{B_y}(\tau, y)|_{(\tau, y) = (t, x^*(t))} \cdot [x^*(t - \Delta t) - x^*(t)]$$
$$- V_{B_\tau}(\tau, y)|_{(\tau, y) = (t, x^*(t))} \cdot \Delta t \qquad (14.19)$$

If we assume that u^* is continuous (x^* is a continuously differentiable function) then for sufficiently small Δt,

$$x^*(t - \Delta t) \cong x^*(t) - f(x^*(t), u^*(t), t) \cdot \Delta t \qquad (14.20)$$

If u^* is continuous $F(x^*(s), u^*(s), s)$ is a continuous function of s at t. Then, for sufficiently small Δt,

$$V_B(t, x^*(t)) - V_B(t - \Delta t, x^*(t - \Delta t)) = \int_{t - \Delta t}^t F(x^*(s), u^*(s), s) ds \cong F(x^*(t), u^*(t), t) \Delta t$$
$$\qquad (14.21)$$

From (14.17) and (14.19) to (14.21) it follows that

$$V_{B_\tau}(\tau, y)|_{(\tau, y) = (t, x^*(t))} \Delta t = [F(x^*(t), u^*(t), t) + \lambda(t) \cdot f(x^*(t), u^*(t), t)] \Delta t + o(\Delta t)$$

Dividing by Δt on both sides and then taking the limit as $\Delta t \to 0$:

$$V_{B_\tau}(\tau, y)|_{(\tau, y) = (t, x^*(t))} = F(x^*(t), u^*(t), t) + \lambda(t) \cdot f(x^*(t), u^*(t), t) \qquad (14.22)$$

Given that $V_{B_\tau}(\tau, y)|_{(\tau, y) = (t_1^*, x^*(t_1^*))} = 0$, (6.4) proves the following transversality condition (v.e) for $\lambda_0 = 1$.

- In fact, *the statement of the Maximum Principle for the new problem is the same as that for Problem (A) except for the addition of the transversality condition:*

  (v.e) $H(x^*(t_1^*), u^*(t_1^*), t_1^*, \lambda(t_1^*)) = 0$

Sufficient Conditions: The following is a special case of a result due to Seierstad (1984).

Suppose for every $t_1 \in (t_0, \infty)$ there exists an admissible pair for Problem (A) which satisfies the Arrow sufficiency conditions for Problem (A). Let, for every $t_1 \in (t_0, \infty)$, $(x_{t_1}^*, u_{t_1}^*)$ denote this admissible pair and let λ_{t_1} denote the associated co-state variable with continuous trajectory which appears in the Arrow sufficiency conditions. It follows that we can then define functions $\hat{u} : (t_0, \infty) \to U$, $\hat{x}: (t_0, \infty) \to R^n$ and $\hat{\lambda}: (t_0, \infty) \to R^n$ such that for all $t \in (t_0, \infty)$, $\hat{u}(t) = u_t^*(t)$, $\hat{x}(t) = x_t^*(t)$ and $\hat{\lambda}(t) = \lambda_t(t)$.

Suppose $\bigcup_{t_1 \in (t_0, \infty)} R(u_{t_1}^*)$ is bounded, $\hat{x}$ is continuous and $R(\hat{\lambda})$ is bounded.

Then, for any given $t_1^* \in (t_0, \infty)$ it is true that $(x_{t_1^*}^*, u_{t_1^*}^*)$ is an optimal control for the new problem if:

$$(v.f) \quad F(\hat{x}(t), \hat{u}(t), t,) + \hat{\lambda}(t).f(\hat{x}(t), \hat{u}(t), t) \geq 0, \text{ if } t < t_1^*$$
$$\leq 0, \text{ if } t > t_1^*$$

Refer back to the optimization exercise considered in discussing necessary conditions and note that $\hat{x}(\tau) = x^1$ for all $\tau \in (t_0, \infty)$. A sufficient condition for $\tau = t_1^*$ to be a solution is that the function $V_{B_\tau}(\tau, y)|_{(\tau, y)=(t_1^*, \hat{x}(t_1))} + V_{B_y}(\tau, y)|_{(\tau, y)=(t_1, \hat{x}(t_1))}..\hat{x}(t_1)$ is non-decreasing for $t_1 < t_1^*$ and is non-increasing for $t_1 > t_1^*$. Given that (6.4) holds for $(x^*(t), u^*(t)) = (x_t^*(t), u_t^*(t))$ for all $t \in (t_0, \infty)$ and given that $.\hat{x}(t_1) = 0$ for all $t_1 \in (t_0, \infty)$, $(v.f)$ follows.

Suppose the decision-making problem has alternative conditions involving the terminal state as discussed earlier. The above result still defines sufficient conditions for optimality provided references to Problem (A) in the statement of the result are replaced with references to the relevant problem with fixed terminal time and appropriate conditions involving the terminal state.

$t_1 \in [T_1, T_2]$, $t_0 < T_1 < T_2$

Necessary Conditions: The same argument as in the case where $t_1 \in (t_0, \infty)$ can be made except that now $t_1 = t_1^*$ must be a solution for the problem:

$$\max_\tau V_B(\tau, x^1)$$
$$\text{s.t. } \tau \in [T_1, T_2]$$

Therefore, $V_B(t_1^*, x^1) \leq 0$ if $t_1^* = T_1$, $V_B(t_1^*, x^1) = 0$ if $t_1^* \in (T_1, T_2)$ and $V_B(t_1^*, x^1) \geq 0$ if $t_1^* = T_2$. Given (14.22), the necessity of the following transversality condition $(v.g)$ is established for $\lambda_0 = 1$.

The statement of the Maximum Principle for the new problem is the same as that for Problem (A) except for the addition of the transversality condition:

$$(v.g) \quad H(x^*(t_1^*), u^*(t_1^*), t_1^*, \lambda(t_1^*)) \leq 0, \text{ if } t_1 = T_1$$
$$= 0, \text{ if } t_1 \in (T_1, T_2)$$
$$\geq 0, \text{ if } t_1 = T_2$$

Sufficient conditions:

> *The result stated in the case where $t_1 \in (t_0, \infty)$ continues to provide sufficient conditions for an optimal control in the present problem if '(t_0, ∞)' is replaced with '$[T_1, T_2]$' in the statement of the result.*

14.6.5 Inequality constraints with control variables[17]

Till now we have assumed that the value of the control chosen by the decision-maker at any time t must lie in a fixed set $U \subseteq R^r$. Therefore, the set from which the decision-maker can pick the vector of values of the control variables for any time t is independent of the value t of the time variable as well as of the values of the state variables at t. Suppose, instead, we assume that for all $t \in [t_0, t_1]$ the decision-maker must choose $u(t)$ such that

$$g_j(x(t), u(t), t) \geq 0, \; j = 1, 2, ..., m$$

where m is a positive integer and $g_j : R^n \times R^r \times R \to R$ for all $j \in \{1, 2, ..., m\}$. We assume that for all $j \in \{1, 2, ..., m\}$, the values of the function g_j are not independent of the values of the control variables, that is,

$$(\exists x)(\exists s)(\exists v_1)(\exists v_2)(((x \in R^n \wedge s \in R) \wedge (v_1 \in R^r \wedge v_2 \in R^r)) \wedge g_j(x, v_1, s) \neq g_j(x, v_2, s))$$

Let $g : R^n \times R^r \times R \to R^n$ be a continuous function where for all $(x, u, s) \in R^n \times R^r \times R$, $g(x, u, s) = (g_1(x, u, s), g_2(x, u, s), ..., g_m(x, u, s))$. Also, let $\partial g_j(x, u, s)/\partial u_k$ and $\partial g_j(x, u, s)/ \partial x_i$ be continuous on $R^n \times R^r \times R$ for all $j \in \{1, 2, ..., m\}$, for all $k \in \{1, 2, ..., r\}$ and for all $i \in \{1, 2, ..., n\}$, where $x = (x_1, x_2, ..., x_n)$ and $u = (u_1, u_2, ..., u_r)$. In this section, we simply note how the necessary and sufficient conditions for an optimal control may change as a result of this alternative specification of the choice set for the values of the control variables. Inequality constraints of the above form, which explicitly involve control variables, are known as **mixed constraints**.

The decision-making problem we consider now is the same as Problem (B) except that the constraint $u(t) \in U$ in the statement of the problem is replaced with $g(x(t), u(t), t) \geq 0.$

$$\textbf{Problem (C):} \quad \max_{\substack{u(t) \\ t \in [t_0, t_1]}} \int_{t_0}^{t_1} F(x(t), u(t), t)dt + S(x(t_1))$$

$$\text{s.t. } \dot{x}(t) = f(x(t), u(t), t)$$

$$x(t_0) = x^0$$

$$x_i(t_1) = x_i^1, \text{ for all } i \in S_E$$

---

[17] Note that the constraint $u(t) \in U$ may itself require $u(t)$ to take values in R^r which satisfy certain inequality constraints, for example, $U = \{u | u \in R^r \wedge u \geq 0\}$.

$$x_i(t_1) \geq x_i^1, \text{ for all } i \in S_{IE}$$

$$(S_E \cup S_{IE} \subseteq \{1, 2, ..., n\}, S_E \cap S_{IE} = \phi)$$

$$g(x(t), u(t), t) \geq 0$$

We will assume that Problem (C) satisfies some additional conditions: All partial derivatives of the functions F and f_i $(i = 1, 2, ..., n)$ with respect to the control variables are continuous on $R^n \times R^r \times R$; moreover, the rank of the matrix

$$\begin{pmatrix} \dfrac{\partial g_1(x,u,s)}{\partial u_1} & \dfrac{\partial g_1(x,u,s)}{\partial u_2} & \cdot & \cdot & \dfrac{\partial g_1(x,u,s)}{\partial u_r} & g_1(x,u,s) & 0 & \cdot & \cdot \\ \dfrac{\partial g_2(x,u,s)}{\partial u_1} & \dfrac{\partial g_2(x,u,s)}{\partial u_2} & \cdot & \cdot & \dfrac{\partial g_2(x,u,s)}{\partial u_r} & 0 & g_2(x,u,s) & 0 & \cdot & 0 \\ \cdot & \cdot & & & & & & \cdot \\ \cdot & \cdot & & & & & & \\ \cdot & \cdot & & & & & & \\ \dfrac{\partial g_m(x,u,s)}{\partial u_1} & \dfrac{\partial g_m(x,u,s)}{\partial u_2} & \cdot & \cdot & \cdot & \dfrac{\partial g_m(x,u,s)}{\partial u_r} & 0 & \cdot & 0 & g_m(x,u,s) \end{pmatrix}$$

is equal to m at all $(x, u, s) \in \{(x, u, s) | (x \in R^n \wedge u \in R^r) \wedge (s \in [t_0, t_1] \wedge g(x, u, s) \geq 0)\}$.

Necessary conditions:

A Maximum Principle for Problem (C):[18] *Suppose (x^*, u^*) is an optimal pair for Problem (C). Then, there exists $\lambda_0 \in R$, a continuous function $\lambda : [t_0, t_1] \to R^n$ and a piece-wise continuous function $\mu : [t_0, t_1] \to R^m$ which is continuous on every interval of continuity of u^*, such that the following conditions hold:*

(i) *$\lambda_0 = 0$ or $\lambda_0 = 1$.*

(ii) *If $\lambda_0 = 0$ then $\lambda(t) \neq 0$ for all $t \in [t_0, t_1]$.*

(iii) *If for all $(x, u, s, \lambda, \mu) \in R^n \times R^r \times R \times \Re^n \times \Re^m$ we define,*
$H(x, u, s, \lambda) = \lambda_0 F(x, u, s) + \lambda . f(x, u, s)$ and $L(x, u, s, \lambda, \mu) = H(x, u, s, \lambda) + \mu . g(x, u, s)$
then, except at points of discontinuity of u^*, for all $t \in [t_0, t_1]$,
$\dot{\lambda}(t) = -L_x(x, u, s, \lambda, \mu)|_{(x,u,s,\lambda) = (x^*(t), u^*(t), t, \lambda(t), \mu(t))}$.

(iv) *If H is defined as in (iii) then, for all $t \in [t_0, t_1]$,*
$H(x^*(t), u, t, \lambda(t))$ is maximized on $\{u | u \in R^r \wedge g(x^*(t), u, t) \geq 0\}$ at $u = u^*(t)$.

(v) *If L is defined as in (iii) then, except at points of discontinuity of u^*, for all $t \in [t_0, t_1]$,*
$L_u(x^*(t), u, t, \lambda(t), \mu(t))|_{u=u^*(t)} = 0$.

(vi) *For all $t \in [t_0, t_1]$, $\mu(t) \geq 0$ and $\mu(t) . g(x^*(t), u^*(t), t) = 0$.*

(vii) *Transversality condition (v.d) is satisfied.*

[18] Russak (1970).

The function $L(x, u, s, \lambda, \mu)$ in the statement of the Maximum Principle is known as the **Generalized Hamiltonian** or **Lagrangian** for the problem. Note that the Maximum Principle for Problem (C) is very similar to the Maximum Principle for Problem (B), that is, the Maximum Principle for Problem (A) augmented with transversality condition $(v.d)$. Conditions (i) and (ii) are the same as (i) and (ii) in the Maximum Principle for Problem (A). Condition (iii) is obtained if we replace the Hamiltonian function in condition (iii) of the Maximum Principle for Problem (A) with the Generalized Hamiltonian function. Condition (iv) is obtained if we replace, in condition (iv) of the Maximum Principle for Problem (A), the choice set U for $u(t)$ with $\{u | u \in R^r \wedge g(x^*(t), u, t) \geq 0\}$, the choice set for $u(t)$ in Problem (C) when the state at t is $x^*(t)$.

In the section on necessary conditions, a heuristic argument was given as to why the conditions in the Maximum Principle for Problem (A) might be necessary conditions for optimality in Problem (A). Here we will do the same for Problem (C).

Recall that we established the following: Suppose, given $t \in (t_0, t_1)$ and a sufficiently small positive number Δt, the decision-maker had to choose a value $v \in U$ such that the control which had: (a) values same as u^* (the optimal control for Problem (A)) on $[t_0, t]$; (b) the value v on $(t, t + \Delta t]$; (c) values on $(t + \Delta t, t_1]$ chosen to maximize net returns on $(t + \Delta t, t_1]$; and (d) the associated state at t_1, $x(t_1) = x^1$ (given $x(t_0) = x^0$), maximized net returns on $[t_0, t_1]$. Then, the optimal choice of value for the decision-maker would then be $u^*(t)$ (more precisely, $\lim_{s \to t+} u^*(s)$, if we allow t to be a point of discontinuity of u^*).

We made certain assumptions (for example, Ω, the set of ordered pairs of initial times and initial states from which the terminal state x^1 could be reached at t_1, being assumed open in $[t_0, t_1] \times \Re^n$) which ensured that, for a sufficiently small Δt, there would exist controls satisfying (a)–(d) for all $v \in U$. Therefore, the choice set for the decision-maker in this problem was U. This led us to the conclusion that the function $H(x^*(t), u, t, \lambda(t))$ is maximized on the entire set U (the choice set for the problem) at $u = u^*(t)$.

Similarly, suppose, for $t \in (t_0, t_1)$ and a sufficiently small positive number Δt, the decision-maker considered the analogous problem of choosing $v \in U$ such that the control, which had: (a) values same as u^* on $(t, t_1]$; (b) the value v on $(t - \Delta t, t]$; (c) values on $[t_0, t - \Delta t]$ chosen to maximize net returns on $[t_0, t - \Delta t]$; and (d) the associated state at t_1, $x(t_0) = x^0$ (given $x(t_1) = x^1$), maximized net returns on $[t_0, t_1]$. The optimal choice of value for the decision-maker would again be $u^*(t)$ ($= \lim_{s \to t-} u^*(s)$). By suitable choice of assumptions it could again be ensured that for all $v \in U$ and for sufficiently small Δt there would exist a control satisfying (a) to (d). Consequently, the choice set for the decision-maker in this problem was U. This could then be used to derive the conclusion that the function $H(x^*(t), u, t, \lambda(t))$ is maximized on U (the choice set for the problem) at $u = u^*(t)$.

Now, suppose that for all $s \geq 0$ the decision-maker faces the single mixed constraint: $g_1(x(s), u(s), s) \geq 0$ instead of the constraint $u(s) \in U$. Consider some

$t \in (t_0, t_1)$ where u^* is continuous and suppose that the decision-maker confronts the first of the two problems with $x^1 = x^*(t_1)$. A value of v for which $g_1(x^*(t), v, t) > 0$ can be chosen on the interval $(t, t + \Delta t]$. Given that the trajectories of the state variables are continuous and the function g_1 is continuous, for sufficiently small Δt we would then have $g_1(x(s), v, s) > 0$ for all $s \in (t, t + \Delta t]$. However, if a value of v is such that $g_1(x^*(t), v, t) = 0$ then that value of v can be chosen on the interval $(t, t + \Delta t]$ only if $\dot{g}_1(x^*(t), v, t) = g_{1x}(x, u, s)|_{(x,u,s)=(x^*(t),v,t)} f(x^*(t), v, t) + g_{1s}(x, u, s)|_{(x,u,s)=(x^*(t),v,t)} \geq 0$ (assuming that all partial derivatives of g_1 exist). The choice set for the first decision-making problem, therefore, becomes $\{v | v \in \mathfrak{R}^r \wedge (g_1(x^*(t), v, t) > 0 \vee ((g_1(x^*(t), v, t) = 0 \wedge \dot{g}_1(x^*(t), v, t) \geq 0))\}$. Consequently, we can, using an argument similar to that used in the earlier section, conclude that $H(x^*(t), u, t, \lambda(t))$ is maximized on the set $\{v | v \in \mathfrak{R}^r \wedge (g_1(x^*(t), v, t) > 0 \vee (g_1(x^*(t), v, t) = 0 \wedge \dot{g}_1(x^*(t), v, t) \geq 0))\}$ at $u = u^*(t)$.

Similarly, if we consider the second of the two decision-making problems with $x^1 = x^*(t_1)$, the choice set for the problem can be seen to be $\{v | v \in \mathfrak{R}^r \wedge (g_1(x^*(t), v, t) > 0 \vee (g_1(x^*(t), v, t) = 0 \wedge \dot{g}_1(x^*(t), v, t) \leq 0))\}$ and we can conclude that $H(x^*(t), u, t, \lambda(t))$ is maximized on this set at $u = u^*(t)$. Taken in conjunction with the preceding statement this leads to the conclusion that $H(x^*(t), u, t, \lambda(t))$ is maximized on $\{v | v \in \mathfrak{R}^r \wedge g_1(x^*(t), v, t) \geq 0\}$ at $u = u^*(t)$.

The above argument can easily be extended to the case of m constraints $g_j(x(s), u(s), s) \geq 0$ $(j = 1, 2, ..., m)$ and provides a heuristic argument for (iv) to be a necessary condition for optimality in Problem (C).

The reader who has gone through the earlier discussion of static optimization under inequality constraints should gauge that the conditions (v) and (vi) follow from condition (iv). This is indeed true because it can be proved that the assumption about the rank of a matrix made in this section (known as the **full rank condition** or the **rank constraint qualification**) ensures that the non-degenerate constraint qualification (see the chapter on Static Optimization) is satisfied for the family of static constrained optimization problems in (iv).

Suppose that $m = 1$ and assume that the constraint $u(s) \in U$ in Problem (A) is replaced with the single mixed constraint $g_1(x, u, s) \geq 0$ for all $s \in [t_0, t_1]$. Given that $u^* : [t_0, t_1] \to \mathfrak{R}^r$ is an optimal control for Problem (C) with associated trajectory $x^* : [t_0, t_1] \to \mathfrak{R}^n$, set $x^1 = x^*(t_1)$ in Problem (A). (x^*, u^*) is obviously an optimal control for this modified version of Problem (A). With appropriate assumptions we can, as in Section 14.3, derive (14.14). Taking the limit on both sides of (14.14) as $\Delta t \to 0+$, we obtain

$$\dot{\lambda}^+(t) = -[F_x(x, u, s)|_{(x,u,s)=(x^*(t),u^*(t),t)} + \lambda(t).f_x(x, u, s)|_{(x,u,s)=(x^*(t),u^*(t),t)}]$$
$$- [F_u(x, u, s)|_{(x,u,s)=(x^*(t),u^*(t),t)}$$
$$+ \lambda(t).f_u(x, u, s)|_{(x,u,s)=(x^*(t),u^*(t),t)}]c_y(\tau, \tau, y)|_{(\tau,\tau,y)=(t,t,x^*(t))} \qquad (14.23)$$

From condition (iv) in the Maximum Principle for Problem (C) it follows that, if $g_1(x^*(t), u^*(t), t) > 0$ then (assuming $\lambda_0 = 1$),

$$F_u(x, u, s)|_{(x,u,s)=(x^*(t),u^*(t),t)} + \lambda(t).f_u(x, u, s)|_{(x,u,s)=(x^*(t),u^*(t),t)} = 0.$$

From condition (*vi*) of the Maximum Principle for Problem (C) we know that $\mu_1(t) = 0$ so that

$$\dot{\lambda}^+(t) = -[F_x(x, u, s) + \lambda(s).f_x(x, u, s) + \mu_1(t)g_{1x}(x, u, s)]_{(x,u,s)=(x^*(t),u^*(t),t)} \qquad (14.24)$$

Suppose instead that $g_1(x^*(t), u^*(t), t) = 0$. Recall that we denoted the value of the optimal control (assumed unique) for Problem (A) $(\tau, y))$ at time $s \in [\tau, t_1]$ by $c(s, \tau, y)$. Suppose we continue with this notation for the modified version of Problem (A) discussed above. Assume that $\partial c_k(s, s, x)/\partial x_j$ $(k = 1, 2, ..., r; j = 1, 2, ..., n)$ is defined at all $(x, s) \in \{(x^*(t), t)|t \in (t_0, t_1)\}$. Let $h(x, s) = g_1(x, c(s, s, x), s)$. Since $g_1(x^*(t), u^*(t), t) = 0$, given the constraint $g_1(x, u, s) \geq 0$ it follows that, the rhs derivative of $h(x, s)$ with respect to x_i at $(x, s) = (x^*(t), t)$ must be non-negative and the lhs derivative must be non-positive. This is only possible if, for all $i \in \{1, 2, ..., n\}$,

$$\left[\frac{\partial h(x, s)}{\partial x_i}\right]_{(x,s)=(x^*(t),t)} = \left[\frac{\partial g_1(x, u, s)}{\partial x_i} + \sum_{k=1}^{r} \frac{\partial g_1(x, u, s)}{\partial u_k}\frac{\partial c_k(s, s, x)}{\partial x_i}\right]_{(x,u,s)=(x^*(t),u^*(t),t)} = 0$$

It follows, therefore, that

$$\mu_1(t)g_{1x}(x, u, s)|_{(x,u,s)=(x^*(t),u^*(t),t)}$$
$$+\mu_1(t)g_{1u}(x, u, s)|_{(x,u,s)=(x^*(t),u^*(t),t)}c_y(\tau, \tau, y)|_{(\tau,\tau,y)=(t,t,x^*(t))} = 0 \qquad (14.25)$$

Condition (*v*) in the Maximum Principle for Problem (C) implies that

$$\mu_1(t)g_{1x}(x, u, s)|_{(x,u,s)=(x^*(t),u^*(t),t)} = -[F_u(x, u, s)|_{(x,u,s)=(x^*(t),u^*(t),t)}|$$
$$+\lambda(t).f_u(x, u, s)|_{(x,u,s)=(x^*(t),u^*(t),t)}] \qquad (14.26)$$

(14.23), (14.25) and (14.26) can be used to show that even in this case (14.24) is satisfied. Proceeding along the line suggested while discussing necessary conditions, we can derive a counterpart of (14.24) in which $\dot{\lambda}^+(t)$ is replaced with $\dot{\lambda}^-(t)$. The above argument can again be extended to the case of m mixed constraints and used to justify condition (*iii*) in the Maximum Principle for Problem (C).

Sufficient Conditions:

Mangasarian Sufficiency Conditions for Problem (C): *Let for all (x, u, s, λ) $\in \Re^n \times \Re^r \times \Re \times \Re^n$, $H(x, u, s, \lambda) = F(x, u, s) + \lambda.f(x, u, s)$ and for all*

$(x, u, s, \lambda, \mu) \in R^n \times R^r \times R \times R^n \times R^m$, $L(x, u, s, \lambda, \mu) = H(x, u, s, \lambda) + \mu.g(x, u, s)$.

Suppose that for all $t \in [t_0, t_1]$,

(i) $H(x, u, t, \lambda(t))$ is jointly concave in (x, u); and
(ii) for all $j \in \{1, 2, ..., m\}$, $g_j(x, u, t)$ is quasi-concave in (x, u).
Suppose (x^*, u^*) is an admissible pair for Problem (C) and suppose there exists a continuous function $\lambda : [t_0, t_1] \to R^n$ and a piece-wise continuous

function $\mu : [t_0, t_1] \to R^m$, continuous at every point of continuity of u^*, such that the following conditions hold:

(iii) Except at points of discontinuity of u^*, for all $t \in [t_0, t_1]$,

$$\dot{\lambda}(t) = -L_x(x, u, s, \lambda, \mu)|_{(x,u,s,\lambda,\mu)=(x^*(t),u^*(t),t,\lambda(t),\mu(t))};$$

(iv) Except at points of discontinuity of u^*, for all $t \in [t_0, t_1]$,

$$L_u(x^*(t), u, t, \lambda(t), \mu(t))|_{u=u^*(t)} = 0;$$

(v) For all $t \in [t_0, t_1]$, $\mu(t) \geqq 0$ and $\mu(t).g(x^*(t), u^*(t), t) = 0$

(vi) Transversality condition (v.d) is satisfied.

Then, (x^*, u^*) is an optimal pair for Problem (C).

In order to establish the sufficiency of the above conditions proceed, using condition (i), as in the derivation of (14.18). Since (iii) is satisfied,

$$H_x(x, u, s, \lambda)|_{(x,u,s,\lambda)=(x^*(s),u^*(s),s,\lambda(s))} = -\dot{\lambda}(s) - \mu(s).g_x(x, u, s)|_{(x,u,s)=(x^*(s),u^*(s),s)}.$$

Also, from (iv),

$$H_u(x, u, s, \lambda)|_{(x,u,s,\lambda)=(x^*(s),u^*(s),s,\lambda(s))} = -\mu(s).g_u(x, u, s)|_{(x,u,s)=(x^*(s),u^*(s),s)}.$$

Conditions (ii) and (v) then ensure that

$$\mu(s).[g_x(x, u, s)|_{(x,u,s)=(x^*(s),u^*(s),s)}.(x^a(s) - x^*(s))$$
$$+ g_u(x, u, s)|_{(x,u,s)=(x^*(s),u^*(s),s)}.(u^a(s) - u^*(s))] \geq 0.$$

Condition 14.18 is, therefore, again satisfied and the use of condition (vi) establishes the optimality of (x^*, u^*). Note that if the Hamiltonian function is strictly concave in (x, u) then (x^*, u^*) is again a unique optimal pair for the problem.

Note that the Mangasarian sufficiency conditions for Problem (C) holds even if the rank constraint qualification is not satisfied.

The Arrow sufficiency conditions are not so easily extended to Problem (C). For Arrow-type sufficiency conditions, the reader is requested to refer to Seierstad and Sydsaeter (1977 and 1987, p. 289). A natural extension of Problem (C) would involve the inclusion of constraints of the form $h_k(x(t), t) \geq 0$ $(k = 1, 2, ..., p)$, which do not involve control variables and are, therefore, known as **pure state constraints**. Their treatment is beyond the scope of the current text and the interested reader may refer to Hartl et al. (1995), pp. 181–218, for a discussion of such problems.

14.7 Infinite Horizon Problems

Suppose, instead of choosing values for the control variables in a time interval beginning at some point t_0 and terminating at some point t_1, the decision-maker has to make choices for all points of time t_0 onwards. That is, suppose the time horizon for the decision-making problem is now $[t_0, \infty)$ instead of $[t_0, t_1]$. Then, corresponding to the set of admissible controls (permissible patterns of choice) in any finite horizon problem considered earlier, we can define a set of admissible controls for a corresponding infinite horizon decision-making problem, by simply replacing restrictions with respect to the interval $[t_0, t_1]$ with the same restrictions with respect to the interval $[t_0, \infty)$ and restrictions with respect to the terminal state at t_1 with the same restrictions with respect to the limiting value of the state $x(t)$ as $t \to \infty$. It would, therefore, appear that corresponding to the basic finite horizon problem (A), we can define a basic infinite horizon optimal control problem, where the set $\mathcal{U}$ of admissible controls is derived from that in Problem (A) and where the decision-maker now aims to choose an admissible control which maximizes the integral sum of instantaneous net returns over the new decision-making horizon $[t_0, \infty)$. However, it is here that we run into a major difficulty in infinite horizon problems which is absent in the case of finite horizon problems.

14.7.1 Definition of an optimal control

The infinite horizon counterpart to the basic finite horizon optimal control problem (A) would appear to be the problem $\max_{u \in \mathcal{U}} \int_{t_0}^{\infty} F(x(t), u(t), t) \, dt$, or written out in greater detail,

$$\textbf{Problem (A}^*\textbf{):} \quad \max_{\substack{u(t) \\ t \in [t_0, \infty]}} \int_{t_0}^{\infty} F(x(t), u(t), t) dt$$

$$\text{s.t. } \dot{x}(t) = f(x(t), u(t), t)$$

$$x(t_0) = x^0, \lim_{t \to \infty} x(t) = x^1$$

$$u(t) \in U$$

Note, however, that the improper integral $\int_{t_0}^{\infty} F(x(t), u(t), t) dt$ is defined as the limit of the definite integral $\int_{t_0}^{T} F(x(t), u(t), t) dt$ as $T \to \infty$. If this limit exists (that is, exists finitely) for every control in $\mathcal{U}$ then the value of the objective functional is defined for all admissible controls and an optimal control can be unambiguously defined as one for which this value is maximum. This is, for example, true, if $F(x, u, t) = G(x, u) e^{-\beta t}$, $\beta > 0$, and G is a bounded function which takes only non-negative or only non-positive values. $\int_{t_0}^{T} F(x(t), u(t), t) dt$ is then a bounded and monotonic function of T.

There is, however, no guarantee that this limit will exist for every admissible control. In particular, in many economic problems, one finds that for more than one admissible control the value of $\int_{t_0}^{T} F(x(t), u(t), t) dt$ diverges to ∞ as $T \to \infty$. If the value of the objective functional is not defined for every admissible control then the definition of an optimal control as 'an admissible control for which the value of the objective functional is not less than that for any other admissible control' becomes meaningless. In such circumstances one has to consider alternative ways of defining an optimal control which are broadly consistent with the stated objective of the decision-maker.

We will now define three alternative criteria for optimality (of admissible controls) which are often used in cases where the objective functional in Problem (A*) is not defined for all admissible controls.

The overtaking criterion

An admissible control u^* is defined to be **optimal according to the overtaking criterion** iff there exists a function $T : \mathcal{U} \to (t_0, \infty)$ such that, given any admissible control u, it is true that for all $T \geq T(u)$,

$$\int_{t_0}^{T} F(x^*(t), u^*(t), t)dt \geq \int_{t_0}^{T} F(x(t), u(t), t)dt$$

where x^* and x are the trajectories associated with u^* and u, respectively. Thus, over a sufficiently long time interval, the integral sum of instantaneous net returns yielded by an admissible control which is optimal by the overtaking criterion, is at least as large as that yielded by any given admissible control.

Note that in a case where the value of the objective functional is defined for every admissible control, a control which is optimal by the overtaking criterion is optimal by the maximum value criterion but the converse is not always true. For example, for every $T > t_0$, the integral sum of instantaneous net returns for an admissible control u^* over a time interval $[t_0, T]$ may be less than that for another control u^{**} by $\delta(T)$, where $\delta : (t_0, \infty) \to \Re_{++}$ and $\lim_{T \to \infty} \delta(T) = 0$. If u^{**} is optimal by the maximum value criterion then u^* is optimal by the maximum value criterion but is not optimal by the overtaking criterion. This suggests that it might be reasonable to use optimality criteria which are less stringent than the overtaking criterion.

The catching-up criterion

An admissible control u^* is defined to be **optimal according to the catching-up criterion** iff there exists a function $T: \mathcal{U} \times \Re_{++} \to [t_0, \infty)$ such that, given any admissible control u and any positive real number ε, for all $T \geq T(u, \varepsilon)$ it is true that,

$$\int_{t_0}^{T} F(x^*(t), u^*(t), t)dt \geq \int_{t_0}^{T} F(x(t), u(t), t)dt - \varepsilon$$

where x^* and x are the trajectories associated with u^* and u, respectively.

In a case where the value of the objective functional is defined for every admissible control it can be proved that an admissible control is optimal by the catching-up criterion iff it is optimal by the maximum value criterion. However, consider a case where there are just two admissible controls u^* and u^{**} such that, for every $T > t_0$, the integral sum of net returns yielded by u^* over the interval $[t_0, T]$ is strictly less than that yielded by u^{**} and the difference is given by $\{1 + \cos(\pi + T - t_0)\} + \delta(T)$ where $\delta : (t_0, \infty) \to \Re_{++}$ and $\lim_{T \to \infty} \delta(T) = 0$. $\{1 + \cos(\pi + T - t_0)\}$ is a periodic function of T with range $[0, 2]$. Therefore, u^{**} is the unique optimal control by the catching-up criterion. Yet, given any positive number, however small, we can always find a time horizon, longer than any arbitrarily specified time horizon, for which the difference between the integral sums of net returns yielded by the two controls is less than the given positive number. If this is sufficient for the decision-maker to consider u^* to be as good as u^{**} for attaining his objective, a less stringent criterion for optimality than the catching-up criterion can be considered.

The sporadically catching-up criterion

An admissible control u^* is defined to be **optimal according to the sporadically catching-up criterion** iff there exists a function $T : \mathcal{U} \times R_{++} \times [t_0, \infty) \to [t_0, \infty)$ such that, given any admissible control u, any positive real number ε and any real number $\bar{T} \geq t_0$, there exists $T \geq T(u, \varepsilon, \bar{T})$ for which it is true that

$$\int_{t_0}^{T} F(x^*(t), u^*(t), t)dt \geq \int_{t_0}^{T} F(x(t), u(t), t)dt - \varepsilon,$$

where x^* and x are the trajectories associated with u^* and u, respectively.

14.7.2 Necessary conditions for optimality

Halkin (1974) extended the Maximum Principle to infinite horizon problems. Note that if u^* is an optimal control for Problem (A*) with associated trajectory x^* then for all intervals $[t_2, t_3]$, $t_3 > t_2 \geq t_0$, (x^*, u^*) must be an optimal pair for the following problem:

$$\max_{\substack{u(t) \\ t \in [t_2, t_3]}} \int_{t_2}^{t_3} F(x(t), u(t), t)dt$$

$$\text{s.t. } \dot{x}(t) = f(x(t), u(t), t)$$

$$x(t_2) = x^*(t_2), x(t_3) = x^*(t_3)$$

$$u(t) \in U$$

This is true if u^* achieves the maximum value for the objective functional on the set of admissible controls but also if the value of the objective functional is not defined for u^* but u^* is optimal according to any of the three criteria for optimality defined in section 7.1. If there exists an interval $[t_2, t_3]$ for which the above is not true then it is possible to define a new admissible pair $(\hat{x}, \hat{u})$ such that for all $T \geq t_3$,

$$\int_{t_0}^{T} F(\hat{x}(t), \hat{u}(t), t)dt - \int_{t_0}^{T} F(x^*(t), u^*(t), t)dt = \delta,$$

$$\text{where } \delta = \int_{t_2}^{t_3} F(\hat{x}(t), \hat{u}(t), t)dt - \int_{t_2}^{t_3} F(x^*(t), u^*(t), t)dt > 0.$$

In that case (whichever the criterion of optimality being used) u^* can never be an optimal control.

Since u^* must be optimal for the above finite horizon problem it follows that u^* must satisfy the Maximum Principle for such a problem for all time intervals $[t_2, t_3]$, $t_3 > t_2 \geq t_0$. Halkin proved that:

> The Maximum Principle for Problem (A*) is the same as the Maximum Principle for Problem (A) except that every reference to the interval $[t_0, t_1]$ in the statement of the latter must be replaced with a reference to the interval $[t_0, \infty)$.

The argument that an optimal control in the infinite horizon problem is (where admissible) optimal for truncated finite horizon versions of the original problem remains true even if we remove the condition $\lim_{t \to \infty} x(t) = x^1$ in Problem (A*) or if we replace this condition by the condition $\lim_{t \to \infty} x(t) \geq x^1$ or any other constraint on the limiting value of the state. The Maximum Principle for the resulting problems is, therefore, the same as for Problem (A*).

In finite horizon problems when there are no restrictions on the terminal value of the i-th state variable $x_i(t_1)$ ($i \in \{1, 2, ..., n\}$), the Maximum Principle for the problem includes the transversality condition $\lambda_i(t_1) = 0$. Similarly, when there is an inequality constraint on the terminal value of the i-th state variable, $x_i(t_1) \geq x_i^1$, the Maximum Principle for the problem includes the transversality condition $\lambda_i(t) \geq 0, x_i(t_1) \geq x_i^1, \lambda_i(t)\left[x_i(t_1) - x_i^1\right] = 0$. However, in infinite horizon problems when there is no constraint on $\lim_{t \to \infty} x_i(t)$ or when there is a constraint $\lim_{t \to \infty} x_i(t) \geq x_i^1$, the analogous transversality conditions $\lim_{t \to \infty} \lambda_i(t) = 0$ and $\lim_{t \to \infty} \lambda_i(t) \geq 0, \lim_{t \to \infty} \lambda_i(t)\left[x_i(t_1) - x_i^1\right] = 0$ do not necessarily hold.[19]

Michel (1982) has, however, shown that if the problem considered is of the following type:

$$\max_{\substack{u(t) \\ t \in [t_0, \infty]}} \int_{t_0}^{\infty} G(x(t), u(t)) e^{-\beta t} dt \quad (\beta > 0)$$

$$\text{s.t. } \dot{x}(t) = f(x(t), u(t))$$

$$x(t_0) = x^0$$

and if the objective functional is defined for every admissible control, then the transversality condition

$$\lim_{t \to \infty} H(x^*(t), u^*(t), t, \lambda(t)) = 0$$

must be satisfied by any optimal pair (x^*, u^*). Michel's proof of the Maximum Principle for the problem together with the above transversality condition assumes that the state is restricted to lie in an open subset of $\Re^n$ and can be extended to problems with constraints of the form $u(t) \in U$ or $g(x(t), u(t)) \geq 0$. Moreover, if G is a non-negative real valued function and if the set $\{f(x^*(t), u) | u \in \Re^n\}$ contains an open neighbourhood of the zero vector in $\Re^n$ for all t which are sufficiently large then the transversality condition in the finite horizon counterpart of the above problem

$$\lim_{t \to \infty} \lambda(t) = 0$$

is also satisfied.

The fact that necessary transversality conditions cannot be specified for infinite horizon problems, in general, implies that the solutions to these problems cannot, in

---

[19] Chiang (1992) has argued for the assumption of such transversality conditions in infinite horizon problems by criticizing some commonly cited examples of problems where these conditions are violated. Caputo (2005) addresses this criticism in detail.

general, be characterized using the Maximum Principle. In many economic applications, therefore, it is the practice to directly impose an appropriate constraint on the behaviour of the state variable as $t \to \infty$. This constraint then serves as a transversality condition which has to be satisfied by a solution to the problem.

14.7.3 Sufficient conditions for optimality

Mangasarian-type sufficiency conditions

Consider Problem (A*). From the discussion of sufficient conditions due to Mangasarian in the case of finite horizon control problems it is obvious that,

> If the objective functional is defined for every admissible control then the Mangasarian sufficiency conditions for Problem (A) (with the time interval $[t_0, t_1]$ replaced with $[t_0, \infty)$) together with the condition that
>
> $\lim_{t \to \infty} \lambda(t).(x(t) - x^*(t)) \geq 0$ for all trajectories x associated with admissible controls,
>
> is sufficient for optimality of the admissible pair (x^*, u^*) in Problem (A*).

These Mangasarian-type sufficiency conditions continue to be sufficient for optimality even in case the constraint $\lim_{t \to \infty} x(t) = x^1$ is not present or when it is replaced by an inequality constraint of the form $\lim_{t \to \infty} x(t) \geq x^1$. Given that these sufficiency conditions hold, we can rewrite (5.1) as

$$\int_{t_0}^{T} [F(x^*(s), u^*(s), s) - F(x^a(s), u^a(s), s)]ds = \lambda(T).(x^a(T) - x^*(T)) - \lambda(t_0).(x^a(t_0) - x^*(t_0))$$

for any $T \geq t_0$ where (x^a, u^a) is any arbitrarily given admissible pair. If $\lim_{t \to \infty} \lambda(t).(x(t) - x^*(t)) \geq 0$ for all trajectories x associated with admissible controls then it follows that for every positive real number ε it is possible to find a real number $T(u^a, \varepsilon)$ such that for all $T \geq T(u^a, \varepsilon)$,

$$\int_{t_0}^{T} [F(x^*(s), u^*(s), s) - F(x^a(s), u^a(s), s)]ds \geq -\varepsilon$$

This implies that:

> Even in cases where the objective functional is not defined for every admissible control, the Mangasarian sufficiency conditions for Problem (A) (with the time interval $[t_0, t_1]$ replaced with $[t_0, \infty)$) together with the condition that
>
> $\lim_{t \to \infty} \lambda(t).(x(t) - x^*(t)) \geq 0$ for all trajectories x associated with admissible controls,
>
> is sufficient for the admissible pair (x^*, u^*) to be optimal in Problem (A*) by the catching-up criterion of optimality.

Arrow-type sufficiency conditions

In a similar vein, the Arrow sufficiency conditions for finite horizon problems can be extended to the infinite horizon case. Thus:

The Arrow sufficiency conditions for Problem (A) (with the time interval $[t_0, t_1]$ replaced with $[t_0, \infty)$) together with the condition that

$\lim_{t \to \infty} \lambda(t).(x(t) - x^*(t)) \geq 0$ *for all trajectories x associated with admissible controls,*

is sufficient for the admissible pair (x^*, u^*) to be optimal in Problem (A^*) by the catching-up criterion of optimality.

Note that for problems in which the objective functional is defined for every admissible control, if an admissible control is optimal by the catching-up criterion then it attains the maximum value for the objective functional on the set of admissible controls. Moreover, the above Arrow-type sufficiency conditions continue to be sufficient for optimality even in cases where the constraint $\lim_{t \to \infty} x(t) = x^1$ is absent or is replaced by an inequality constraint of the form $\lim_{t \to \infty} x(t) \geq x^1$.

The Maximum Principle, the Mangasarian-type sufficiency conditions and the Arrow-type sufficiency conditions for Problem (A^*) remain valid under the weaker set of restrictions on the functions F and f discussed earlier in the section 14.6.3. That is, the conditions hold if the domain of these functions is $R^n \times U \times [t_0, \infty)$, the functions are continuous on this domain and the relevant partial derivatives are continuous in any set D which is open in $R^n \times U \times [t_0, \infty)$ and $\{(x^*(t), u^*(t), t) | t \in [t_0, \infty)\} \subseteq D$.

EXERCISES

Solve the following finite horizon problems:
(In each of them u is the control variable, y is the state variable and T stands for the terminal time.)

1. max $\int_0^4 3y \, dt$ subject to $\dot{y} = y + u$, $y(0) = 5$, $y(4)$ free, $u(t) \in [0, 2]$.

2. max $\int_0^4 3y \, dt$ subject to $\dot{y} = y + u$, $y(0) = 5$, $y(4) \geq 300$, $u(t) \in [0, 2]$.

3. max $\int_0^T -(t^2 + u^2) \, dt$ subject to $\dot{y} = u$, $y(0) = 4$, $y(T) = 5$, T free, u unconstrained.

4. max $\int_0^T -1 \, dt$ subject to $\dot{y} = 2u$, $u \in [-1, 1]$, $y(T) = 0$, $y(0) = 0$.

5. max $\int_0^1 -\frac{1}{2}(y^2 + u^2) dt$ subject to $\dot{y} = u - y$, u unconstrained, $y(0) = 1$, $y(1)$ free.

14.8 Infinite Horizon Problem: An Alternative Aproach

In the preceding sections of this chapter we considered a theory for solving decision-making problems in which the decision-maker has to choose values of control variables at each point belonging to a time interval $[t_0, t_1]$ or $[t_0, \infty)$. In this section we consider the problem of a decision-maker who has to choose values of control variables for an infinite sequence of points or periods of time numbered 0, 1, 2, ... Since our objective in this section is to provide a basic introduction to the theory of dynamic programming which is used to study solutions of such problems, we will confine our discussion to the case where the decision-maker has to choose the values of a single control variable u. Depending on her choices, the decision-maker gets a net return in each period and her objective is to maximize the sum of these returns. There is a single state variable x whose value for any period $t > 0$ is determined by the values of u chosen for preceding periods $0, 1, 2, ...t-1$, and the exogenously given value of x for period 0. The value of x for period $t \geq 0$ (the state in t) determines the set of values of u the decision-maker can choose from for period t, the net return received by the decision-maker in period t from choice of any specific value of u and the value of x for period $t+1$ (the **state** in $t+1$).

We will initially specify the decision-maker's problem in a more general form than that subsequently analysed in this section. Suppose the values of x and u have to lie in the sets $X \subseteq R$ and $U \subseteq R$, respectively. Suppose that the state in any period $t \geq 0$, denoted by $x(t)$, and the chosen value of u for the period t, denoted by $u(t)$, determine the state in the period $t + 1$ according to the rule:

$$x(t + 1) = \phi(x(t), u(t), t)$$

where $\phi : X \times U \times \mathbb{Z}_+ \rightarrow R$. The above equation is referred to as the **transition equation or state equation** for the problem. We shall, in what follows, sometimes refer to $x(t)$ as the **input state** in period t and to $x(t+1)$ as the **output state** in period t.

Suppose also that for any $t \geq 0$, $u(t)$ must lie in the set $\Upsilon(x(t), t)$ where $\Upsilon: X \times \mathbb{Z}_+ \rightarrow 2^U$ is a function and 2^U denotes the set of all subsets of U. Let the net return in any period $t \geq 0$ be denoted by $r(x(t), u(t), t)$ where $r : X \times U \times \mathbb{Z}_+ \rightarrow R$. Then, the decision-maker's problem can be written as

$$\max_{u:\mathbb{Z}_+ \rightarrow U} \sum_{t=0}^{\infty} r(x(t), u(t), t)$$
$$\text{s.t. } x(t + 1) = \phi(x(t), u(t), t)$$
$$x(t + 1) \in X$$
$$u(t) \in \Upsilon(x(t), t)$$
$$x(0) = x^0$$

where $x^0 \in X$.

We begin our analysis of solutions to this problem by assuming that the values of the transition function ϕ and the function Υ are independent of the value of t. Also assume that there exists a function $\bar{r} : X \times U \rightarrow R$ and a number $\beta \geq 0$ such that $r(x, u, t) = \beta^t \bar{r}(x, u)$ for all $(x, u, t) \in X \times U \times \mathbb{Z}_+$. $\bar{r}(x(t), u(t))$ can often be interpreted as the current net return in period t and β the constant **discount factor** employed by the decision-maker to get the discounted value $\beta^t \bar{r}(x(t), u(t))$ of that net return in period 0. Note that these assumptions signify that the initial value of the variable t does not matter for the decision-maker. That is, if we consider a second decision-making problem which is identical to the first in all respects except that periods $0, 1, 2, ...$ are replaced respectively by periods $h, h+1, h+2, ..., h$ being an integer chosen at random, then the function $u^* : \{0, 1, 2, ...\} \to U$ is an optimal control for the first problem iff the function $u^{**} : \{h, h+1, h+2, ...\} \to U$ is an optimal control for the second problem, where for all $t \geq 0$, $u^*(t) = u^{**}(t+h)$. Further, if $x^* : \{0, 1, 2, ...\} \to X$ and $x^{**} : \{h, h+1, h+2, ...\} \to X$ are the trajectories of the state variable associated with u^* and u^{**}, respectively then $\beta^h \sum_{t=0}^{\infty} \beta^t r(x^*(t), u^*(t)) = \sum_{t=h}^{\infty} \beta^t r(x^{**}(t), u^{**}(t), t)$. That is, the maximum value of the sum of net returns attainable in the problem with initial time h is β^h times that attainable in the problem with initial time 0.

Note also that our assumptions imply that we can define a function $\Gamma : X \to 2^X$ such that the first three constraints in the above problem can be replaced by the constraint $x(t+1) \in \Gamma(x(t))$. That is, the input state in period t completely determines the set of feasible output states in period t (the set of feasible states in $t+1$). Let $A = \{(x_I, x_o) | x_I \in X \wedge x_o \in \Gamma(x_I)\}$ be the set of all feasible pairs of input states and output states. We shall refer to A as the **graph** of Γ.

Finally, suppose that the transition equation is such that $u(t)$ can be written explicitly as a function of $x(t)$ and $x(t+1)$. Substituting this function for $u(t)$ in the current net return function $\bar{r}$ it becomes possible to express the current net return in period $t \geq 0$ simply as a function of $x(t)$ and $x(t+1)$. Let $F : A \to \Re$ be this **current net return function**. We can now rewrite the decision-maker's problem as follows:

$$\text{Problem (D):} \quad \max_{x:\mathbb{Z}_+ \to X} \sum_{t=0}^{\infty} \beta^t F(x(t), x(t+1))$$

$$\text{s.t. } x(t+1) \in \Gamma(x(t))$$

$$x(0) = x^0$$

The decision-making problem is now stated not in terms of choice of values of the control variable at $0, 1, 2, ...$ but equivalently in terms of a choice of states at $1, 2, 3, ...$ Any infinite sequence of states $\{x(t)\}_{t=0}^{\infty}$ is a **feasible trajectory** or plan for the state variable in the above problem iff $x(0) = x^0$ and $x(t+1) \in \Gamma(x(t))$ for all $t \geq 0$. If we denote by $\Pi(x^0)$ **the set of feasible plans** in the above problem for any initial state $x^0 \in X$, then the problem is simply one of choosing an element in $\Pi(x^0)$ such that the value of the sum of net returns is maximized.

From our discussion of optimal control problems with an infinite time horizon, we know that the choice of an optimal plan for the state variable may be complicated by the fact that the sum of net returns may not be finite for some elements in $\Pi(x^0)$, so that comparison between different feasible plans may not always be possible. In our subsequent discussion, we will obviate this possibility by assuming that

F is a bounded and continuous function and $\beta \in (0, 1)$.

14.8.1 The value function and the Bellman equation

In Problem (D) we assumed that the decision-maker is given a specific initial state $x^0 \in X$. For every $x_I \in X$ let us denote by $D(x_I)$ the decision-making problem obtained when the initial state x^0 is replaced in Problem (D) with x_I. Suppose for every $x_I \in X$ there exists a solution to Problem $(D(x_I))$. Then, we can define a function $V : X \to \Re$ such that for all $x_I \in X$, $V(x_I)$ is the maximum value of the sum of net returns attainable in Problem $(D(x_I))$. Therefore, the function V, if it exists, gives, for every $x_I \in X$, the maximum value of the sum of net returns that the decision-maker can attain when her problem remains unchanged but the initial state is x_I instead of x^0. V will be referred to as the **value function** for Problem (D).

Suppose the value function for Problem (D) exists and is given by $V : X \to R$. Consider any $x_I \in X$ and let $\{x^*(t)\}_{t=0}^{\infty}$ be an optimal plan for Problem $(D(x_I))$. Consider any $x_o \in \Gamma(x_I)$. The maximum value for the sum of net returns attainable in Problem $(D(x_I))$ by choosing a feasible plan for which $x(1) = x_o$ is the sum of the net return in period 0, $F(x_I, x_o)$ and the maximum value for the sum of net returns attainable in the problem

$$\max_{x:\mathbb{N}\to X} \sum_{t=1}^{\infty} \beta^t F(x(t), x(t+1))$$
$$\text{s.t. } x(t+1) \in \Gamma(x(t))$$
$$x(1) = x_o$$

Remember that the decision-maker's problem is such that a change in the initial time from 0 to an integer h implies that the maximum value attainable in the new problem is β^h times that attainable in the old problem. It follows that the maximum value attainable in the above problem is β times the maximum value attainable in Problem $(D(x_o))$. Therefore, the maximum value attainable for the sum of net returns in Problem $(D(x_I))$ by choosing a feasible plan for which $x(1) = x_o$ is $F(x_I, x_o) + \beta V(x_o)$.

Since $\{x^*(t)\}_{t=0}^{\infty}$ is an optimal plan for Problem $(D(x_I))$, therefore, it must give the maximum value attainable for the sum of net returns in Problem $(D(x_I))$ if the decision-maker is constrained to choose a feasible plan for which $x(1) = x^*(1)$. Therefore,

$$V(x_I) = \sum_{t=0}^{\infty} \beta^t F(x^*(t), x^*(t+1)) = F(x_I, x^*(1)) + \beta V(x^*(1))$$

Given that $\{x^*(t)\}_{t=0}^{\infty}$ is an optimal plan for Problem $(D(x_I))$, it follows that for all $x_o \in \Gamma(x_I)$,

$$F(x_I, x^*(1)) + \beta V(x^*(1)) \geq F(x_I, x_o) + \beta V(x_o)$$

Therefore, for all $x_I \in X$,

$$V(x_I) = \max_{x_o \in \Gamma(x_I)} [F(x_I, x_o) + \beta V(x_o)]$$

We can, therefore, conclude that if V is the value function for Problem (D) then $v = V$ must be a solution to the equation

$$v(x_I) = \max_{x_o \in \Gamma(x_I)} [F(x_I, x_o) + \beta v(x_o)], \text{ for all } x_I \in X \tag{14.27}$$

Since the unknown v in equation (14.27) is a function, (14.27) is often referred to as the **functional equation** for Problem (D). This equation is also referred to as the **Bellman equation** (14.27) for Problem (D) after Richard Bellman who pioneered the dynamic programming approach to optimization. The basic approach uses the interconnections between solutions to problems like Problem (D) and solutions to equations like (14.27) and solves a single optimization problem like Problem (D) involving a large number of variables by solving a sequence of smaller optimization problems similar to that on the rhs of (14.27).

We know that, if the value function for Problem (D) exists then we can find it by solving the Bellman equation for Problem (D). However, the Bellman equation might have more than one solution and we need to sieve through these solutions to find the value function. Since we have assumed that F is a bounded function and $\beta \in (0, 1)$ it follows that V, if it exists, must be a bounded function. Therefore, we need to confine our search only to bounded functions which solve (14.27). Suppose $v = v^*$ is a bounded solution to the functional equation (14.27). Then, for all $x_I \in X$, there must be a solution to the optimization problem

$$\textbf{Problem (E } (x_I))\textbf{: } \max_{x_o \in \Gamma(x_I)} \left[F(x_I, x_o) + \beta v^*(x_o) \right]$$

Let $y \in X$. Let x_1^* be a solution to Problem (**E**(y)). Further, let x_{t+1}^* be a solution to Problem (E(x_t^*)) for all $t \geq 1$. Then, for all $T \geq 1$,

$$v^*(y) = F(y, x_1^*) + \sum_{t=1}^{T-1} \beta^t F(x_t^*, x_{t+1}^*) + \beta^T v^*(x_T^*)$$

Given that v^* is a bounded function and $\beta \in (0, 1)$, taking the limit on both sides of the above equation as $T \to \infty$, we get,

$$v^*(y) = F(y, x_1^*) + \sum_{t=1}^{\infty} \beta^t F(x_t^*, x_{t+1}^*) \tag{14.28}$$

Suppose $\{\tilde{x}(t)\}_{t=0}^{\infty} \in \Pi(y)$, the set of feasible plans for Problem (D(y)). We know that

$$v^*(y) \geq F(y, \tilde{x}(1)) + \beta v^*(\tilde{x}(1)) \tag{14.29}$$

Since $v = v^*$ is a solution to (14.27), for all $t \geq 1$,

$$v^*(\tilde{x}(t)) \geq F(\tilde{x}(t), \tilde{x}(t+1)) + \beta v^*(\tilde{x}(t+1)) \tag{14.30}$$

From (14.29) and (14.30) it follows that, for all $T \geq 1$,

$$v^*(y) \geq F(y, \tilde{x}(1)) + \sum_{t=1}^{T-1} \beta^t F(\tilde{x}(t), \tilde{x}(t+1)) + \beta^T v^*(\tilde{x}(T))$$

Since v^* is a bounded function and $\beta \in (0, 1)$ it follows that

$$v^*(y) \geq F(y, \tilde{x}(1)) + \sum_{t=1}^{\infty} \beta^t F(\tilde{x}(t), \tilde{x}(t+1)) \qquad (14.31)$$

(14.28) and (14.31) imply that for any arbitrarily chosen $\{\tilde{x}(t)\}_{t=0}^{\infty} \in \Pi(y)$,

$$v^*(y) = F(y, x_1^*) + \sum_{t=1}^{\infty} \beta^t F(x_t^*, x_{t+1}^*) \geq F(y, \tilde{x}(1)) + \sum_{t=1}^{\infty} \beta^t F(\tilde{x}(t), \tilde{x}(t+1))$$

Note that $\{x_t^*\}_{t=0}^{\infty}$ with $x_0^* = y$ is a feasible plan for Problem (D(y)). Therefore, it follows that $v^*(y)$ is equal to the maximum value of the sum of net returns attainable in Problem (D(y)). Since y is an arbitrarily chosen element of X it follows that v^* must be the value function for Problem (D). We will consider the conditions under which a bounded solution exists for (14.27) in the next section.

Note that the above argument implies that for all $x_I \in X$ if $\{x_t^*\}_{t=0}^{\infty}$ is a plan such that $x_0^* = x_I$ and such that x_{t+1}^* is a solution to Problem (E(x_t^*)) for all $t \geq 0$ then it is an optimal plan for Problem (D(x_I)). Moreover, if there is a unique solution to Problem (E(x_I)) for all $x_I \in X$ then one can define a function $g : X \to X$ such that for all $x_I \in X$, the optimal value in Problem (D(x_I)) of $x(t+1) = g(x(t))$ for all $t \geq 0$. The function g then provides to the decision-maker the optimal value of the output state in any period for which the input state is known and is referred to as the decision-maker's **policy function**. The existence of a policy function for Problem (D), therefore, implies that there is a unique optimal plan for the problem.

Suppose we know that the value function for Problem (D) exists. The following argument then suggests a way to compute the value function. Consider a function $h : X \to X$ such that for all $x_I \in X$, $h(x_I) \in \Gamma(x_I)$. Consider a succession of hypothetical decision-making regimes numbered 0, 1, 2, ... In regime 0, whatever the value of $x(0)$, the decision-maker is forced to choose the output state in any period $t \geq 0$ by the rule $x(t+1) = h(x(t))$. The sum of net returns that the decision-maker can attain for any $x(0) \in X$ is, therefore, fixed. Let the function $v_0 : X \to R$ be such that $v_0(x_I)$ gives the (maximum) value of the sum of net returns when $x(0) = x_I$, for all $x_I \in X$. Now, consider regime 1. In this regime, whatever the value of $x(0)$, the decision-maker is allowed to choose any output state belonging to $\Gamma(x(0))$ in period 0. However, corresponding to any choice of output state in period 0 she has to choose the output state in any period $t \geq 1$ by the rule $x(t+1) = h(x(t))$. Consider any $x_I \in X$. Since the decision-maker is interested in maximizing the sum of net returns she should, if $x(0) = x_I$, choose $x(1)$ by solving the problem

$$\max_{x_o \in \Gamma(x_I)} [F(x_I, x_o) + \beta v_0(x_o)]$$

If a solution exists for all $x_I \in X$, a function $v_1 : X \to \Re$ can be defined such that, for all $x_I \in X$, $v_1(x_I)$ gives the maximum value of the sum of net returns in regime 1 when $x(0) = x_I$. Then, it follows that, for all $x_I \in X$,

$$v_1(x_I) = \max_{x_o \in \Gamma(x_I)} [F(x_I, x_o) + \beta v_0(x_o)] \geq F(x_I, h(x_I)) + \beta v_0(h(x_I)) = v_0(x_I)$$

Now, consider regime 2. In regime 2, irrespective of the value of $x(0)$, the decision-maker is allowed to choose any output state belonging to $\Gamma(x(t))$ in period t for all $t < 2$. However, corresponding to any such choice she has to choose the output state in any period $t \geq 2$ by the rule $x(t + 1) = h(x(t))$. Consider any $x_I \in X$. If $x(0) = x_I$ the decision-maker maximizes the sum of net returns by solving the problem

$$\max_{x_o \in \Gamma(x_I)} [F(x_I, x_o) + \beta v_1(x_o)]$$

If a solution exists for all $x_I \in X$, there exists a function $v_2 : X \to R$ such that, for all $x_I \in X$, $v_2(x_I)$ gives the maximum value of the sum of net returns in regime 2 when $x(0) = x_I$. Then, since $v_1(x_I) \geq v_0(x_I)$ for all $x_I \in X$, therefore, for all $x_I \in X$,

$$v_2(x_I) = \max_{x_o \in \Gamma(x_I)} [F(x_I, x_o) + \beta v_1(x_o)] \geq \max_{x_o \in \Gamma(x_I)} [F(x_I, x_o) + \beta v_0(x_o)] = v_1(x_I)$$

We can now generalize the argument to the case of any decision-making regime $\tau \geq 1$. The decision-making regime $\tau \geq 0$ can be defined as that under which, given $x(0)$, the decision-maker is allowed to choose any output state belonging to $\Gamma(x(t))$ in period t for $t < \tau$. However, for all $t \geq \tau$, the decision-maker has to choose the output state in period t by the rule $x(t + 1) = h(x(t))$. Note that as τ increases in value, the decision-making regime τ comes closer and closer to the conditions governing the decision-maker's choice in Problem (D) where the decision-maker is free to choose any output state belonging to $\Gamma(x(t))$ in any period t. Assuming that the solution to the maximization problem

$$\max_{x_o \in \Gamma(x_I)} [F(x_I, x_o) + \beta v_{\tau-1}(x_o)]$$

exists for all for all $x_I \in X$ and for all $\tau \geq 1$, we can define, for all $\tau \geq 1$, $v_\tau : X \to R$ as the function for which,

$$v_\tau(x_I) = \max_{x_o \in \Gamma(x_I)} [F(x_I, x_o) + \beta v_{\tau-1}(x_o)], \text{ for all } x_I \in X \tag{14.32}$$

v_τ can be referred to as the value function under regime τ.

It follows that if $v_\tau(x_I) \geq v_{\tau-1}(x_I)$, for all $x_I \in X$, then

$$v_{\tau+1}(x_I) = \max_{x_o \in \Gamma(x_I)} [F(x_I, x_o) + \beta v_\tau(x_o)] \geq \max_{x_o \in \Gamma(x_I)} [F(x_I, x_o) + \beta v_{\tau-1}(x_o)] = v_\tau(x_I)$$

for all $x_I \in X$.

Since $v_1(x_I) \geq v_0(x_I)$, for all $x_I \in X$, by induction it follows that for all $\tau \geq 0$, $v_{\tau+1}(x_I) \geq v_\tau(x_I)$. There is, of course, nothing surprising in this result because, given any initial state, for all $\tau \geq 0$, the set of feasible plans of the decision-maker in regime $\tau + 1$ contains the set of feasible plans in regime τ. However, because the decision-making regime τ comes closer and closer to the decision-making regime underlying Problem (D) as τ increases in value, we can expect that the value function v_τ under regime τ will approach the value function V for Problem (D) as $\tau \to \infty$. This raises the possibility that beginning from an arbitrary function $v_0 : X \to R$ corresponding to any arbitrary choice of rule $h : X \to X$ we can through repeated iterations on the recursive equation (RE) (equation 14.32) arrive at the value function V. The conditions under which this might be true will be considered in the next section.

14.8.2 The existence of the value function

In this section we discuss how sufficient conditions for the existence of a bounded solution to (14.27) can be obtained through the use of an important fixed point theorem known as the Contraction Mapping Theorem. These conditions will, of course, also be sufficient for the existence of the value function for Problem (D). The proof of existence provided in this section is, however, not a complete proof because that would require us to introduce mathematical concepts which are outside the scope of this text. The idea is to indicate the approach and major stages in the argument.

14.8.3 Some properties of bounded continuous functions on X

Given a non-empty set M and a function $d : M \times M \to R$, the ordered pair (M, d) is a **metric space** iff the following sentences are true:

(i) $(\forall x)(\forall y)(x \in M \wedge y \in M \to d(x, y) \geq 0)$
(ii) $(\forall x)(\forall y)(x \in M \wedge y \in M \to (d(x, y) = 0 \leftrightarrow x = y))$
(iii) $(\forall x)(\forall y)(x \in M \wedge y \in M \to d(x, y) = d(y, x))$
(iv) $(\forall x)(\forall y)(\forall z)(x \in M \wedge y \in M \wedge z \in M \to d(x, y) + d(y, z) \geq d(x, z))$

Suppose (M, d) is a metric space. The function d is known as the metric or **distance function** for the metric space (M, d). When the metric being used in a context is implicitly understood, M itself is often referred to as a metric space. The elements of M are known as the **points** of the metric space. Note that the conditions (i) to (iv) imply that the value of the function d corresponding to any two points of the metric space satisfies the intuitive notion of a distance between two points along a straight line or on a plane. Condition (iv), for example, is usually referred to as the **Triangle Inequality** (in Euclidean geometry, the sum of the lengths of two sides of a triangle can never be less than the length of its third side). Not surprisingly, for all $n \in \mathbb{N}$, the Euclidean space of n dimensions is a metric space with distance function $d_{E^n}(x, y) = \|x - y\| = \sqrt{(x - y) \cdot (x - y)}$. That (i) to (iii) are satisfied is obvious and (iv) follows from the Cauchy-Schwartz inequality.

In this section, the particular M we are interested in is the set of all real-valued, bounded, and continuous functions defined on the set X. It is the set or space of functions in which we will seek a bounded solution to (14.27). Let us denote this set by $C(X)$. Let $B(X)$ be the set of real-valued bounded functions with domain X. Let

$d_X : B(X) \times B(X) \to R$ be the function given by

$$(\forall x)(\forall y)(x \in B(X) \wedge y \in B(X) \to d_X(x, y) = \sup_{s \in X} |x(s) - y(s)|)$$

Theorem 14.8.1 $(C(X), d_X)$ *is a metric space.*

Proof. $(C(X), d_X)$ clearly satisfies (i) to (iii). Suppose it does not satisfy (iv). Let a, b and c belong to $C(X)$ and let $\varepsilon \in R_{++}$ such that

$$\sup_{s \in X} |a(s) - c(s)| - \varepsilon = \sup_{s \in X} |a(s) - b(s)| + \sup_{s \in X} |b(s) - c(s)|)$$

It then follows from the definition of supremum (or ℓub) of a set that there exists $s' \in X$ such that

$$\left| a(s') - b(s') \right| + \left| b(s') - c(s') \right| \geq \left| a(s') - c(s') \right| > \sup_{s \in X} |a(s) - b(s)| + \sup_{s \in X} |b(s) - c(s)|)$$

The impossibility of the above inequality then implies that $(C(X), d_X)$ satisfies (iv). Therefore, $(C(X), d_X)$ is a metric space.

An infinite **sequence in a metric space** is an infinite sequence whose terms are points of the metric space. In the remainder of this section we shall not need to refer to finite sequences and, therefore, any infinite sequence will simply be referred to as a sequence. A sequence with successive terms denoted as $a_0, a_1, a_2, \dots$ will be denoted by $\{a_n\}_{n=0}^{\infty}$. Given a metric space (M, d), a sequence $\{a_n\}_{n=0}^{\infty}$ in (M, d) and a point $a \in M$, the sequence $\{a_n\}_{n=0}^{\infty}$ has the **limit** a in (M, d) iff

$$(\forall \varepsilon)(\varepsilon \in R_{++} \to (\exists m)(m \in \mathbb{N} \wedge (\forall n)(n \in \mathbb{N} \wedge n \geq m \to d(a_n, a) < \varepsilon)))$$

If $\{a_n\}_{n=0}^{\infty}$ has the limit a in (M, d) we denote this by $\lim_{n \to \infty} a_n = a$. Any sequence with a limit in (M, d) is defined to be a **convergent sequence** in (M, d). The sequence $\left\{ \frac{1}{n+1} \right\}_{n=0}^{\infty}$ has the limit 0 in one-dimensional Euclidean space R. Suppose $\varepsilon' \in \Re_{++}$. For all $n \in \mathbb{N}$ such that $n > \frac{1}{\varepsilon'} - 1$ it follows that $\left| \frac{1}{n+1} - 0 \right| = \frac{1}{n+1} < \varepsilon'$.

A sequence cannot have more than one limit in a metric space. For a proof, suppose that $l_1 \in M$ and $l_2 \in M$ are both limits of $\{a_n\}_{n=0}^{\infty}$ in (M, d) and $d(l_1, l_2) = \delta$, where $\delta \in R_{++}$. Since $\lim_{n \to \infty} a_n = l_1$, let $m_1 \in \mathbb{N}$ such that $(\forall n)(n \in \mathbb{N} \wedge n \geq m_1 \to d(a_n, l_1) < \delta/2)$. Since $\lim_{n \to \infty} a_n = l_2$, let $m_2 \in \mathbb{N}$ such that $(\forall n)(n \in \mathbb{N} \wedge n \geq m_2 \to d(a_n, l_2) < \delta/2)$. Let $m_3 = \max(m_1, m_2)$. Then, $d(l_1, l_2) \leq d(l_1, m_3) + d(l_2, m_3) < \delta = d(l_1, l_2)$ and our supposition cannot be true.

Given a metric space (M, d) and a sequence $\{a_n\}_{n=0}^{\infty}$ in (M, d), $\{a_n\}_{n=0}^{\infty}$ is a **Cauchy sequence** in (M, d) iff:

$$(\forall \varepsilon)(\varepsilon \in R_{++} \to (\exists k)(k \in \mathbb{N} \wedge (\forall m)(\forall n)$$
$$(m \in \mathbb{N} \wedge n \in \mathbb{N} \wedge m \geq k \wedge n \geq k \to d(a_m, a_n) < \varepsilon)))$$

The sequence $\{\frac{1}{n+1}\}_{n=0}^{\infty}$ is a Cauchy sequence in E. Suppose $\varepsilon' \in R_{++}$. Let $k' \in \mathbb{N}$ such that $k' > \frac{2}{\varepsilon'} - 1$. Then, for all $m \in \mathbb{N}$ and $n \in \mathbb{N}$ such that $m \geq k'$ and $n \geq k'$, $\left|\frac{1}{m+1} - \frac{1}{n+1}\right| \leq \frac{1}{m+1} + \frac{1}{n+1} < \varepsilon'$.

> *Every convergent sequence in a metric space is a Cauchy sequence in that metric space but the converse is not true.* For a proof, suppose (M, d) is a metric space, $\{a_n\}_{n=0}^{\infty}$ is a sequence in (M, d) and $a \in M$ such that $\lim_{n \to \infty} a_n = a$. Let $\varepsilon' \in R_{++}$. Let $k' \in \mathbb{N}$ such that $(\forall n)(n \in \mathbb{N} \wedge n \geq k' \to d(a_n, a) < \frac{\varepsilon'}{2})$. Then, $(\forall m)(\forall n)(m \in \mathbb{N} \wedge n \in \mathbb{N} \wedge m \geq k' \wedge n \geq k' \to d(a_m, a_n) < \varepsilon')$. Thus, $\{a_n\}_{n=0}^{\infty}$ is a Cauchy sequence in (M, d). Note that $\left\{\frac{1}{n+1}\right\}_{n=0}^{\infty}$ is a Cauchy sequence in $((0, 1], d_E)$ where $d_E(x, y) = |x - y|$, for all $x \in (0, 1]$ and for all $y \in (0, 1]$, but is not a convergent sequence in the same metric space.

A metric space is defined to be a **complete metric space** iff every Cauchy sequence in a metric space is a convergent sequence in that metric space. In Section 4.1, we noted that a sequence is a convergent sequence in E iff it satisfies the Cauchy criterion. This implies that *E is a complete metric space.*

Theorem 14.8.2 $(C(X), d_X)$ *is a complete metric space.*

Proof From 14.8.1, we know that $(C(X), d_X)$ is a metric space.

Let $\{f_n\}_{n=0}^{\infty}$ be a Cauchy sequence in $(C(X), d_X)$. Let $a \in X$ and $\varepsilon' \in R_{++}$. Let $k' \in \mathbb{N}$ such that

$$(\forall m)(\forall n)(m \in \mathbb{N} \wedge n \in \mathbb{N} \wedge m \geq k' \wedge n \geq k' \to d_X(f_m, f_n) = \sup_{s \in X} |f_m(s) - f_n(s)| < \varepsilon')$$

Therefore, $k' \in \mathbb{N}$ such that

$$(\forall m)(\forall n)(m \in \mathbb{N} \wedge n \in \mathbb{N} \wedge m \geq k' \wedge n \geq k' \to |f_m(a) - f_n(a)| < \varepsilon')$$

Thus, for all $y \in X$, $\{f_n(y)\}_{n=0}^{\infty}$ is a Cauchy sequence in E. Since E is a complete metric space let the function $f^* : X \to \Re$ be such that $(\forall y)(y \in X \to f^*(y) = \lim_{n \to \infty} f_n(y))$.

We first prove that f^* is a bounded function. Let $\varepsilon' \in R_{++}$. Since $\{f_n\}_{n=0}^{\infty}$ is a Cauchy sequence let $k' \in \mathbb{N}$ such that $(\forall n)(n \in \mathbb{N} \wedge n \geq k' \to d_X(f_n, f_{k'}) < \frac{\varepsilon'}{2})$. Let $a \in X$. Since $\lim_{n \to \infty} f_n(a) = f^*(a)$, let $n' \in \mathbb{N}$ such that $n' \geq k'$ and $|f_{n'}(a) - f^*(a)| < \frac{\varepsilon'}{2}$. Then, $|f_{k'}(a) - f^*(a)| \leq |f_{k'}(a) - f_{n'}(a)| + |f_{n'}(a) - f^*(a)| < \varepsilon'$. Since a is an arbitrarily chosen element of X, therefore, $(\forall y)(y \in X \to |f_{k'}(y) - f^*(y)| < \varepsilon')$. Since $f_{k'}$ is a bounded function, therefore, f^* must be a bounded function.

Next, we prove that $\lim_{n \to \infty} f_n = f^*$ in the metric space $(B(X), d_X)$. Suppose $\lim_{n \to \infty} f_n \neq f^*$. Then, there exists $\varepsilon' \in R_{++}$ such that

$$(\forall n)(n \in \mathbb{N} \to (\exists m)(m \in \mathbb{N} \wedge m \geq n \wedge d_X(f_m, f^*) > \varepsilon'))$$

Since $\{f_n\}_{n=0}^{\infty}$ is a Cauchy sequence, let $k' \in \mathbb{N}$ such that

$$(\forall m)(\forall n)\left(m \in \mathbb{N} \wedge n \in \mathbb{N} \wedge m \geq k' \wedge n \geq k' \to d_X(f_m, f_n) < \frac{\varepsilon'}{2}\right)$$

It follows that

$$(\forall m)(\forall n)\left(m \in \mathbb{N} \wedge n \in \mathbb{N} \wedge m \geq k' \wedge n \geq k' \rightarrow (\forall y)\left(y \in X \rightarrow |f_m(y) - f_n(y)| < \frac{\varepsilon'}{2}\right)\right)$$

Let $n_1 \in \mathbb{N}$ such that $m' \geq k'$ and $d_X(f_{n_1}, f^*) > \varepsilon'$. Let $a \in X$ such that $\left|f_{n_1}(a) - f^*(a)\right| > \varepsilon'$. Since $\lim_{n \to \infty} f_n(a) = f^*(a)$, let $n_2 \in \mathbb{N}$ such that $n_2 \geq k'$ and $\left|f_{n_2}(a) - f^*(a)\right| < \frac{\varepsilon'}{2}$. Then, $\varepsilon' < \left|f_{n_1}(a) - f^*(a)\right| \leq \left|f_{n_1}(a) - f_{n_2}(a)\right| + \left|f_{n_2}(a) - f^*(a)\right|$ $< \frac{\varepsilon'}{2} + \frac{\varepsilon'}{2} = \varepsilon'$. Therefore, our supposition cannot be true and $\lim_{n \to \infty} f_n = f^*$.

Finally, we prove that f^* is continuous on X. It would then follow that $\lim_{n \to \infty} f_n = f^*$ in $(C(X), d_X)$.

Let $a \in X$ and $\varepsilon' \in R_{++}$. Since $\lim_{n \to \infty} f_n = f^*$ in $(B(X), d_X)$, let $n' \in \mathbb{N}$ such that $d_X(f_{n'}, f^*) < \frac{\varepsilon'}{3}$. Since $f_{n'}$ is continuous on X, let $\delta \in R_{++}$ such that $(\forall y)(y \in X \wedge |y - a| < \delta \rightarrow |f_{n'}(y) - f_{n'}(a)| < \frac{\varepsilon'}{3})$. It follows that

$$(\forall y)(y \in X \wedge |y - a| < \delta \rightarrow |f^*(y) - f^*(a)|$$
$$\leq |f^*(y) - f_{n'}(y)| + |f_{n'}(y) - f_{n'}(a)| + |f_{n'}(a) - f^*(a)| < \varepsilon')$$

Since ε' an arbitrarily chosen positive real number f^* is continuous at a. Since a is an arbitrary chosen element of X, therefore, f is continuous on X and $\lim_{n \to \infty} f_n = f^*$ in $(C(X), d_X)$.

Therefore, $\{f_n\}_{n=0}^{\infty}$ is a convergent sequence in $(C(X), d_X)$. Since $\{f_n\}_{n=0}^{\infty}$ is an arbitrarily chosen Cauchy sequence in $(C(X), d_X)$ it follows that $(C(X), d_X)$ is a complete metric space. ●

14.8.4 Restrictions on Γ and application of the Contraction Mapping Theorem

Let us assume:

For all $x_I \in X$, $\Gamma(x_I) \neq \phi$ and $\Gamma(x_I)$ is a compact set in E.

Then, given that F is a continuous function, by Weirstrass' Theorem it follows that for all $f \in C(X)$ and for all $x_I \in X$ there exists a solution to the problem

$$\max_{x_o \in \Gamma(x_I)} [F(x_I, x_o) + \beta f(x_o)]$$

Thus, given $f \in C(X)$, we can define a function $f_{\max} : X \rightarrow R$ such that

$$(\forall x_I)(x_I \in X \rightarrow f_{\max}(x_I) = \max_{x_o \in \Gamma(x_I)} [F(x_I, x_o) + \beta f(x_o)])$$

Since F and f are bounded functions it follows that $f_{\max} \in B(X)$.

The implication of this is that we can define a function $T_{\max} : C(X) \rightarrow B(X)$ such that

$$(\forall f)(f \in C(X) \rightarrow (\forall x_I)(x_I \in X \rightarrow T_{\max} f(x_I) = \max_{x_o \in \Gamma(x_I)} [F(x_I, x_o) + \beta f(x_o)])$$

where $T_{\max} f$ denotes the value of the function $T_{\max}$ at f.

Suppose the following additional restriction is imposed in Problem (D):

A, the graph of Γ, is a closed and convex set in E^2.

Then, it can be proved that,[20] for all $f \in C(X)$, $T_{\max} f \in C(X)$ so that we can write $T_{\max} : C(X) \to C(X)$.

We will use the restrictions on Γ and the corresponding properties of the function $T_{\max}$ to prove the existence of a bounded solution to (14.27). However, before doing so, we will define a special kind of function known as a contraction mapping, discuss a set of sufficient conditions for any function defined on a set of bounded functions to qualify as a contraction mapping and finally, prove a very important fixed point theorem for contraction mappings.

Given a metric space (M, d), a function $T:M \to M$ is a **contraction mapping** in (M, d) with **modulus** γ iff $\gamma \in (0, 1)$ and

$$(\forall y)(\forall z)(y \in M \land z \in M \to d(Ty, Tz) \leq \gamma d(y, z))$$

A simple example of a contraction mapping is the function $f : E \to E$ where $f(y) = by$, $b \in (0, 1)$, for all $y \in E$. Note that for all $y \in E$ and for all $z \in E$, $|f(y) - f(z)| = b|y - z|$.

Given any set S and any two functions $f_1 : S \to R$ and $f_2 : S \to R$ we will write $f_1 \geq f_2$ iff $(\forall y)(y \in S \to f_1(y) \geq f_2(y))$. The following theorem, due to Blackwell (1965), provides a set of sufficient conditions for any given function defined on a set of bounded functions to be a contraction mapping.

Theorem 14.8.3 Let (M, d_S) be a metric space, where $M \subseteq B(S)$, the set of all real-valued functions with domain S, and $d_S: B(S) \times B(S) \to R$ such that $(\forall y)(\forall z)(y \in B(S) \land z \in B(S) \to d_S(y, z) = \sup_{s \in S} |y(s) - z(s)|)$. Let, for all $\varepsilon \in R_+$, $\bar{f}_{S\varepsilon} : S \to R$ be the function such that $(\forall s)(s \in S \to \bar{f}_{S\varepsilon}(s) = \varepsilon)$ and let $(\forall y)(\forall \varepsilon)(y \in M \land \varepsilon \in R_+ \to y + \bar{f}_{S\varepsilon} \in M)$. A function $T : M \to M$ is a contraction mapping in (M, d_S) if it satisfies the following conditions:

  (i) Monotonicity: $(\forall y)(\forall z)(y \in M \land z \in M \to y \geq z \to Ty \geq Tz)$
  (ii) Discounting: $(\exists \gamma)(\gamma \in (0, 1) \land (\forall y)(\forall \varepsilon)(y \in M \land \varepsilon \in R_+ \to T(y + \bar{f}_{S\varepsilon}) \leq Ty + \gamma \bar{f}_{S\varepsilon}))$.

Proof Let $f_1 \in M$ and $f_2 \in M$ such that $\delta = d_S(f_1, f_2)$. Then, $f_1 + \bar{f}_{S\delta} \geq f_2$. If $T : M \to M$ satisfies monotonicity then $T(f_1 + \bar{f}_{S\delta}) \geq Tf_2$. If $T : M \to M$ satisfies the discounting condition then there exists $\gamma \in (0, 1)$ such that $Tf_1 + \gamma \bar{f}_{S\delta} \geq T(f_1 + \bar{f}_{S\delta})$. It follows that $Tf_1 + \gamma \bar{f}_{S\delta} \geq Tf_2$. Similarly, since $f_2 + \bar{f}_{S\delta} \geq f_1$, it follows that $Tf_2 + \gamma \bar{f}_{S\delta} \geq Tf_1$. This implies that $d_S(Tf_1, Tf_2) \leq \gamma \delta = \gamma d_S(f_1, f_2)$. Therefore, since f_1 and f_2 were arbitrarily chosen, T is a contraction mapping in (M, d_S) with modulus γ. ●

The above theorem implies that $T_{\max} : C(X) \to C(X)$ is a contraction mapping in $(C(X), d_X)$ with modulus β. Note that, for all $f_1 \in C(X)$ and for all $f_2 \in C(X)$, if

---

[20] See Stokey and Lucas with Prescott (1989).

$f_1 \geq f_2$ then

$$(\forall x_I)\,(x_I \in X \to T_{\max} f_1(x_I) = \max_{x_o \in X} [F(x_I, x_o) + \beta f_1(x_o)]$$

$$\geq \max_{x_o \in X} [F(x_I, x_o) + \beta f_2(x_o)] = T_{\max} f_2(x_I))$$

Therefore, $T_{\max}$ satisfies monotonicity. Further, for all $\varepsilon \in R_+$ and for all $f \in C(X)$, $f + \bar{f}_{X\varepsilon} \in C(X)$ and

$$(\forall x_I)\,(x_I \in X \to T_{\max}(f + \bar{f}_{X\varepsilon})(x_I) = \max_{x_o \in X} [F(x_I, x_o) + \beta \{f(x_o) + \varepsilon\}]$$

$$= \max_{x_o \in X} [F(x_I, x_o) + \beta f(x_o)] + \beta \varepsilon = T_{\max} f(x_I) + \beta \varepsilon)$$

where $\beta \in (0, 1)$. This implies that $T_{\max}$ satisfies the discounting condition.

Once we have established that $T_{\max}$ is a contraction mapping in $(C(X), d_X)$ with modulus β the following theorem can be used to obtain a solution for (14.27) in $C(X)$.

Theorem 14.8.4 (The Contraction Mapping Theorem): *Given a metric space (M, d) and a function $T : M \to M$, if (M, d) is a complete metric space and T is a contraction mapping in (M, d) with modulus γ then,*

(i) $(\exists y)(y \in M \land Ty = y \land (\forall z)(z \in M \land Tz = z \to z = y))$ *(T has a unique fixed point), and*

(ii) $(\forall y)(\forall z)(y \in M \land z \in M \land Ty = y \to \lim_{n \to \infty} T^n z = y)$,
where $(\forall y)(y \in M \to T^1 y = Ty \land (\forall n)(n \in \mathbb{N} - \{1\} \to T^n y = TT^{n-1}y))$.

Proof Suppose (M, d) is a complete metric space and T is a contraction mapping in (M, d) with modulus γ. Let $f_0 \in M$ and let $(\forall n)(n \in \mathbb{N} \to f_n = T^n f_0)$ where $(\forall n)(n \in \mathbb{N} \to T^n f_0 = Tf_{n-1} \in M)$. Then, $(\forall n)(n \in \mathbb{N} \to d(f_n, f_{n+1}) \leq \gamma d(f_{n-1}, f_n) \leq \gamma^2(f_{n-2}, f_{n-1}) \leq \ldots \leq \gamma^n d(f_0, f_1))$. Therefore, using the Triangle Inequality, we get

$$(\forall m)(\forall n)(m \in \mathbb{N} \land n \in \mathbb{N} \land m < n \to d(f_m, f_n) \leq \sum_{k=m}^{n-1} d(f_k, f_{k+1})$$

$$\leq d(f_0, f_1) \sum_{k=m}^{n-1} \gamma^k < \frac{\gamma^m}{1 - \gamma} d(f_0, f_1))$$

It follows from above that $\{f_n\}_{n=0}^{\infty}$ is a Cauchy sequence in (M, d). Therefore, since (M, d) is a complete metric space, $\{f_n\}_{n=0}^{\infty}$ is a convergent sequence in (M, d).

Let $f \in M$ such that $\lim_{n \to \infty} f_n = f$. Since T is a contraction mapping in the metric space (M, d), $(\forall n)(n \in \mathbb{N} \to d(f, Tf) \leq d(f, T^n f_0) + d(T^n f_0, Tf) \leq d(f, T^n f_0) + \beta d(T^{n-1} f_0, f))$. Further, since $\lim_{n \to \infty} f_n = f$, it follows that

$$d(f, Tf) = \lim_{n \to \infty} d(f, Tf) \leq \lim_{n \to \infty} d(f, T^n f_0) + \beta \lim_{n \to \infty} d(T^{n-1} f_0, f) = 0$$

Therefore, $Tf = f$ and f is a fixed point of T.

We next prove that f is the only fixed point of T. Suppose $\tilde{f}$ is a fixed point of T and $\tilde{f} \neq f$. Then, $d(\tilde{f}, f) = d(T\tilde{f}, Tf) \leq \gamma d(\tilde{f}, f)$. Since $\gamma \in (0, 1)$, this implies that, $d(\tilde{f}, f) = 0$ and $f = \tilde{f}$. Therefore, our supposition cannot be true and f must be the only fixed point of T. Further, since f_0 was an arbitrarily chosen element of M, it follows that $(\forall y)(y \in M \to \lim_{n \to \infty} T^n y = f)$ where $(\forall y)(y \in M \to T^1 y = T y \wedge (\forall n)(n \in \mathbb{N} - \{1\} \to T^n y = T T^{n-1} y))$. •

Since $(C(X), d_X)$ is a complete metric space and $T_{\max} : C(X) \to C(X)$ is a contraction mapping in $(C(X), d_X)$ with modulus β, the Contraction Mapping Theorem implies that:

(i) $(\exists y)(y \in C(X) \wedge y = T_{\max} y \wedge (\forall z)(z \in C(X) \wedge T_{\max} z = z \to z = y))$, and

(ii) $(\forall y)(\forall z)(y \in C(X) \wedge z \in C(X) \wedge T y = y \to \lim_{n \to \infty} T_{\max}^n z = y)$,

where $(\forall y)(y \in C(X) \to T_{\max}^1 y = T_{\max} y \wedge (\forall n)(n \in \mathbb{N} - \{1\} \to T_{\max}^n y = T_{\max} T_{\max}^{n-1} y))$.

(i) implies that there exists a unique bounded and continuous solution to the functional equation (14.27). We know that this must also be the value function V for Problem (D). (ii) implies that beginning from any bounded and continuous function on X, v_0, if we generate a sequence of functions $\{v_n\}_{n=0}^{\infty}$ through repeated iterations on equation (14.32), then this sequence has the value function V as its limit in the sense that as $n \to \infty$, $d_X(v_n, V) \to 0$. This provides a way to actually calculate the value function for Problem (D). However, except in a few rare cases, it is not possible to calculate the exact function analytically and one has to rely on numerical methods to arrive at approximations. Since the few special cases where analytical solutions do apply are also cases where F, contrary to the assumptions of this chapter, is an unbounded function, we ask the interested reader to refer to Stokey and Lucas with Prescott (1989) for the applicable theory as well as examples of analytical solutions.

14.8.5 Differentiability of the value function and the Euler equation

Stokey and Lucas with Prescott (1989) proceed to derive conditions under which the value function for Problem (D) satisfies additional properties. The broad approach is that of the previous section which establishes the existence of a continuous value function under certain conditions on F and Γ. Thus, if F satisfies the following additional condition (which is stronger than the concavity of F but weaker than the strict concavity of F):

$$(\forall x_I')(\forall x_o')(\forall x_I'')(\forall x_o'')(\forall \theta)((x_I', x_o') \in A \wedge (x_I'', x_o'') \in A \wedge \theta \in [0, 1] \to (x_I' = x_I'' \to$$
$$F(\theta x_I' + (1 - \theta)x_I'', \theta x_o' + (1 - \theta)x_o'') \geq \theta F(x_I', x_o') + (1 - \theta) F(x_I'', x_o''))$$

$$\wedge (x_I' \neq x_I'' \to F(\theta x_I' + (1 - \theta)x_I'', \theta x_o' + (1 - \theta)x_o'') > \theta F(x_I', x_o') + (1 - \theta) F(x_I'', x_o''))),$$

then V is a strictly concave function and the policy function g for Problem (D) exists and is continuous. Further, beginning from any bounded, continuous, and concave function v_0, if $\{g_n\}_{n=0}^{\infty}$ is a sequence of functions on X such that $(\forall n)(n \in \mathbb{N} \to$

$(\forall x_I)(x_I \in X \rightarrow g_n(x_I) = \arg\max_{x_o \in X}[F(x_I, x_o) + \beta T_{\max}^n v_0]))$ then $(\forall x_I)(x_I \in X \rightarrow \lim_{n \to \infty} g_n(x_I) = g(x_I))$. This result then provides a way to compute the policy function for Problem (D).

If, in addition,

$F(x_I, x_o)$ is continuously differentiable with respect to x_I in the interior of A, then,

$$(\forall x_I)\left(x_I \in \text{int} X \wedge g(x_I) \in \text{int}\Gamma(x_I) \rightarrow V'(x_I) = \left[\frac{\partial F(y, z)}{\partial y}\right]_{(y,z)=(x_I, g(x_I))}\right)$$

Thus, the above restrictions on the current net return function F in Problem (D) are sufficient to ensure that the value function is continuously differentiable in the interior of X given that, for all $x_I \in int\ X$, $g(x_I) \in int\ \Gamma(x_I)$.

Note also that if $\{x^*(t)\}_{t=0}^{\infty}$ is an optimal plan for Problem (D) then, for all $t \geq 0$, $x^*(t+1)$ must be a solution to the problem

$$\max_y \left[F(x^*(t), y) + \beta F(y, x^*(t+2))\right]$$

$$\text{s.t. } y \in \Gamma(x^*(t))$$

$$x^*(t+2) \in \Gamma(y)$$

Let $\Omega(x^*(t+2)) = \{y | y \in X \wedge x^*(t+2) \in \Gamma(y)\}$ for all $t \geq 0$. Under all the above assumptions about F and Γ it follows that if $x^*(0) = x^0 \in int\ X$, $x^*(t+1) \in int\ \Gamma(x^*(t))$ and $x^*(t+1) \in int\ \Omega(x^*(t+2))$ for all $t \geq 0$ then, a necessary condition for optimality of $\{x^*(t)\}_{t=0}^{\infty}$ is that for all $t \geq 0$,

$$\left[\frac{\partial F(x_I, x_o)}{\partial x_o}\right]_{(x_I, x_o)=(x^*(t), x^*(t+1))} + \beta \left[\frac{\partial F(x_I, x_o)}{\partial x_I}\right]_{(x_I, x_o)=(x^*(t+1), x^*(t+2))} = 0$$

This second-order difference equation is known as the **Euler equation** for Problem (D).

Suppose $X \subseteq \Re_+$ and $F(x_I, x_o)$ is increasing in x_I. Now if $\{x^*(t)\}_{t=0}^{\infty}$ is a feasible plan for the decision-maker satisfying the Euler equation then a sufficient condition[21] for $\{x^*(t)\}_{t=0}^{\infty}$ to be an optimal plan for the decision-maker is that

$$\lim_{t \to \infty} \beta^t \left[\frac{\partial F(x_I, x_o)}{\partial x_I}\right]_{(x_I, x_o)=(x^*(t), x^*(t+1))} x^*(t) = 0$$

This condition involving the limiting value of the state variable as $t \to \infty$ is analogous to the transversality conditions discussed in the theory of optimal control. Note that $\beta^t \left[\frac{\partial F(x_I, x_o)}{\partial x_I}\right]_{(x_I, x_o)=(x^*(t), x^*(t+1))}$ can be interpreted as the marginal contribution of the state variable to the net return at time t. However, from the equation defining the derivative of the value function, we know that this is also equal to $\beta^t V'(x^*(t))$.

---

[21] See Stokey and Lucas with Prescott (1989), pp. 98–9.

The latter measures the marginal contribution of the state variable at time t to the maximum value attainable time t onwards (remember that, given an initial state, the maximum value attainable for the decision-making problem with initial time t is β^t times that attainable in the decision-making problem with initial time 0). If we, therefore, interpret this quantity as the value to the decision-maker of a unit of the state variable at time t, the transversality condition can be interpreted as saying that, along an optimal plan, the total value of the quantity of the state variable must approach zero over time.

KEY TERMS

Abnormal Problems
Admissible Control
Admissible Pair
Alternative Conditions
Arrow Sufficiency Conditions
Backward Value Function
Bellman Equation
Catching-up Criterion
Contraction Mapping Theorem
Control Region
Control Variables
Current Net Return Function
Euler Equation
Feasible Trajectory
Full Rank Condition or Rank
    Constraint Qualification
Generalized Hamiltonian or
    Lagrangian
Hamiltonian Function
Infinite Horizon Problems
Initial State
Mangasarian Sufficiency Conditions
Maximum Principle
Mixed Constraints

Necessary Conditions
Necessary Conditions for Optimality
Objective Functional
Optimal Control
Overtaking Criterion
Output State
Discount Factor
Piecewise Continuous Function
Pontryagin's Maximum Principle
Pure State Constraints
Salvage Value Function
Shadow Price
Sporadically Catching-up Criterion
State Equations
State Variables
Sufficient Conditions
Sufficient Conditions for Optimality
Terminal State
Theory of Optimal Control
Transition Equation
Input State
Transversality Conditions
Value Function
Variable Terminal Time

USEFUL WEB LINKS

http://www.fiu.edu/~thompsop/modeling/modeling_chapter2.pdf (accessed on 2 August 2010): Optimal Control Theory

http://ocw.mit.edu/courses/economics/14-451-dynamic-optimization-methods-with-applications-fall-2009/lecture-notes/MIT14_451F09_lec10.pdf (accessed on 2 August 2010): dynamic optimization methods http://hassler-j.iies.su.se/Courses/MacroII/Notes/book.pdf (accessed on 2 August 2010); http://ocw.mit.edu/courses/economics/14-451-dynamic-optimization-methods-with-applications-fall-2009/lecture-notes/ (accessed on 2 August 2010);

http://www.fiu.edu/~thompsop/modeling/modeling_chapter4.pdf (accessed on 2 August 2010); http://www.ssc.upenn.edu/~rwright/courses/app-dp.pdf (accessed on 2 August 2010); www.econ.iastate.edu/tesfatsi/dpintro.cooper.pdf (accessed on 2 August 2010): dynamic programming

CHAPTER FIFTEEN

Economic Applications III

15.1 Introduction

Traditionally, dynamic problems appear in many areas of economic theory. We shall pick up some areas to provide a sample of the type of queries and the provided answers. Naturally this will not be an exhaustive account. The first area we choose is the stability problem of competitive equilibrium. The second area we choose is the theory of optimal economic growth. Both are vast and important enough to justify books being written about them and hence we shall just take up some central questions and exhibit how the mathematical theory developed earlier can help us to understand these problems. We shall discuss some examples of continuous time processes as well as some discrete time processes: the differences in conclusions will perhaps be somewhat disconcerting.

The problem appears to be that when one is telling an economic story, one is by force made to use the discrete paradigm; yet much of economic theory used is based on continuous time processes; the use of discrete time processes started relatively recently; results should alert us to the speciality of the results claimed on the basis of continuous time processes used earlier. Often these differences are neglected by theorists as well as policy makers for the sake of expediency: it does not look very smart or knowledgeable to say that one does not know or that one cannot make specific predictions; it is so much more forceful to appear all knowing and make point predictions. But then is it surprising that so many of these policies fail or that the predictions go wrong? One then finds convenient scapegoats: the vagaries of the weather or some such extraneous factor.

Consider the problem of stability of competitive equilibrium.

15.2 The Stability of Competitive Equilibrium

From the time of Adam Smith, there was a belief that the so-called Invisible Hand, the market force, should be successful in attaining equilibrium or a matching of demand and supply. That this force was believed to be as mysterious and powerful as the Force which Luke Skywalker is supposed to have battled the odds with, is perhaps not very surprising. What is surprising, however, is the almost complete neglect of this area by economic theorists.

Briefly the situation was as follows: if demand and supply did not match, it was entirely expected that prices would adjust in the direction of excess demand (which is demand – supply) and would thus remove any vestige of mismatch. Even to this day, a sudden spurt of demand is associated with a rise in prices. The intuitive appeal of this story is, however, far stronger than its actual effectiveness.

To motivate our analysis, we consider first of all, the case of two goods so that there is a single market. In a single market, the notion of a locally stable equilibrium is identified with the slope of the excess demand curve being negative at the equilibrium. But there does not seem to be any easy identification of global stability with any property even in such a simple set-up.

Consider the following example. There are two persons **A** and **B** with utility functions defined over commodities (x, y) as follows: $U_A(x, y) = \min(x, 2y)$ and $U_B(x, y) = \min(2x, y)$; their endowments are specified by $w_A = (1, 0)$, $w_B = (0, 1)$; routine computations lead to the excess demand function of the first good (x), $Z(p)$, where p is the relative price of good x:

$$Z(p) = \frac{p - 1}{(p + 2)(2p + 1)}$$

Thus the unique **interior** equilibrium is given by $p = 1$; now note that if the adjustment on prices is given by

$$\dot{p} = Z(p) \tag{15.1}$$

Let us take $Z(p)$ as being continuously differentiable so that the solution to (15.1) say $p_t(p^o)$ is well defined for any initial point $p^o > 0$. Note too that the interior equilibrium is unstable. In a single market case, the derivative of $Z(p)$ at equilibrium is enough to clinch matters and as one may check that the derivative is positive. See Figure 15.1.

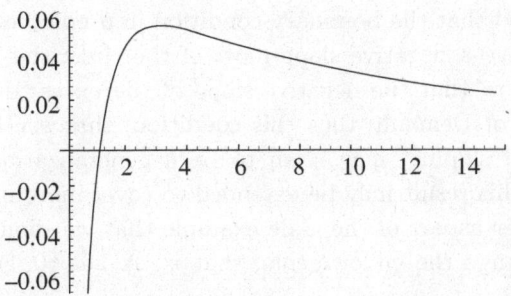

FIGURE 15.1   Excess Demand—The Gale Example

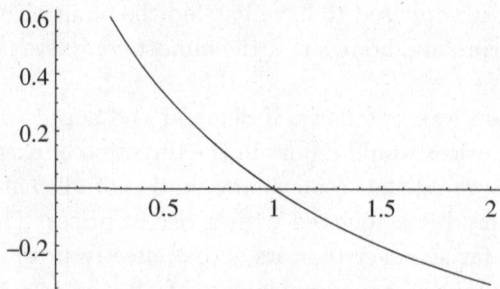

FIGURE 15.2   Excess Demand—A Stable Equilibrium

Alternatively, a look at Figure 15.1 should convince us that the unique interior equilibrium is unstable under the usual adjustment of prices. How do we rule out such situations?

As an illustration of our analysis below, consider the following assumptions for the single market (two-good case):[1]

α: $Z_x(p)$, $Z_y(p)$ are continuous functions of p for all $p > 0$ and satisfy Walras Law, that is, $pZ_x(p) + Z_y(p) = 0 \forall p > 0$; further $\lim_{p \to 0} Z_x(p) > 0$; similarly, $\lim_{p \to \infty} Z_y(p) > 0$.

It is easy to check that condition α implies that the set of equilibria $E = \{p > 0 : Z_x(p) = 0\}$ is non-empty. Moreover, if $Z_x(p)$ is differentiable, then there must exist **at least one** $p^\star \in E$ such that $Z_x'(p^\star) < 0$. That is, there must be **at least one** locally stable equilibrium. For global stability considerations, however, we must proceed further.

β: $p \in E. \Rightarrow Z_x'(p) \neq 0$.

With conditions α and β, however, we can show the following: first, there are only a finite number of points in E and second, for any $p^o > 0$, $p_t(p^o) \to p^\star \in E$ for some $p^\star \in E$: this may be considered to be the counterpart of global stability when there are multiple equilibria. When there is a single equilibrium the situation will be as in Figure 15.2. **Thus, condition α implies that overall the excess demand curve is downward sloping: since it begins from somewhere near the vertical axis and ends up below the horizontal axis.**

It should be noted that the boundary condition in α enforces the fact that excess demand functions have a negative slope most of the time; this is enough for global stability. We also know that the negative slope of the excess demand function may be related to a Law of Demand; thus this condition, that excess demand functions be *mainly* downward sloping, may seem to be a generalization. One can actually investigate whether this result may be extended to cover more markets.

There is one other aspect of the Gale example that we should note; suppose, for example, we interchange the endowments, that is, **A** has $(0, 1)$ while **B** has $(1, 0)$;

---

[1] We shall write $Z_x(p)$, $Z_y(p)$ as the excess demand functions for the two goods x, y, respectively, with p being the relative price of good x.

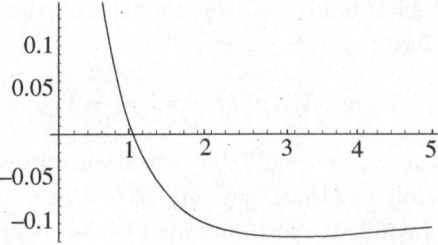

FIGURE 15.3 Excess Demand—The Gale Example with Endowments Interchanged

recomputing excess demand functions, we note that the situation is as in Figure 15.3: the unique interior equilibrium is now globally stable. One may, therefore, say that we had instability of the interior equilibrium because the distribution of endowments had not been *right* previously. With the new pattern of endowments, both α, β are met and the excess demand curve becomes downward sloping.

15.2.1 Gross substitutes and the weak axiom of revealed preference

A competitive equilibrium is known to be stable under the assumption that goods are gross substitutes of one another. This is one of the main assumptions under which stability is obtained. Another assumption which has been employed is the assumption that the weak axiom of revealed preference (WARP) holds in the aggregate. Both these assumptions are strong requirements. We use the notation introduced earlier that $Z(P)$ denotes excess demand. Recall that $Z(P) = X(P) - W$, where $X(p) = \sum_i x^i(p)$, and $W = \sum_i w^i$; further for each i, $x^i(p)$ uniquely solves $\max U^i(x)$ subject to $p.x \leq p.w^i$. Thus we shall confine attention to exchange economies. The extension to cover production will not be difficult. Also recall the following properties of the excess demand function: $Z(\lambda.p) = Z(p)$ for all $\lambda > 0$ and that $p.Z(p) = 0$ for all $p > 0$, that is, homogeneity of degree zero in the prices and Walras Law, respectively.

Consider WARP, first: If $p^\star$ is an equilibrium price configuration and $p \neq \lambda p^\star$ for any scalar λ, and if for all such p, we have $p^\star.Z(p) > 0$, then we shall say that WARP holds. Note that had there been only one person, WARP would always hold. Since there is only one person, it follows that $Z(p) = x^i(p) - w^i$; at the equilibrium of course, then $x^i(p^\star) = w^i$. Thus, if $p \neq \lambda p^\star$, it follows that $p.x^i(p) = p.w^i \Rightarrow x^i(p)$ is revealed preferred to $x^i(p^\star)$. Consequently, when $x^i(p^\star)$ is chosen, rationality demands that $x^i(p)$ is not available or that $p^\star.[x^i(p) - w^i] > 0$ that is, $p^\star.Z(p) > 0$: which is WARP.

WARP is also seen to hold if there are many persons but the distribution of endowments $\{w^i\}$ is Pareto Optimal. It follows then that the only equilibrium possible is the price $p^\star$ at which $x^i(p^\star) = w^i$ for all i; thus we are back in the environment of the last paragraph, and the entire argument goes through exactly as above.

Suppose then that WARP holds and we assume that the adjustment on prices in disequilibrium takes the form:

$$\dot{p}_j = k_j Z_j(p) \forall j \neq n, \ p_n = 1 \tag{15.2}$$

where $k_j > 0 \forall j$ are constants; we recall that given an initial price configuration p^o, we are assured of a solution to (15.2), say $p(t, p^o)$, which is continuous and unique with respect to the initial configuration, provided the excess demand functions $Z_j(p)$ are continuously differentiable. We shall take that this is so. Also we need to ensure that the solution or trajectory stays within the positive orthant given that $p^o > 0$. The following boundary restriction on excess demand functions achieves this:

Boundary Restriction

For any configuration of prices $p, p_n = 1$, *for any* $i \neq n$, *there is some* ϵ_i *such that* $p_i \leq \epsilon_i \Rightarrow Z_i(p) > 0$.

Under the above mentioned restrictions, we have the following:

Proposition 15.1   The unique equilibrium $p^\star$ *is globally stable under* (15.2).

Proof: First of all, that WARP implies the equilibrium is unique, is straightforward. Suppose to the contrary $\bar{p} \neq \lambda p^\star$ are both equilibria; then, by WARP, $\bar{p}.Z(p^\star) > 0$: which implies that $Z(p^\star) \neq 0 \Rightarrow p^\star$ cannot be an equilibrium. This proves the first part of the claim. For the remaining part of the claim, we shall show that $V(p(t, p^o)) \equiv V(t) = \sum_i^{(n-1)} (p_i(t, p^o) - p_i^\star)^2 / k_i$ is a Liapunov function. Clearly $V(t) > 0$; also $\dot{V}(t) = 2\sum_i^{(n-1)} (p_i(t, p^0) - p_i^\star) Z_i(p(t, p^o)) = 2[-Z_n(p(t, p^o)) - \sum_i^{(n-1)} p_i^\star Z_i(p(t, p^o))] < 0$ unless $p(t, p^o) = p^\star$. The above also proves that $V(t) < V(0)$ for $t > 0$ and hence that the trajectory lies in a bounded region of the positive orthant: the last bit follows by virtue of the boundary restriction. Hence the claim. •

Thus WARP, together with the boundary restriction, ensures that the unique equilibrium is globally stable. In other words, beginning with any price configuration, price revision in the direction of excess demand leads to the equilibrium.

We next consider the case of gross substitution (**GS**): this is an assumption which is employed on the nature of excess demand functions:

$$\frac{\partial Z_i}{\partial p_j} > 0 \text{ for } i \neq j$$

This assumption has some implications and it has to be employed with care because of other more basic requirements placed on excess demand functions. First of all, recall a basic requirement on each excess demand function: homogeneity of degree zero in prices; this implies that for each i

$$\sum_{i=1}^{n} \frac{\partial Z_i}{\partial p_j} p_j = 0 \forall p \geq 0, \ p \neq 0$$

Together with the GS restriction, this implies that

$$\frac{\partial Z_i}{\partial p_i} < 0 \forall i \; \forall p \geq 0, \, p \neq 0 \tag{15.3}$$

Consider a price configuration $p = (0, p_2, ..., p_{n-1}, 1)$: now consider $\lambda > 1$, then at λp, $Z_1(\lambda p) = Z_1(p)$ from homogeneity of degree zero in prices; however since all **other** prices have gone up, with own price fixed, by GS, we must have $Z_1(\lambda p) > Z_1(p)$. To resolve this kind of conflict, we shall employ GS only over strictly positive prices or in the interior of the non-negative orthant. We note:

Claim 15.2.1 Under GS, the equilibrium is unique up to scalar multiples. In other words, $\bar{p}, \hat{p} > 0$ both equilibria implies that $\bar{p} = \lambda \hat{p}$ for some $\lambda > 0$.

Proof: Consider $\mu = \min_{1 \leq i \leq n} \bar{p}_i / \hat{p}_i = \bar{p}_k / \hat{p}_k$ say ; then $\mu \hat{p}_i \leq \bar{p}_i \forall i$ with equality for $i = k$; we assume that there is at least one strict inequality (otherwise there is nothing to prove) and necessarily this inequality will occur for some $i \neq k$. Hence using GS, we can conclude that $Z_k(\mu \hat{p}) < Z_k(\bar{p}) = 0$ but $Z_k(\hat{p}) = Z_k(\mu \hat{p}) < 0 \Rightarrow \hat{p}$ cannot be an equilibrium. Thus there can be no inequality and the two equilibria are scalar multiples of one another. •

We shall provide two different demonstrations: the first a diagrammatic one is a useful method to analyse differential equations when there are only two variables, that is, for motion on the plane (See Figure 15.4). The second is the method we discussed earlier: the construction of a Liapunov function in the general case. Consider the simpler case when $n = 3$; by virtue of Walras Law we know that we can restrict attention to two markets only and by means of homogeneity, we measure prices relative to the third good, the numeraire. Thus there are actually only two markets: in market 1 good 1 is exchanged for the numeraire whereas in market 2, good 2 is exchanged against the numeraire good 3. Accordingly, in disequilibrium, the relative prices of goods 1 and 2 are adjusted as follows:

$$\dot{p}_i = \phi_i(Z_i(p)) \; i = 1, 2 \tag{15.4}$$

where ϕ_i are sign-preserving functions of $Z_i(p))$, that is, $\phi_i(Z_i(p))$ has the same sign as $Z_i(p)$. We shall analyse the solution of this system by means of Figure 15.4.

In Figure 15.4, in the $p_1 - p_2$ plane, the curves $Z_i(p_1, p_2, 1) = 0$ have been drawn; note that the two curves are upward rising, by virtue of the assumption of GS; for example, consider $Z_1(p_1, p_2) = 0$; along this curve we must have

$$\frac{\partial Z_1}{\partial p_1} dp_1 + \frac{\partial Z_1}{\partial p_2} dp_2 = 0$$

hence we must have:

$$\frac{dp_2}{dp_1}\Big|_{Z_1 = 0} = -\frac{\dfrac{\partial Z_1}{\partial p_1}}{\dfrac{\partial Z_1}{\partial p_2}} > 0$$

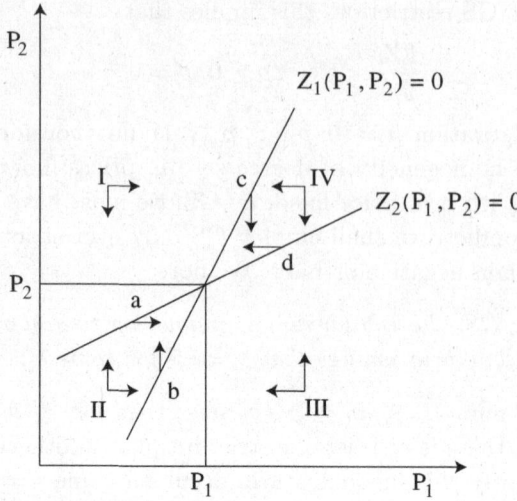

FIGURE 15.4 Phase Diagram with Gross Substitutes

The slope of the other curve can be similarly obtained. We have already noted that there can only be a single intersection since the point of intersection is an equilibrium and equilibrium is unique under GS. Also we have drawn the curves so that $Z_1 = 0$ is steeper at the point of intersection. This follows on account of GS and homogeneity of degree zero in prices; the latter implies

$$Z_{i1}p_1 + Z_{i2}p_2 = -Z_{i3} < 0 \text{ for } i = 1, 2.$$

where we write:

$$Z_{ij} = \frac{\partial Z_i}{\partial p_j}$$

Hence it follows that

$$\frac{-Z_{11}}{Z_{12}} > \frac{p_2}{p_1} > \frac{-Z_{21}}{Z_{11}}.$$

Thus the curves divide up the plane into four zones marked I–IV. Each zone has a different configuration of excess demands. To pin down these, we note that everywhere above $Z_1 = 0$, $Z_1 > 0$ while below the curve $Z_1 < 0$; for the $Z_2 = 0$ curve, we have $Z_2 > 0$ everywhere to its right and $Z_2 < 0$ everywhere to the left. Hence, the following patterns of excess demand exist in the four zones:

$I : Z_1 > 0, Z_2 < 0; \quad II : Z_1 > 0, Z_2 > 0; \quad III : Z_1 < 0, Z_2 > 0; \quad IV : Z_1 < 0, Z_2 < 0.$

The arrows in each zone represent the direction of price adjustment given the equation (15.4). Note that from regions I and III, such an adjustment would lead to points such as a, b, c, or d; from points such as a, b price revision leads to the zone II; from which price adjustment cannot leave; similarly from the points such as c, d price adjustment leads to zone IV and once again, from this zone there is no escape. Zones II and IV are called absorbing states. In zones II and IV, price adjustment will lead to equilibrium. Thus, the only possibility is that the orbits will approach the equilibrium: this is the global stability result that we had mentioned at the outset.

We proceed next to consider an analytical argument to arrive at the same conclusion when there are many goods; but still, we employ the assumption of GS. For the general case of n-goods the only method is to construct a Liapunov function. For the general form of the equation (15.4), the interested reader is referred to the paper by McKenzie (1960). We shall not consider the details of that demonstration, but merely indicate the direction of argument.

Let $P(p) = \{i : Z_i(p) \geq 0\}$: the set of goods in positive excess demand. Further define $V(p) = \sum_{i \in P(p)} p_i Z_i(p)$: the value of goods in positive excess demand. McKenzie (1960) shows that $V(p)$ has the properties of a Liapunov function, when goods satisfy weak GS, that is, the inequality in (15.3) is weak.

15.2.2 Scarf example

Next, we consider the Scarf (1960) example. Consider an exchange model where there are three individuals $h = 1, 2, 3$ and three goods $j = 1, 2, 3$. The utility functions and endowments are as follows:

$$U^1(q_1, q_2, q_3) = min(q_1, q_2); \ w^1 = (1, 0, 0)$$

$$U^2(q_1, q_2, q_3) = min(q_2, q_3); \ w^2 = (0, 1, 0)$$

$$U^3(q_1, q_2, q_3) = min(q_1, q_3); \ w^3 = (0, 0, 1)$$

Routine calculations lead to the following excess demand functions, where good 3 is treated as numeraire (that is, $p_3 = 1$):

$$Z_1(p_1, p_2) = \frac{p_1(1 - p_2)}{(1 + p_1)(p_1 + p_2)}$$

$$Z_2(p_1, p_2) = \frac{p_2(p_1 - 1)}{(1 + p_2)(p_1 + p_2)}$$

$$Z_3(p_1, p_2) = \frac{p_2 - p_1}{(1 + p_1)(1 + p_2)}$$

and the tatonnement process, for this example is given by

$$\dot{p}_i = Z_i(p_1, p_2) \ i = 1, 2 \tag{15.5}$$

Note that equilibrium for this exchange model (and for the process defined above) is given by $p_1 = 1, p_2 = 1$. It would be helpful to transform variables by setting

$x_i = p_i - 1$ for $i = 1, 2$. With this change in variables, our process becomes

$$\dot{x}_1 = -\frac{x_2(1 + x_1)}{(x_1 + 2)(x_1 + x_2 + 2)} \ , \ \dot{x}_2 = \frac{x_1(1 + x_2)}{(x_2 + 2)(x_1 + x_2 + 2)} \tag{15.6}$$

In what follows, we shall analyse the answer to the following question: given an arbitrary $x^o = (x_1^o, x_2^o)$, how does the solution $x(t, x^o)$ to (15.6) behave as $t \to \infty$?

We introduce the function $v : R \to R$ by

$$v(x) = \frac{x^2}{2} + x - \ln(1 + x)$$

which is continuously differentiable for all x such that $1 + x > 0$. One may show that

Claim 15.2.2 $v(x) > 0$ if $x > -1$, $x \neq 0$; $v(0) = 0$.

Proof: Note that $v(0) = 0$, and for $x > -1$ we have $v'(x) = \frac{x(x + 2)}{1 + x}$ and $v''(x) = 1 + \frac{1}{1 + x^2}$. Thus for $x > -1$, $v(x)$ is strictly convex with $v'(0) = 0$; hence $x = 0$ yields a global minimum for $v(x)$ for all $x > -1$. •

Next define $V(x_1, x_2) = v(x_1) + v(x_2)$. We then have:

Claim 15.2.3 Along the solution $x(t, x^o)$ to (15.6), $\dot{V} = 0$ provided $x_i(t, x^o) > -1$ for $i = 1, 2$.

Proof: Note that

$$\dot{V} = v'(x_1).\dot{x}_1 + v'(x_2).\dot{x}_2 = 0$$

•

Next, we may claim:

Claim 15.2.4 Given $x^o = (x_1^o, x_2^o)$, $x_i^o > -1$, $i = 1, 2$ the solution $x(t, x^o)$ to (15.6) is such that $\exists a_i, b_i$ such that $-1 < a_i < b_i$ and $x(t, x^o) \in [a_1, b_1] \times [a_2, b_2] \ \forall t > 0$.

Proof: Follows from the last two claims. •

For local stability, it may be of some interest to note the following:

Claim 15.2.5 For x small, $v(x) \approx x^2$.

Proof: This follows since for x small one may use the following approximation:

$$\ln(1 + x) \approx x - \frac{x^2}{2}$$

•

The above may be used to classify the solution to (15.6) when the initial point x^o is close to the equilibrium, that is, the origin. It is approximately a circle with the centre origin and passing through x^o. In the general case, the nature of the orbit is provided by Figure 15.5: the cyclical behaviour of prices around equilibrium is revealed; however, since the figure cannot be taken as a demonstration, we provide such a demonstration, next.

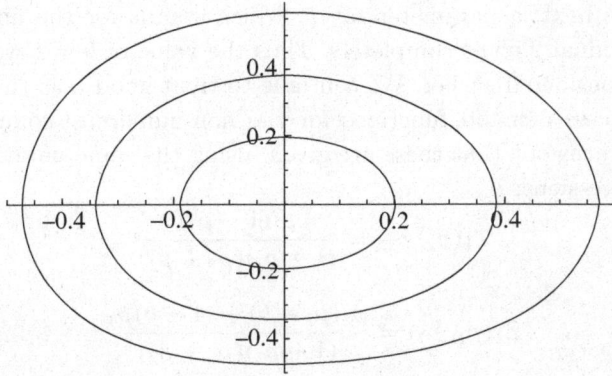

FIGURE 15.5 Orbits of the Scarf Example

First of all, note that since $V(t) = V(x_1(t), x_2(t)) = V(x_1^o, x_2^o)$ for all t, it follows that the solution or trajectory $x(t, x^o) = (x_1(t), x_2(t))$ is bounded and each $x_i(t)$ is bounded away from -1: since if either of these conditions is violated, $V(t)$ would tend to $+\infty$. Hence the ω-limit set corresponding to x^o, $L_\omega(x^o)$, is non-empty and compact; also, $(0, 0) \notin L_\omega(x^o)$ if $x^0 \neq (0, 0)$ (remember, $(0, 0)$ is the equilibrium for the system) hence by the Poincaré-Bendixson Theorem[2] $L_\omega(x^o)$ must be a closed orbit. This means that either we have a *limit cycle* or the trajectory $x(t, x^o)$ itself is a closed orbit.

If there is a limit cycle $\mathcal{L}$, then by virtue of Claim 15.2.3, it follows that for any $y \in \mathcal{L}$, $V(y) = V(x^o)$; further, in such circumstances, there would be a neighbourhood $\mathcal{N}$ of x^o such that for any solution $x(t, y)$ originating from any $y \in \mathcal{N}$, $x(t, y) \to \mathcal{L}$.[3] Consequently, we must have $V(y) = V(x^o) \forall y \in \mathcal{N}$: this is, of course, not possible, since the function V cannot be constant on an open set. Hence, no such limit cycle exists. And the solution $x(t, x^o)$ must be a closed orbit. Thus, we have shown the following to be true:

Claim 15.2.6 For any initial configuration x^o, the solution to (15.6), $x(t, x^o)$ is a closed orbit around the equilibrium $(0, 0)$.

For the Scarf example, the solution is a closed curve given by Remark 15.1 (b); consequently, the set of limit points L coincides with this curve; note that the Poincaré-Bendixson Theorem stated that so long as the set of limit points does not contain an equilibrium, again guaranteed by the point (b) noted above, a cycle is the only possible alternative. As we hope to show, there are some more interesting features of the Scarf example.

---

[2] See, Proposition 13.4.

[3] The argument is similar to the one used in the proof of Claim 13.1.5.

We introduce next, a parameter say b, which stands for the amount of second good which individual 2 owns completely. Thus the value of $b = 1$ would revert back to the example considered earlier. We continue to treat good 3 as the numeraire and then compute excess demand functions for the non-numeraire commodities for the case at hand; it turns out that these are given, using the same notation as above, by the following expressions:

$$Z_1(p_1, p_2) = \frac{p_1(1 - p_2)}{(1 + p_1)(p_1 + p_2)}$$

$$Z_2(p_1, p_2) = \frac{p_2(p_1 - b) + (1 - b)p_1}{(1 + p_2)(p_1 + p_2)}$$

Consequently, the system (15.5) now takes the form:

$$\dot{p_1} = \frac{p_1(1 - p_2)}{(1 + p_1)(p_1 + p_2)} \text{ and } \dot{p_2} = \frac{p_2(p_1 - b) + (1 - b)p_1}{(1 + p_2)(p_1 + p_2)} \tag{15.7}$$

Once more standard computations ensure that the **unique equilibrium** is given by

$$p_1^* = \frac{b}{2 - b} = \theta \text{ say, } p_2^* = 1$$

Thus, it may be noted that our choice of parameter places a restriction on its magnitude

$$0 < b < 2;$$

and we shall take it that this is met. Note also that when $b = 1$, $\theta = 1$ too, and we have the earlier situation. That there have been some changes to the stability property of equilibrium is contained in the next claim:

Claim 15.2.7 For the process (15.7), $(\theta, 1)$ is a locally asymptotically stable equilibrium iff $b < 1$; for $b > 1$, the equilibrium is locally unstable.

Proof: The characteristic roots of the Jacobian of the system (15.5) evaluated at the equilibrium are given by:

$$\frac{1}{8}(-b + b^2 \pm \sqrt{b}\sqrt{(-32 + 49b - 26b^2 + 5b^3)})$$

and it should be noted that for $0 < b < 1.5$ approximately, the characteristic roots are imaginary; moreover, the real part, that is, $-b + b^2 < 0 \Leftrightarrow b < 1$ and the claim follows. •

We are now ready to show that:

Claim 15.2.8 For the system (15.7), the unique equilibrium $(\theta, 1)$ is globally stable whenever $b < 1$. When $b > 1$ any solution with an arbitrary non-equilibrium initial point is unbounded.

Proof: Consider the function:

$$W(p_1, p_2) = 2(1 - b)p_1 + (2 - b)p_1^2/2 - b \log p_1 + p_2^2/2 - \log p_2$$

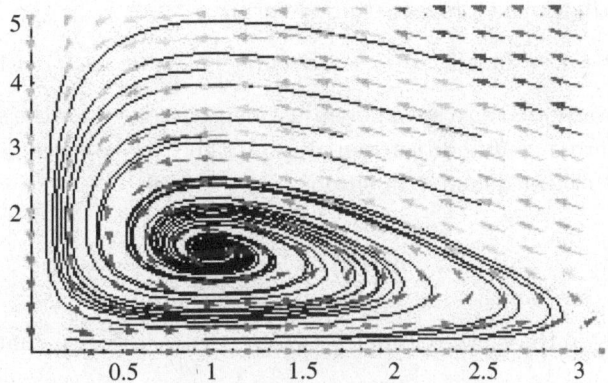

FIGURE 15.6   Scarf Example Stabilized

Then consider the derivative of the function $W(.,.)$ along any solution to the system (15.7), we have:

$$\dot{W} = \{((2-b)p_1 - b)(1 + p_1)\}\frac{\dot{p}_1}{p_1} + (p_2^2 - 1)\frac{\dot{p}_2}{p_2}$$

$$= -(1 - p_2)^2 \frac{p_1(1 - b)}{p_2(p_1 + p_2)} < 0$$

whenever $b < 1$ and $p_2 \neq 1$. We may now conclude that the function $W(p_1, p_2)$ is a Liapunov function for the system and the first part of the claim follows. For the remaining part note that whenever $b > 1$, $\dot{W} \geq 0$ along any solution. The main point of interest about the function $W(p_1, p_2)$ is that it is a **strictly convex function** with an absolute minimum at (p_1^*, p_2^*). Suppose then that some solution remains bounded and hence, limit points exist; consequently, along such a solution, $W(t)$ will be mono-tonically non-decreasing and bounded too and hence convergent and thus $\dot{W}$ must converge to zero; one may conclude that any limit point for the bounded solution must be the equilibrium and consequently, $W(t)$ is **non-decreasing and converges** to its minimum value: thus $W(t)$ must be constant and the only possibility for a bounded solution is that it must begin from the equilibrium, as claimed (See Figure 15.6).●

Thus, an easy stability condition for the Scarf example is that $b < 1$; just as, for a meaningful equilibrium to exist, we need to have $b < 2$, a more stringent requirement has to be placed on the magnitude of b to ensure global stability. More importantly, it is clearly demonstrated that income effects need not necessarily be the villain of the piece. In the Scarf example, there are no substitution effects, yet it is possible to have global convergence.

15.2.3 Discrete price adjustment: preliminary difficulties

Among economists, the problems that discreteness, *per se* introduces, it seems to me, are not well appreciated. To exhibit the problem, we consider first of all, a discrete version of the tatonnement process studied earlier:

$$p_j(t+1) = \max\{0, p_j(t) + \gamma Z_j(p(t))\} \ \forall j \neq n \ , \ p_n(t) = 1 \forall t \qquad (15.8)$$

where $\gamma > 0$ is interpreted as a speed of adjustment, assumed for the sake of simplicity, to be constant across commodities/markets. Further, to ease our demonstration, we assume that the excess demand functions $Z_j(.)$ exhibit the property of GS for all $p\Re^n_{++}$, that is,

$$\frac{\partial Z_j(p)}{\partial p_k} > 0 \ \forall j \neq k$$

and for all $p \in R^n_{++}$. Under the property of GS, it is well-known that

  (i) There is a unique $p* = (p*_1, \cdots, p*_{n-1}, 1) \in R^n_{++}$ such that $Z(p*) = 0$.
  (ii) For every $p \in R^n_{++}, p \neq p*$, $p*.Z(p) > 0$ (WARP holds).

The continuous version of the process (15.8) converges to the unique equilibrium and the demonstration is relatively straightforward, using the fact that the function $V(p) = \|p - p*\|^2 = \sum_j (p_j - p*_j)^2$ is a Liapunov function. It would be instructive to try to see what we can say about the iterates of the process (15.8).

One may show the following:

Claim 15.2.9 *If γ is sufficiently small, then for any $\epsilon > 0$, there is an integer t^o such that $V(p(t+1)) \leq V(p(t))$ for $0 \leq t < t^o$ and $V(p(t)) \leq \epsilon$ for $t \geq t^o$.*

Proof: Now consider the following:

$$\left. \begin{array}{l} \sum_{j \neq n} p_j^2(t+1) \leq \sum_{j \neq n} p_j^2(t) + 2\gamma \sum_{j \neq n} p_j(t) Z_j(p(t)) + \gamma^2 \sum_{j \neq n} Z_j^2(p(t)) \\ -2 \sum_{j \neq n} p*_j \, p_j(t+1) \leq -2 \sum_{j \neq n} p*_j \, p_j(t) - 2\gamma \sum_{j \neq n} p*_j \, Z_j(p(t)) \end{array} \right\} (15.9)$$

Also note that by definition, we have

$$V(p(t+1)) = \sum_{j \neq n} p_j^2(t+1) - 2 \sum_{j \neq n} p_j(t+1) p*_j + \sum_{j \neq n} p*_j^2$$

so that using the equation (15.9) and using Walras Law, we have:

$$V(p(t+1)) \leq V(p(t)) + \gamma[\gamma \sum_{j \neq n} Z_j^2(p(t)) - \sum_{j=1}^{n} p*_j Z_j(p(t))$$

Next let, for any $\epsilon > 0$, $K = \max[\epsilon, V(p^o)]$ where p^o denotes the initial price; and further choose:

$$\gamma < \min\{\min_{p \in B_\epsilon} \frac{p*.Z(p)}{\sum_{j \neq n} Z_j(p)^2}, \ \min_{p \in C_\epsilon} \sqrt{\frac{\epsilon/2}{\sum_{j \neq n} |Z_j(p)|}}\}$$

where $B_\epsilon = \{p \in R^n_{++} : K \geq V(p) \geq \epsilon/2\}$ and $C_\epsilon = \{p \in R^n_{++} : V(p) \leq \epsilon/2\}$. Note that such a choice of γ can be made and taken to be positive. This follows, by the continuity of the functions and the sets being compact over which positive minimum may be shown to exist. Next note that for such a choice of γ, we have:

(a) if $p(t) \in B_\epsilon$ then $V(p(t+1)) < V(p(t))$

(b) if $p(t) \in C_\epsilon$ then $V(p(t+1)) \leq \epsilon$.

Since $V(p^o) \leq K$, it follows that $V(p(t)) \leq K, \forall t$ and hence the sequence of prices $p(t)$ is bounded and limit points exist: let $\hat{p}$ be a limit point such that $V(\hat{p}) \leq V(\bar{p})$ for any other limit point $\bar{p}$. It may be shown that $V(\hat{p}) \leq \epsilon/2$; to see this, let $p(t_s) \to \hat{p}$ while $p(t_s+1) \to \bar{p}$ say; then it follows that $\bar{p}_j = \max\{0, \hat{p}_j + \gamma Z_j(\hat{p})\}$ for all $j \neq n$; thus $V(\hat{p}) > \epsilon/2 \Rightarrow V(\bar{p}) < V(\hat{p})$; hence we must have $V(\hat{p}) \leq \epsilon/2$; hence the claim follows. ●

It should be easy to see that choosing the appropriate γ is a difficult task indeed; for that we need to have pretty exhaustive information about the price space; in fact, it would appear that we must know the equilibrium itself. In that case, we need not worry about processes such as the one constructed above. If knowledge about the equilibrium and the price space is less than perfect, and if we may only be able to guess what the speed of adjustment should be, what happens to the convergence question? We discuss this and related questions in the next section.

There is another point which is perhaps noteworthy about the example considered below. The fact that non-linear discrete processes may give rise to problems has not been properly appreciated by economists (as we mentioned above), has to do with the fact that the examples considered were mainly the logistic equation mentioned in Section 3.1. What we show below is that similar situations can be found within models which are very familiar to economists.

15.2.4 Bifurcation and complex dynamics in a discrete tatonnement

Consider an exchange model involving two individuals A and B, and two goods x and y. A's preferences are given by $x^\alpha y^{1-\alpha}$ where $0 < \alpha < 1$ and B's preferences are given by $x^\beta y^{1-\beta}$ where $0 < \beta < 1$; further A has the endowment $(x^\circ, 0)$ and B has the endowment $(0, y^\circ)$ where x°, y° are both assumed to be positive. To continue with our analysis, standard calculations lead to the following excess demand function for good x, $Z(p)$, where p is the price of x relative to y (see Figure 15.7)

$$Z(p) = \frac{\beta y^\circ}{p} - (1-\alpha)x^\circ \tag{15.10}$$

Equilibrium $p^\star$ is then uniquely determined by

$$p^\star = \frac{\beta y^\circ}{(1-\alpha)x^\circ}. \tag{15.11}$$

Consider the standard adjustment on prices in disequilibrium, the tatonnement, for the model described above:

$$p(t+1) = p(t) + \gamma Z(p(t)) \tag{15.12}$$

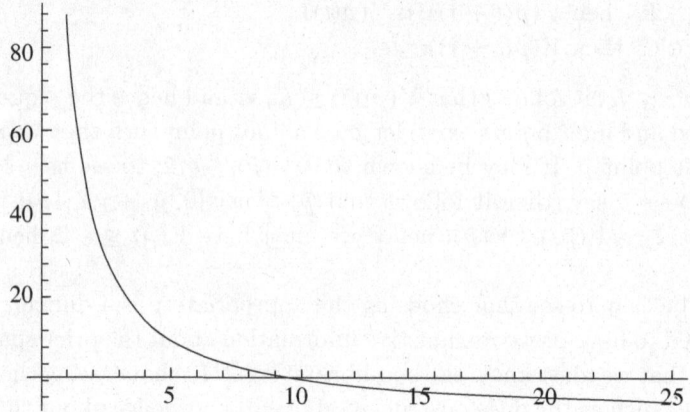

FIGURE 15.7   The Excess Demand Function

where $\gamma > 0$ is some speed of adjustment, assumed constant. We can rewrite the equation (15.12) as $p(t+1) = f(p(t))$ where

$$f(p) = p + \gamma Z(p) = p + \gamma(\frac{\beta y^\circ}{p} - (1-\alpha)x^\circ);$$

note that at $p = \overline{p} = \sqrt{\gamma \beta y^\circ}$, $f(.)$ attains a minimum value which is given by:

$$f(\overline{p}) = 2\sqrt{\gamma \beta y^\circ} - \gamma(1-\alpha)x^\circ$$

and for the adjustment process to be well defined for all values of p, we shall require that the parameter values are so defined that $f(p) > 0$ for all p; this is ensured by requiring that the minimum value, $f(\overline{p})$, is positive, that is,

$$4 > \frac{\gamma((1-\alpha)x^\circ)^2}{\beta y^\circ} \qquad (15.13)$$

Let us, for future reference, define

$$K = \frac{\gamma((1-\alpha)x^\circ)^2}{\beta y^\circ}$$

It follows that

$$K = \frac{(\overline{p})^2}{(p^\star)^2}$$

Consequently, $\overline{p} \le p^\star \Leftrightarrow K \le 1$. Figure 15.8 depicts the nature of the function $f(p)$ when $K > 1$.

We first note that the following.

Claim 15.2.10   There is an interval $I = [a,b]$ such that $f : I \to I$.

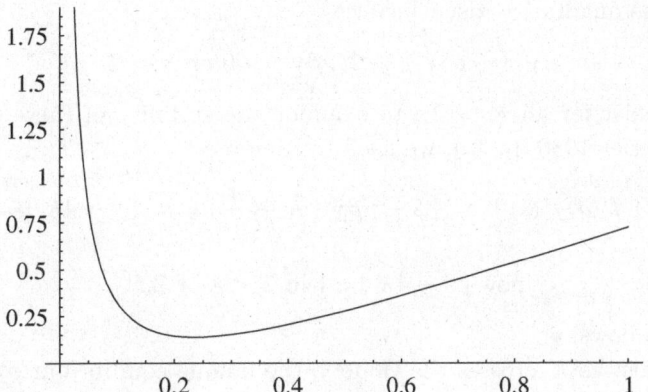

FIGURE 15.8   The f-Map with K > 1

Proof: Consider the case when $K > 1$; that is, $\overline{p} > p^\star$. Let $b = Max[f(f(\overline{p})), \overline{p}]$; $a = f(\overline{p})$ Since $\overline{p} > p^\star$, and $f'(p) < 0$ $\forall p < \overline{p}$, note that

$$a = f(\overline{p}) < f(p^\star) = p^\star < \overline{p} \leq b$$

Consider $I = [a,b]$; clearly, $p^\star \in I$; consider $p \in I$ and suppose, if possible, that $f(p) \notin I$.

$$\text{Since} f(p) \geq f(\overline{p}) = a; f(p) \notin I \Rightarrow f(p) > b.$$

That is, we must have $Z(p) > 0$ or $f(p) > p$ or $a = f(\overline{p}) \leq p < p^\star < \overline{p}$; hence $b \geq f(f(\overline{p})) \geq f(p) > b$: a contradiction. Hence, no such p can exist and the claim is established. A similar construction for I may be provided when $K \leq 1$. •

Since $f'(p^\star) = 1 - K$, we have the following:

Claim 15.2.11 $K < 2 \Rightarrow p^\star$ *is locally stable for the process (15.12); $p^\star$ is locally stable for the process (15.12)* $\Rightarrow K \leq 2$.

Given above, we shall investigate what happens when $2 < K < 4$.
It should be next pointed out that:

Claim 15.2.12 *For $2 < K < 2.5$ there exists a stable 2-cycle.*

Proof: For a 2-cycle to exist, we need to show that there are q, s such that $q \neq s$ satisfying $q = f(s)$ and $s = f(q)$, that is,

$$q = s + \gamma(\frac{\beta y^\circ}{s} - (1 - \alpha)x^\circ) \text{ and } s = q + \gamma(\frac{\beta y^\circ}{q} - (1 - \alpha)x^\circ)$$

Thus $2sq = \gamma\beta y^\circ$ and $s + q = \gamma(1 - \alpha)x^\circ$ or s, q are roots of the quadratic equation

$$z^2 - \gamma(1 - \alpha)x^\circ z + \frac{\gamma\beta y^\circ}{2} = 0$$

Real roots of the quadratic exist whenever

$$(\gamma(1-\alpha)x^\circ)^2 - 2(\gamma\beta y^\circ) > 0 \text{ or } K > 2$$

Thus 2-cycles exist for all $K > 2$; to examine the stability of these cycles (see, for example, Lauwerier 1986, p. 40), we need to examine

$$\mid f'(q)f'(s) \mid \; = \; \mid 5 - \gamma\beta y^\circ(\frac{1}{s^2} + \frac{1}{q^2}) \mid \; = \; \mid 9 - 4K \mid;$$

$$\text{now} \mid 9 - 4K \mid < 1 \text{ iff } 2 < K < 2.5$$

and the claim follows. •

Note that just as K crosses the value 2, the unique equilibrium p^* loses stability and a stable 2-cycle is born; the 2-cycle loses stability just as K crosses the value 2.5 and a stable 4-cycle may be shown to exist. To analyse the behaviour of the attractors for different values of K, we fix the values of all the parameters except γ:

$$\beta y^\circ = 1 \text{ and } (1-\alpha)x^\circ = 6;$$

so that $K = 36\gamma$ and our difference equation (15.12) takes the particular form

$$p(t+1) = p(t) + (\frac{1}{p(t)} - 6)K/36 \tag{15.14}$$

Let a_n denote the critical value of K where a 2^n cycle is born; then $a_1 = 2$ and $a_2 = 2.5$. It is known that the sequence a_n converges to a_∞ as n becomes large and further, for any one dimensional iterate, with a single parameter, there is a constant, F (the Feigenbaum constant = 4.669202...) which allows us to approximate the value of a_∞ by means of the following (See Lauwerier 1986, p. 44–5.):

$$a_\infty \approx \frac{Fa_{n+1} - a_n}{F - 1}$$

Using the above formula, one may see that for the example we have been discussing,

$$a_\infty \approx 2.6362694$$

To properly appreciate the nature of the dynamics, please see the bifurcation diagram (Figure 15.9).

In Figure 15.9, on the x-axis we have values of the parameter K; on the vertical axis, we have plotted the iterates $f^n(x)$, say for $n = 200$ to $n = 300$ for a point chosen at random from the corresponding [a,b]. This experimental method generates the period doubling phenomenon mentioned above; beginning with a 2-period cycle, bifurcating into 4-period cycles and so on, till a_∞; beyond this value, apart from small gaps the picture is diffused; within the gaps, we have stable cycles. Outside these regions of stable cycles, we may conclude that either there is no stable cycle or that there are cycles with very long periods: the reason we are not able to capture these may be attributed to the fact that we have looked at only a few iterates. This is only to be expected from running such experiments. The above exercise reveals the

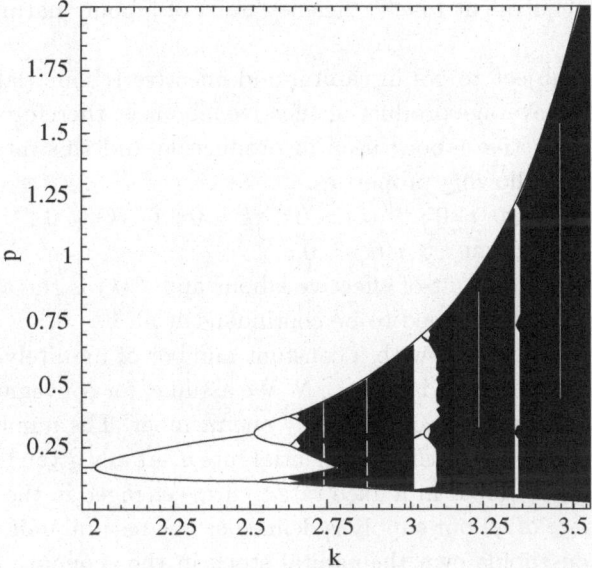

FIGURE 15.9 The Bifurcation Diagram

complexity in discrete price adjustment processes. For a more detailed consideration, the interested reader may consult Mukherji (1999).

15.3 Optimal Economic Growth: Ramsey-Cass-Koopmans Model

The Ramsey-Cass-Koopmans growth model is a decentralized version of the optimal growth model due to Ramsey (1928), Cass (1965), and Koopmans (1965).

15.3.1 The outlines of the model

Consider a closed economy with a single producible good and labour. The good can either be consumed or stocks of the good, constituting the capital stock of the economy, can be used in production. Stocks of the good are subject to depreciation at a constant exponential rate δ. Labour is homogenous in quality at any given point of time. The government plays no role (at least explicitly) as regards economic activity.

The single good in the economy is produced using labour and existing stocks of the good itself (capital). Production of the single good is subject to purely labour-augmenting technical progress at a constant exponential rate μ. This means that the nature of technical progress is such that the effect of using one natural (physical) unit of labour at any time $t \geq 0$ is the same as that of using $e^{\mu t}$ natural units of labour at the initial time 0. The latter is, therefore, taken as a measure of the efficiency of a natural unit of labour in production at time t and each natural unit of labour at time

t is assumed to be equivalent to $e^{\mu t}$ efficiency units of labour (natural units of labour at time 0).

Production is subject to crs in capital and effective labour (labour measured in efficiency units). The average product of effective labour is, therefore, a function of the ratio of capital to effective labour used in production and this intensive production function satisfies the following properties:

$\forall k \geq 0 : f(k) \geq 0$; $\forall k > 0 : f'(k) > 0$; $\forall k > 0 : f''(k) < 0$;
$\lim_{k \to 0} f'(k) = \infty$; $\lim_{k \to \infty} f'(k) = 0$;

where, k is capital per unit of effective labour and $f(k)$ is the average product of effective labour. $f''(k)$ is assumed to be continuous at all $k > 0$.

The economy is populated with a constant number of infinitely-lived households. Let the households be numbered 1, 2, ..., N. We assume, for convenience of exposition, that at time 0 each household contains only one member. The number of members of each household grows at a constant exponential rate n. At any given time, each member of a household supplies labour at a fixed rate to firms engaged in the production of the single good. This rate of labour supply is defined as one natural unit of labour per unit time. Similarly, households own the capital stock in the economy and lease out the entire stock to firms for use in production. Firms aim to maximize their profits at each point in time. Note that the production technology for the good is such that capital and labour can be combined in any proportion and both capital and labour have positive but diminishing marginal products. Therefore, assuming that factor markets are perfectly competitive (households and firms act as price takers), equilibrium factor rentals equal respective marginal products and factors are fully employed. Moreover, given that production of the single good is subject to crs, in equilibrium, every firm produces using the same ratio of capital to effective labour.

Let $c(t)$ denote the rate of consumption per efficiency unit of labour and $k(t)$ the amount of capital per efficiency unit of labour in the economy at time $t \geq 0$. The initial amount of capital per unit of effective labour in the economy $k(0)$ is positive. Let $w(t)$ denote the wage per efficiency unit of labour and $r(t)$ the rental price of capital net of depreciation at time $t \geq 0$. Then, for any $t \geq 0$, $w(t) = f(k(t)) - k(t).f'(k(t))$ and $r(t) = f'(k(t)) - \delta$.

At any point in time households can borrow units of the good from other households against their future wage earnings. Households have perfect foresight about the rate of growth in the size of each household and about the time paths of the wage rate and the rate of interest in the economy. In this economy, at any point in time the present discounted value of the stream of future wage earnings of any household is assumed to be a finite sum. Therefore, households can calculate the maximum amount of net liabilities which any other household can accumulate, if that household has to, eventually repay its debts out of its future earnings and cannot indulge in Ponzi games (use chain finance schemes to roll over its debt indefinitely). Households also have perfect knowledge about the existing net liabilities of other households. Under these conditions, any given household will lend to another household only if it is certain to receive repayment of its principal with interest. Therefore households will be indifferent between holding their accumulated savings as capital or as outstanding loans to other households. In equilibrium, therefore, the rate of interest in the economy will always be equal to the net rate of rental for capital. Therefore, at any time $t \geq 0$, the rate of interest in the economy is equal to $r(t)$. The present discounted value of the stream of future wage earnings of any household at time t per efficiency unit of labour supplied by the household at time t is denoted by $W(t)$ and is equal to $\int_t^\infty w(\tau) e^{(n+\mu)(\tau-t)} e^{-\int_t^\tau r(v) dv} d\tau$.

Let $c_i(t)$ and $a_i(t)$, respectively, denote the rate of consumption per efficiency unit of labour and the amount of wealth per efficiency unit of labour of household i ($i = 1, 2, ..., N$) at time $t \geq 0$. Wealth (or net worth) of a household is equal to the amount of capital owned by it less its net outstanding debts. Note that $c(t) = (1/N) \sum_{i=1}^N c_i(t)$ and $k(t) = (1/N) \sum_{i=1}^N a_i(t)$ for all $t \geq 0$.

Given that a household will lend to another household only if the present discounted value of the latter's future wage earnings is greater than its existing net liabilities, it follows that for all $t \geq 0$,

$$W(t) \geq -a_i(t) \tag{15.15}$$

The wealth of household i at any time $t \geq 0$ is equal to $a_i(t).e^{(n+\mu)t}$. Since $r(t)$ is the rate of return on the assets of the household (capital and outstanding loans given to other households) and the rate of payment on its liabilities (outstanding debts) therefore $r(t).a_i(t)e^{(n+\mu)t}$ is the household's rate of earnings from its wealth at time t. Its rate of wage earnings and its rate of consumption at time t are respectively $w(t).e^{(n+\mu)t}$ and $c_i(t).e^{(n+\mu)t}$. Since, the rate of change of wealth for the household at any time is equal to its rate of saving, therefore, it follows that for all $t \geq 0$,

$$\dot{a}_i(t) = w(t) + \{r(t) - n - \mu\} a_i(t) - c_i(t)$$

Therefore, for all $\tau \geq 0$,

$$\int_0^\tau \frac{d}{dt} \{a_i(t).e^{-\int_0^t \{r(v)-n-\mu\} dv}\} dt = \int_0^\tau w(t) e^{-\int_0^t \{r(v)-n-\mu\} dv} dt - \int_0^\tau c_i(t) e^{-\int_0^t \{r(v)-n-\mu\} dv} dt$$

That is, for all $\tau \geq 0$,

$$\int_0^\tau c_i(t) e^{-\int_0^t \{r(v)-n-\mu\} dv} dt = a_i(0) - a_i(\tau).e^{-\int_0^\tau \{r(v)-n-\mu\} dv} + \int_0^\tau w(t) e^{-\int_0^t \{r(v)-n-\mu\} dv} dt \tag{15.16}$$

Given (15.15) it, therefore, follows that for all $\tau \geq 0$,

$$\int_0^\tau c_i(t) e^{-\int_0^t \{r(v)-n-\mu\} dv} dt \leq a_i(0) + \int_0^\tau w(t) e^{-\int_0^t \{r(v)-n-\mu\} dv} dt + W(\tau) e^{-\int_0^\tau \{r(v)-n-\mu\} dv}$$

That is, for all $\tau \geq 0$,

$$\int_0^\tau c_i(t) e^{-\int_0^t \{r(v)-n-\mu\} dv} dt \leq a_i(0) + W(0) \tag{15.17}$$

Since the lhs in the inequality is a monotonic function of τ which is bounded above, therefore, $\lim_{\tau \to \infty} \int_0^\tau c_i(t) e^{-\int_0^t \{r(v)-n-\mu\}dv} dt$ exists. From (8.2) it therefore follows that

$$\lim_{\tau \to \infty} a_i(\tau).e^{-\int_0^\tau \{r(v)-n-\mu\}dv} = a_i(0) + W(0) - \int_0^\infty c_i(t) e^{-\int_0^t \{r(v)-n-\mu\}dv} dt \qquad (15.18)$$

and from (15.17) it follows that

$$\int_0^\infty c_i(t) e^{-\int_0^t \{r(v)-n-\mu\}dv} dt \le a_i(0) + W(0)$$

Hence, it follows that

$$\lim_{\tau \to \infty} a_i(\tau).e^{-\int_0^\tau \{r(v)-n-\mu\}dv} \ge 0$$

The above inequality will be referred to as the No Ponzi game condition.[4]

The rate of utility enjoyed by household i $(i = 1, 2, ..., N)$ at instant $t \ge 0$ is a function of the rate of consumption per member of the household $c_i(t).e^{\mu t}$. The instantaneous utility function remains unchanged over time and is characterized by a constant elasticity of marginal utility (constant coefficient of relative risk aversion).[5] More specifically, the instantaneous utility function takes one of the following forms:

$$u(c_i(t)e^{\mu t}) = \frac{\{c_i(t)e^{\mu t}\}^{1-\theta}}{1-\theta} (\theta > 0, \theta \ne 1),$$

$$\text{or, } u(c_i(t)e^{\mu t}) = \ln\{c_i(t)e^{\mu t}\}$$

The elasticity of marginal utility for the two forms of the instantaneous utility function are θ and 1, respectively.

At any time $t \ge 0$, household i $(i = 1, 2, ..., N)$ decides on its rates of consumption and saving by formulating a consumption plan for the household time t onwards. The consumption plan is chosen to maximize the present discounted value at time t of the stream of instantaneous utility received by the household time t onwards. The rate of utility obtained at future instants is discounted by the household at a constant exponential rate $\rho > 0$. It is assumed that $\rho > (1-\theta)\mu$. In choosing its consumption plan the household is constrained by its existing wealth at time t and by the current and future values of the wage rate and the rate of interest in the economy. Given that at every instant the rate of change in the wealth of the household is equal to its rate of earnings less its rate of consumption, any feasible consumption plan of the household must be associated with a time path of household wealth which satisfies the No Ponzi game condition. Therefore, for every $i \in \{1, 2, ..., N\}$ and for every $\tau \ge 0$, household i

---

[4] The derivation of the above condition has been carried out on the assumption that $w(t), r(t)$, and $c_i(t)$ are continuous functions and $a_i(t)$ is a continuously differentiable function. However, the same condition can also be derived if $c_i(t)$ is a piece-wise continuous function and the differential equation defining $\dot{a}_i(t)$ is valid only at points of continuity of $c_i(t)$.

[5] This is necessary for a steady state growth path to exist for the model economy.

solves the following optimization problem at time τ:

$$\max_{c_i(t)t\in[\tau,\infty)} \int_\tau^\infty \frac{\{c_i(t)e^{\mu t}\}^{1-\theta}}{1-\theta}e^{-\rho(t-\tau)}dt$$

$$\text{s.t. } \dot{a}_i(t) = w(t) + \{r(t) - n - \mu\}a_i(t) - c_i(t)$$

$$a_i(\tau)\text{given} \lim_{t\to\infty} a_i(t).e^{-\int_\tau^t \{r(v)-n-\mu\}dv} \geq 0$$

$$c_i(t) \geq 0$$

In order to be able to apply the theory of optimal control to the above problem we will assume that any feasible plan for household consumption per efficiency unit of labour $c_i : [\tau, \infty) \to \Re_+$ must be a piece-wise continuous function of time and must be associated with a unique continuous and piece-wise continuously differentiable time path of household wealth per efficiency unit of labour $a_i : [\tau, \infty) \to \Re$. We will assume that $w(t)$ and $r(t)$ are continuous functions and the reader can verify that $\{ce^{\mu s}\}^{1-\theta}e^{-\rho(s-\tau)}/(1-\theta)$ and $w(s) + \{r(s) - n - \mu\}a - c$ are continuous functions of (a, c, s) on $\Re \times \Re_+ \times [0, \infty)$ with continuous partial derivatives with respect to a. The resulting optimization problem then assumes the features of an optimal control problem where the household's rate of consumption per efficiency unit of labour c_i is the control variable, the household's wealth per efficiency unit of labour a_i is the state variable, any consumption plan for the household $c_i : [\tau, \infty) \to \Re_+$ is a control for the household and the associated time path for household wealth per efficiency unit of labour $a_i : [\tau, \infty) \to \Re$ is the associated trajectory for the state variable.

Since $\{ce^{\mu s}\}^{1-\theta}/(1-\theta)$ is not a bounded function of (c, s) on $\Re_+ \times [0, \infty)$ it is difficult to verify directly whether the objective functional is defined for every admissible control. We will, therefore, initially proceed under the assumption that the objective functional may not be defined for every admissible control. Accordingly, we will assume that every household chooses its optimal consumption plan using the catching-up criterion for optimality. Any optimal control selected by this criterion yields the maximum value of the objective functional when the improper integral corresponding to the value of the objective functional either exists for all admissible controls or exists for some admissible controls and diverges to $-\infty$ for some others.

Note that in stating the household optimization problem, we have used the general form of an instantaneous utility function with constant elasticity of marginal utility where the elasticity θ is not equal to unity. Also, since we allow for the possibility that $c_i(t) = 0$ for any $t \geq 0$ it is necessary to assume that $\theta < 1$. Otherwise the instantaneous utility function is incompletely specified since the rate of utility is undefined for a zero rate of consumption. If we allow $\theta > 1$ then the constraint $c_i(t) \geq 0$ in the problem must be replaced with $c_i(t) > 0$. However, the subsequent analysis of the aggregate dynamics of the economy remains exactly the same if $\theta > 1$ or if the constant elasticity of marginal utility is equal to unity provided that in the latter case we replace θ, wherever it occurs, with 1.

Suppose (a_i^*, c_i^*) is an optimal pair for household i's ($i = 1, , ..., N$) optimization problem at time 0. Note that for the household's optimization problem at any time $\tau > 0$ with $a_i(\tau) = a_i^*(\tau)$, $c_i^* : [\tau, \infty) \to R_+$ must be an optimal consumption plan with the associated time path of wealth $a_i^* : [\tau, \infty) \to R$.[6] This implies that optimal consumption plans chosen by the household at different points in time will be consistent and the actual time paths of consumption and wealth for the household will be identical with the optimal plans for consumption and wealth chosen by the household at time 0. Therefore, in order to determine the time paths of consumption and wealth for household i we simply need to solve the household's optimization problem at time 0. This is given by:

$$\max_{c_i(t)t\in[0,\infty)} \int_0^\infty \frac{\{c_i(t)e^{\mu t}\}^{1-\theta}}{1-\theta} e^{-\rho t} dt$$
$$\text{s.t. } \dot{a}_i(t) = w(t) + \{r(t) - n - \mu\}a_i(t) - c_i(t)$$
$$a_i(\tau) = a_i^0$$
$$\lim_{t\to\infty} a_i(t).e^{-\int_0^t \{r(v)-n-\mu\}dv} \geq 0$$
$$c_i(t) \geq 0$$

where a_i^0 is the given initial wealth per efficiency unit of labour of household i. We shall see later that a necessary condition for an optimal consumption plan to exist for all households is that $a_i^0 + W(0) > 0$ for all $i \in \{1, 2, ..., N\}$. The implication is that the initial level of indebtedness of no household should be such that the repayment of initially existing debts exhausts the entire future wage earnings of the household.

15.3.2 Solution to the optimal control problem for the household (using the maximum principle for infinite horizon problems)

Suppose $c_i^* : [0, \infty) \to R_+$ is an optimal consumption plan for household i according to the catching-up criterion for optimality with associated trajectory for household wealth per efficiency unit of labour $a_i^* : [0, \infty) \to R$. Then, the Maximum Principle tells us that there must exist a positive constant λ_0 and a continuous function $\lambda_i : [0, \infty) \to R$ such that

$$\lambda_0 = 0 \text{ or } \lambda_0 = 1 \tag{15.19}$$

$$\text{If } \lambda_0 = 0, \text{ then for all } t \geq 0, \lambda_i(t) \neq 0 \tag{15.20}$$

---

[6] In particular, this is due to the assumption of a constant rate of time preference ρ which ensures that the value of the objective functional of the problem at time τ for any feasible consumption plan is proportional to the contribution that would be made by that plan to the value of the objective functional at time 0.

Except at points of discontinuity of c_i^*, for all $t \geq 0$,

$$\dot{\lambda}_i(t) = -\left[\frac{\partial H(a, c, s, \lambda)}{\partial a}\right]_{(a,c,s,\lambda)=(a_i^*(t),c_i^*(t),t,\lambda_i(t))} \tag{15.21}$$

For all $t \geq 0$ and for all $c \geq 0$,

$$H\left(a_i^*(t), c_i^*(t), t, \lambda_i(t)\right) \geq H\left(a_i^*(t), c, t, \lambda_i(t)\right) \tag{15.22}$$

where $H(a, c, s, \lambda) = \lambda_0\left[\{ce^{\mu s}\}^{1-\theta}e^{-\rho s} / (1-\theta)\right] + \lambda\left[w(s) + \{r(s) - n - \mu\}a - c\right]$

From (15.22), it follows that, if $\lambda_0 = 0$, then for all $t \geq 0$, $\lambda_i(t)[w(t) + \{r(t) - n - \mu\}a_i^*(t) - c]$ is maximized on $[0, \infty)$ at $c = c_i^*(t)$.

Note that for any $t \geq 0$ this implies that:

If $\lambda_0 = 0$ and $\lambda_i(t) < 0$ then (15.22) cannot be satisfied for any value of $c_i^*(t)$

$$\tag{15.23}$$

If $\lambda_0 = 0$ and $\lambda_i(t) > 0$ then (15.22) is satisfied for $c_i^*(t) = 0$. $\tag{15.24}$

It follows from (15.20) and (15.22) to (15.24) that

$$\text{If } \lambda_0 = 0 \text{ then } c_i^*(t) = 0 \text{ for all } t \geq 0. \tag{15.25}$$

We shall now demonstrate that $c_i(t) = 0$, for all $t \geq 0$, cannot be an optimal consumption plan for the household. For any admissible consumption plan $c_i : [0, \infty) \to \Re_+$ and its associated trajectory $a_i : [0, \infty) \to R$ we know that the following conditions hold: Except at points of discontinuity of c_i, for all $t \geq 0$,

$$\dot{a}_i(t) = w(t) + \{r(t) - n - \mu\}a_i(t) - c_i(t) \tag{15.26}$$

$$a_i(0) = a_i^0 \tag{15.27}$$

$$\lim_{t \to \infty} a_i(t) e^{-\int_0^t \{r(v)-n-\mu\}dv} \geq 0 \tag{15.28}$$

From (15.26), given that c_i is a piece-wise continuous function and a_i is a continuous function, it can be shown that (15.16) is satisfied. From (15.27) and (15.28) it then follows that

$$\int_0^\infty c_i(t) e^{-\int_0^t \{r(v)-n-\mu\}dv} dt = a_i^0 - \lim_{t \to \infty} a_i(t) e^{-\int_0^t \{r(v)-n-\mu\}dv} + W(0) \tag{15.29}$$

Consider a household consumption plan $\hat{c}_i(t) = \beta\left(a_i^0 + W(0)\right) e^{\int_0^t \{r(v)-n-\mu-\beta\}dv}$, for all $t \geq 0$, where β is a positive constant. Define a trajectory $\hat{a}_i : [0, \infty) \to R$ for household wealth per efficiency unit of labour from (15.26) and (15.27) by setting $c_i(t) = \hat{c}_i(t)$, for all $t \geq 0$. Both $\hat{c}_i$ and $\hat{a}_i$ are continuous on $[0, \infty)$. It can be checked that the pair $(\hat{a}, \hat{c})$ satisfies (15.26) to (15.28) with $\lim_{t \to \infty} \hat{a}_i(t) e^{-\int_0^t \{r(v)-n-\mu\}dv} = 0$. $(\hat{a}, \hat{c})$ is, therefore, an admissible pair for the household's optimization problem.

Given that $a_i^0 + W(0) > 0$, it follows that $\hat{c}_i(t) > 0$, for all $t \geq 0$.

Therefore, $\int_0^T [\{\hat{c}_i(t) e^{\mu t}\}^{1-\theta} e^{-\rho t} / (1-\theta)]dt$ is a positive and strictly increasing function of T for all $T > 0$. If $c_i^*(t) = 0$ for all $t \geq 0$ then $\int_0^T [\{c_i^*(t) e^{\mu t}\}e^{-\rho t} / (1-\theta)]dt =$

0 for all $T \geq 0$. Then, given that $\hat{c}_i$ is an admissible consumption plan, c_i^* cannot be optimal by the catching-up criterion.

Therefore, from (15.25) it follows that $\lambda_0 \neq 0$. From (15.20) it follows that $\lambda_0 = 1$. Therefore, the Hamiltonian for the household's optimization problem is given by

$$H(a, c, s, \lambda) = \left[\{ce^{\mu s}\}^{1-\theta} e^{-\rho s} / (1 - \theta) \right] + \lambda \left[w(s) + \{r(s) - n - \mu\}a - c \right].$$

We have seen that it cannot be the case that $c_i^*(t) = 0$, for all $t \geq 0$. We shall now prove that for all $t \geq 0$, $c_i^*(t) > 0$.

Suppose this is not true. Then there exists $\bar{t} \geq 0$ such that $c_i^*(\bar{t}) = 0$. Then it follows that $H\left(a_i^*(\bar{t}), c_i^*(\bar{t}), \bar{t}, \lambda_i(\bar{t})\right) = 0 + \lambda_i(\bar{t})\left[w(\bar{t}) + \{r(\bar{t}) - n - \mu\}a_i^*(\bar{t}) - 0\right]$.

Clearly, if $\lambda_i(\bar{t}) \leq 0$ then for any $c > 0$, $H\left(a_i^*(\bar{t}), c, \bar{t}, \lambda_i(\bar{t})\right) = \left[\{ce^{\mu \bar{t}}\}^{1-\theta} e^{-\rho \bar{t}} / (1 - \theta)\right] + \lambda_i(\bar{t})\left[w(\bar{t}) + \{r(\bar{t}) - n - \mu\}a_i^*(\bar{t}) - c\right] > H(a_i^*(\bar{t}), c_i^*(\bar{t}), \bar{t}, \lambda_i(\bar{t}))$.

Then (15.22) is violated.

Suppose $\lambda_i(\bar{t}) > 0$.

Then, $H\left(a_i^*(\bar{t}), c, \bar{t}, \lambda_i(\bar{t})\right) - H\left(a_i^*(\bar{t}), c_i^*(\bar{t}), \bar{t}, \lambda_i(\bar{t})\right) = \left[\{ce^{\mu \bar{t}}\}^{1-\theta} e^{-\rho \bar{t}} / (1 - \theta)\right] - \lambda_i(\bar{t})c$.

Note that $c^{-\theta} \to \infty$ as $c \to 0$. Therefore, for sufficiently small positive c, $c^{1-\theta}/c > (1 - \theta)\lambda_i(\bar{t})e^{\{\rho - (1-\theta)\mu\}\bar{t}}$. Hence, (15.22) is again violated. It follows that there cannot exist $t \geq 0$ such that $c_i^*(t) = 0$.

Hence, for all $t \geq 0$,

$$c_i^*(t) > 0 \tag{15.30}$$

Since the rates of consumption in the optimal consumption plan must always lie in the interior of the set $\{c | c \geq 0\}$ (the control region for the household's optimization problem) and since the functions $\{ce^{\mu s}\}^{1-\theta} e^{-\rho s} / (1 - \theta)$ and $\lambda[w(s) + \{r(s) - n - \mu\} a - c]$ have partial derivatives with respect to c in $\{c | c > 0\}$, (15.22) implies that for all $t \geq 0$,

$$\left[\frac{\partial H(a, c, s, \lambda)}{\partial c}\right]_{(a,c,s,\lambda) = \left(a_i^*(t), c_i^*(t), t, \lambda_i(t)\right)} = 0 \tag{15.31}$$

From (15.31) it follows that $\lambda_i(t) = c_i^*(t)^{-\theta} e^{\{(1-\theta)\mu - \rho\}t}$, for all $t \geq 0$. That is, $c_i^*(t) = \lambda_i(t)^{-1/\theta} e^{(1/\theta)\{(1-\theta)\mu - \rho\}t}$, for all $t \geq 0$. Note that $c_i^*(t) > 0$ for all $t \geq 0$ and $\lambda_i(t)$ is a continuous function. Therefore, $\lambda_i(t) > 0$ for all $t \geq 0$ and $c_i^*(t)$ is a continuous function. Hence, from (15.21) and (15.26) it follows that for all $t \geq 0$,

$$\dot{a}_i^*(t) = w(t) + \{r(t) - n - \mu\}a_i^*(t) - c_i^*(t) \tag{15.32}$$

$$\dot{\lambda}_i(t) = -\lambda_i(t)\{r(t) - n - \mu\} \tag{15.33}$$

Substituting for $\lambda_i(t)$ in (15.33) it follows that for all $t \geq 0$,

$$\frac{\dot{c}_i^*(t)}{c_i^*(t)} = \frac{1}{\theta}\{r(t) - \rho - n - \theta \mu\} \tag{15.34}$$

Therefore, for all $t \geq 0$,

$$c_i^*(t) = c_i^*(0) e^{(1/\theta) \int_0^t \{r(v) - \rho - n - \theta \mu\} dv} \tag{15.35}$$

Note that if c_i^* is an optimal consumption plan then it must also be an admissible consumption plan satisfying (15.26) to (15.28) and, therefore, (15.29). From (15.29) and (15.35) it can be shown that for all $t \geq 0$,

$$c_i^*(t) = \frac{a_i^0 - \lim_{t \to \infty} a_i^*(t)\, e^{-\int_0^t \{r(v)-n-\mu\}dv} + W(0)}{\int_0^\infty e^{\int_0^t [\{(1/\theta)-1\}\{r(v)-n\}-(\rho/\theta)]dv}\, dt}\, e^{(1/\theta)\int_0^t \{r(v)-\rho-n-\theta\mu\}dv}$$

Since c_i^* is an admissible consumption plan the value of $\lim_{t \to \infty} a_i(t)\, e^{-\int_0^t \{r(v)-n-\mu\}dv}$ must lie in the interval $[0, a_i^0 + W(0)]$. Note that the smaller is the value of

$$lim_{t \to \infty} a_i(t)\, e^{-\int_0^t \{r(v)-n-\mu\}dv}$$

the greater is the value of $c_i^*(t)$ for any $t \geq 0$. Given the household's instantaneous utility function, the marginal instantaneous utility of consumption is always positive. Clearly, if c_i^* is the optimal consumption plan then

$$\lim_{t \to \infty} a_i^*(t)\, e^{-\int_0^t \{r(v)-n-\mu\}dv} = 0 \tag{15.36}$$

We have proved that if an optimal consumption plan $c_i^* : [0, \infty) \to R_+$ exists for household i then it must be given by:

$$c_i^*(t) = \frac{a_i^0 + W(0)}{\int_0^\infty e^{\int_0^t [\{(1/\theta)-1\}\{r(v)-n\}-(\rho/\theta)]dv}\, dt}\, e^{(1/\theta)\int_0^t \{r(v)-\rho-n-\theta\mu\}dv}, \quad \text{for all } t \geq 0 \tag{15.37}$$

Note that a unique consumption plan for the household is determined by using the Maximum Principle for infinite horizon problems even though the Maximum Principle does not provide us with any transversality condition. This is because, given the identity relating to the rate of change in household wealth and given the assumption of positive marginal utility of consumption, the No Ponzi game condition translates into the boundary condition (15.36) on the state variable.

15.3.3 The solution using Arrow-type sufficiency conditions

We know that for any optimal consumption plan of household i ($i = 1, 2, ..., n$), $c_i(t) > 0$ for all $t \geq 0$. Therefore, for any optimal consumption plan of the household, $\{(a_i(t), c_i(t), t) | t \in [0, \infty)\} \subseteq R \times (0, \infty) \times [0, \infty)$. The control region for the household's optimization problem is $[0, \infty)$ and $R \times (0, \infty) \times [0, \infty)$ is an open set in $R \times [0, \infty) \times [0, \infty)$. The functions $(ce^{\mu s})^{1-\theta} e^{-\rho s}$ and $w(s) + \{r(s) - n - \mu\}a - c$ are continuous at all $(a, c, s) \in \Re \times [0, \infty) \times [0, \infty)$ and the partial derivatives of these functions with respect to a and c are continuous at all $(a, c, s) \in R \times (0, \infty) \times [0, \infty)$. The Arrow-type sufficiency conditions are, therefore, applicable for the household's optimization problem (see discussion of the Arrow-type sufficiency conditions).

From (15.32) it follows that the trajectory $a_i^* : [0, \infty) \to R$ associated with the control c_i^*, defined by (15.37) is given by

$$a_i^*(t) e^{-\int_0^t \{r(v)-n-\mu\}dv} = a_i^0 + \int_0^t \{w(\tau) - c_i^*(\tau)\} e^{-\int_0^\tau \{r(v)-n-\mu\}dv} d\tau, \text{ for all } t \geq 0.$$

(15.38)

Also, we know that there exists a continuous function $\lambda_i : [0, \infty) \to R$ such that

$$\lambda_i(t) = c_i^*(t)^{-\theta} e^{\{(1-\theta)\mu-\rho\}t} \text{ for all } t \geq 0. \tag{15.39}$$

If $H(a, c, s, \lambda) = [\{ce^{\mu s}\}^{1-\theta} e^{-\rho s} / (1-\theta)] + \lambda[w(s) + r(s)a - c]$ then for all $t \geq 0$, (15.21) and (15.31) hold.

Also, for all $c > 0$ and for all $t \geq 0$,

$$\left[\frac{\partial^2 H(a, c, s, \lambda)}{\partial c^2} \right]_{(a,c,t,\lambda)=(a_i^*(t),c,t,\lambda_i(t))} = -\theta c^{-\theta-1} e^{\{(1-\theta)\mu-\rho\}t} < 0$$

Therefore, given that (15.31) holds, for all $t \geq 0$, $H\left(a_i^*(t), c, t, \lambda_i(t)\right)$ is maximized on $[0, \infty)$ at $c = c_i^*(t)$, that is, (15.22) is satisfied. Note that in the previous section it was proved that $H(a_i^*(t), c, t, \lambda_i(t))$ cannot be maximized on $[0, \infty)$ at $c = 0$ for any $t \geq 0$.

Since (15.21) is satisfied, for all $t \geq 0$, $\dot{\lambda}_i(t) / \lambda_i(t) = -\{r(t) - n - \mu\}$.

Therefore, $\lambda_i(t) = \lambda_i(0) e^{-\int_0^t \{r(v)-n-\mu\}dv}$, for all $t \geq 0$.

We know that $\lambda_i(t) > 0$, for all $t \geq 0$, $\lim_{t\to\infty} a_i(t) e^{-\int_0^t \{r(v)-n-\mu\}dv} \geq 0$ for any admissible pair (a_i, c_i) in the household's optimization problem and $\lim_{t\to\infty} a_i^*(t) e^{-\int_0^t \{r(v)-n-\mu\}dv} = 0$.

Therefore, for any trajectory $a_i : [0, \infty) \to R$ associated with an admissible control,

$$\lim_{t\to\infty} \lambda_i(t) \left[a_i(t) - a^*(t)\right] = \lim_{t\to\infty} \lambda_i(0) \left[a(t) - a^*(t)\right] e^{-\int_0^t \{r(v)-n-\mu\}dv} \geq 0 \tag{15.40}$$

For all $(a, c, t) \in R \times [0, \infty) \times [0, \infty)$,
$H(a, c, t, \lambda_i(t)) = \{(ce^{\mu t})^{1-\theta} e^{-\rho t} / (1-\theta)\} + \lambda_i(t)[w(t) + r(t)a - c]$.

Given that $\lim_{c\to 0} c^{-\theta} = \infty$, it follows that for all $a \in R$ and for all $t \geq 0$, any value of c which maximizes $H(a, c, t, \lambda_i(t))$ on $[0, \infty)$ lies in the open interval $(0, \infty)$.

For all $a \in R$ and for all $t \geq 0$, given that $\lambda_i(t) > 0$, $\partial H(a, c, t, \lambda_i(t))/\partial c = 0$ implies that $c = \lambda_i(t)^{-1/\theta} e^{(1/\theta)\{(1-\theta)\mu-\rho\}t} > 0$ and $\partial^2 H(a, c, t, \lambda_i(t))/\partial c^2 = -\theta c^{-\theta-1} e^{\{(1-\theta)\mu-\rho\}t} < 0$ for all $c > 0$.

Therefore, for all $a \in R$ and for all $t \geq 0$, $H(a, c, t, \lambda_i(t))$ is maximized on $[0, \infty)$ at $c = \lambda_i(t)^{-1/\theta} e^{(1/\theta)\{(1-\theta)\mu-\rho\}t}$

Let, for all $a \in \Re$ and for all $t \geq 0$, $H^0(a, t, \lambda_i(t)) = \max_{c\in[0,\infty)} H(a, c, t, \lambda_i(t)) = [\{\lambda_i(t)^{-1/\theta} e^{(1/\theta)\{(1-\theta)\mu-\rho\}t} e^{\mu t}\}^{1-\theta} e^{-\rho t} / (1-\theta)] + \lambda_i(t)[w(t) + \{r(t) - n - \mu\}a - \lambda_i(t)^{-1/\theta} e^{(1/\theta)\{(1-\theta)\mu-\rho\}t}]$.

Note that for all $a \in R$ and for all $t \geq 0$, $\partial^2 H^0(a, t, \lambda_i(t))/\partial a^2 = 0$.

Therefore, for all $(a, t) \in R \times [0, \infty)$,

$$H^0(a, t, \lambda_i(t)) = \max_{c \in [0, \infty)} H(a, c, t, \lambda_i(t)) \text{ exists and is concave in } a. \qquad (15.41)$$

Thus, (a_i^*, c_i^*), defined by (15.37) and (15.38), is an admissible pair for household i's optimization problem satisfying (15.26) to (15.28) and $\lambda_i : [0, \infty) \to \Re$, defined by (15.39), is a continuous function such that (15.21), (15.22), (15.40), and (15.41) are satisfied. Hence, c_i^* satisfies the Arrow-type sufficiency conditions for optimality according to the catching-up criterion in household i's optimization problem (see Section 14.7.1). Note that c_i^* is the unique optimal consumption plan for household i because it is also the only admissible consumption plan which satisfies the Maximum Principle for household i's optimization problem.

15.3.4 Aggregate dynamics in the model

From (15.30), (15.32), (15.34), and (15.36) it follows that the time paths of consumption per efficiency unit of labour $c : [0, \infty) \to \Re_+$ and capital stock per efficiency unit of labour $k : [0, \infty) \to \Re$ satisfy the following conditions for all $t \geq 0$:

$$\frac{\dot{c}(t)}{c(t)} = \frac{1}{\theta}\{r(t) - \rho - n - \theta\mu\}$$

$$\dot{k}(t) = w(t) + \{r(t) - n - \mu\}k(t) - c(t)$$

$$\lim_{t \to \infty} k(t)\, e^{-\int_0^t \{r(v) - n - \mu\}dv} = 0$$

$$c(t) > 0 \qquad (15.42)$$

Since, for all $t \geq 0$, $w(t) = f(k(t)) - k(t)f'(k(t))$ and $r(t) = f'(k(t))$ therefore for all $t \geq 0$,

$$\dot{c}(t) = \frac{c(t)}{\theta}\{f'(k(t)) - \delta - \rho - n - \theta\mu\} \qquad (15.43)$$

$$\dot{k}(t) = f(k(t)) - (\delta + n + \mu)k(t) - c(t) \qquad (15.44)$$

$$\lim_{t \to \infty} k(t)\, e^{-\int_0^t \{f'(k(v)) - \delta - n - \mu\}dv} = 0 \qquad (15.45)$$

From (15.43) and (15.44) it follows that $k = k(t)$, $c = c(t)$ must be a solution on $[0, \infty)$ to the following two-dimensional system of differential equations in (k, c):

$$\dot{k} = f(k) - (\delta + n + \mu)k - c$$

$$\dot{c} = \frac{c}{\theta}\{f'(k) - \delta - \rho - n - \theta\mu\}$$

$$(k > 0, \ c \geq 0)$$

Moreover, (15.42) and (15.45) are satisfied. Also, $k : [0, \infty) \to (0, \infty)$ must be such that $w(t) = f(k(t)) - k(t)f'(k(t))$ and $r(t) = f'(k(t)) - \delta$ are continuous functions and the improper integral $W(0)$ exists (the present discounted value of the stream of household wage earnings is a finite sum). Note that there cannot exist $t \geq 0$ such

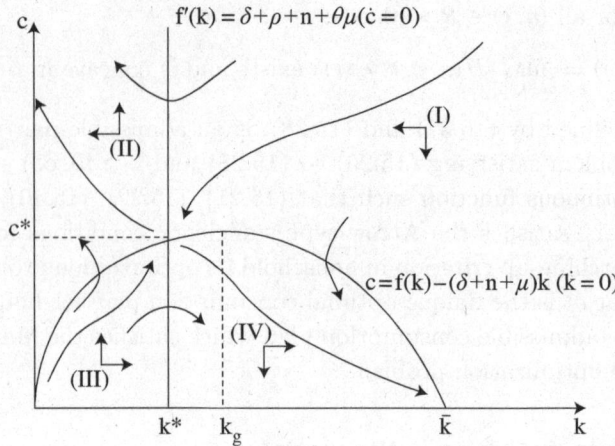

FIGURE 15.10   The Ramsey-Cass-Koopmans Model

that $k(t) = 0$ since $f'(k(t))$ must be a continuous function of t and $\lim_{k \to 0} f'(k)$ $= \infty$.[7]

In Figure 15.10, the hump-shaped curve is the graph of the equation $c = f(k) -$ $(\delta + n + \mu)k$ and represents on the k-c plane the locus of all possible combinations of k and c for which $\dot{k} = 0$. The shape of the curve is determined by the properties of the intensive production function f.[8] Given the assumption that $\rho > (1 - \theta)\mu$, these properties also imply that there exist unique values $k_g > 0$ and $k^* > 0$ such that $f'(k_g) = \delta + n + \mu$, $f'(k^*) = \delta + \rho + n + \theta\mu$ and $k_g > k^*$. The curve $c = f(k) - (\delta + n + \mu)k$, therefore, attains its maximum height at $k = k_g$ and the graph of the equation $f'(k) = \delta + \rho + n + \theta\mu$ is given by the vertical line at $k = k^*$. Note that the part of the graph for $f'(k) = \delta + \rho + n + \theta\mu$ lying on or above the horizontal axis ($c \geq 0$) and the part of the graph for the equation $c = 0$ lying to the right of the origin ($k > 0$) give the locus on the k-c plane of all possible combinations of k and c for which $\dot{c} = 0$. From the diagram it is evident that there are only two points on the k-c plane at which both $\dot{k} = 0$ and $\dot{c} = 0$. These are the points (k^*, c^*), where $c^* = f(k^*) - (\delta + n + \mu)k^*$, and $(\bar{k}, 0)$, where $f(\bar{k}) = (\delta + n + \mu)\bar{k}$. These correspond to the two equilibria for the above dynamic system.

---

[7] Note also that even if we assume $\lim_{k \to 0} f'(k) = f'(0)$ is finite (and greater than $\delta + \rho + n + \theta\mu$), if there exists a value of t such that $k(t) = 0$ and $f(k(t)) = 0$ then a function $c : [0, \infty) \to R_+$ satisfying (15.46) cannot represent a feasible time-path of consumption in the model.

[8] These properties also imply $f(0) = 0$ as evident from the fact that the curve passes through the origin. See Barro and Sala-i-Martin (2003), p. 77.

The Jacobian for the system when evaluated at (k^*, c^*) and $(\bar{k}, 0)$ is respectively equal to

$$\begin{pmatrix} \rho - (1 - \theta)\mu & -1 \\ (c^*/\theta)f''(k^*) & 0 \end{pmatrix}$$

and

$$\begin{pmatrix} f'(\bar{k}) - (\delta + n + \mu) & -1 \\ 0 & (1/\theta)\{f'(\bar{k}) - \delta - \rho - n - \theta\mu\} \end{pmatrix}$$

Given that $c^* > 0$ and $f''(k^*) < 0$, it follows that the determinant of the Jacobian evaluated at (k^*, c^*) is negative. Therefore, the characteristic roots of the Jacobian evaluated at (k^*, c^*) are real and of opposite sign and the equilibrium (k^*, c^*) is an SP. Since $k^* < k_g < \bar{k}$, therefore, $f'(\bar{k}) < \delta + n + \mu < \delta + \rho + n + \theta\mu$. Thus, the trace and determinant of the Jacobian evaluated at $(\bar{k}, 0)$ are negative and positive respectively and it is easily shown that the square of the trace is greater than four times the value of the determinant. This implies that the characteristic roots of the Jacobian evaluated at $(\bar{k}, 0)$ are real and negative and the equilibrium $(\bar{k}, 0)$ is a stable node.

At any point on the k-c plane with $k > 0$ and $c \geq 0$, $\dot{k} \lesseqgtr 0$ accordingly as $c \gtreqless f(k) - (\delta + n + \mu)k$. Also, at any point on the k-c plane with $k > 0$ and $c > 0$, $\dot{c} \lesseqgtr 0$ accordingly as $k \gtreqless k^*$. In Figure 15.10, the first quadrant of the k-c plane is, therefore, divided into four regions (I) to (IV) by the curve $c = f(k) - (\delta + n + \mu)k$ along which $\dot{k} = 0$ and the line $k = k^*$ along which $\dot{c} = 0$. Within each region, the directions of motion of k and c (the signs of $\dot{k}$ and $\dot{c}$) are the same at every point. The directions of motion of k and c in each region are represented by the direction of the horizontal and vertical arrows in the region. Along the curves separating any two regions, the variable which has the same direction of motion in both regions maintains its direction, while the variable which has different directions of motion in the two regions has a zero rate of change with respect to time. In the figure, the curve with an arrowhead passing through any given point represents the time path of (k, c) if from the given point we move either forward or backward through time. The part of the curve which, beginning from the given point, ends in the arrowhead represents the time path of (k, c) as we move forward through time. These directed curves are known as the **phase paths** of the dynamic system. The growth path of the economy in the model will, therefore, be described by the phase path passing through $(k(0), c(0))$.

Since the equilibrium (k^*, c^*) is an SP it can be shown that there is exactly one phase path in (I) and exactly one phase path in (III) along which (k, c) converges asymptotically to (k^*, c^*). All other phase paths in the plane diverge away from (k^*, c^*). These two phase paths are known as the **stable branches or arms of the SP** (k^*, c^*). Let us denote the equations for these two phase paths as $c = \psi_1(k)$, $k^* < k < \infty$, and $c = \psi_2(k)$, $0 < k < k^*$. From the theory of differential equations we know that because the first order partial derivatives of the functions $f(k) - (\delta + n + \mu)k - c$ and $(c/\theta)\{f'(k) - \delta - \rho - n - \theta\mu\}$ are continuous at all $k > 0$ and $c \geq 0$, there cannot be two phase paths passing through any given point. Since $c = 0$ is the phase path passing through any point $(k, 0)$ such that $k > 0$, it follows that for all $k > 0$, $\psi_2(k) > 0$.

Note that if the initial ratio of capital to effective labour $k(0)$ is equal to k^* then $c(0)$ may be less than, equal to or greater than c^*. If $c(0) = c^*$ then because (k^*, c^*) is an equilibrium for the dynamic system, it must be the case that $k(t) = k^*$ and $c(t) = c^*$ for all $t \geq 0$. This implies that the values of all endogenous aggregate quantity variables (aggregate capital stock, aggregate consumption, aggregate output) per unit of efficiency labour remains constant in the economy. This implies that the values of all such variables will be growing at the constant rate $n + \mu$ which is the rate of growth of the labour force measured in efficiency units and the per capita values of these variables will be growing at the constant rate μ, the rate of labour-augmenting technical progress. The wage per efficiency unit of labour and the rental price of capital will be constant along this growth path so that the distribution of income between capital and labour will also remain constant. The growth path of the economy will, therefore, be a steady state or balanced growth path. In this case, given that $\rho > (1 - \theta)\mu$, all the conditions (15.42) to (15.45) will be satisfied by $c : [0, \infty) \to R_+$ and $k : [0, \infty) \to R$ as required. $w(t)$ and $r(t)$ will be constant functions and the improper integral $W(0)$ will exist. If $c(0) < c^*$ then the phase path passing through $(k(0), c(0))$ will enter and remain in region (IV) and if $c(0) > c^*$ then the phase path passing through $(k(0), c(0))$ will enter and remain in region (II).

If $k(0) > k^*$ then $c(0)$ may be less than, equal to or greater than $\psi_1(k(0))$. If $c(0) = \psi_1(k(0))$ then $(k(t), c(t))$ must converge asymptotically to (k^*, c^*) along the stable branch of the SP in (I). That is, the economy will converge asymptotically to the steady state growth path with capital-effective-labour ratio k^*. Given $\rho > (1 - \theta)\mu$ it is again possible to verify that all the conditions (15.42)–(15.45) are satisfied by the time paths of consumption per efficiency unit of labour and capital per efficiency unit of labour $c : [0, \infty) \to R_+$ and $k : [0, \infty) \to R$ as required. Again, $w(t)$ and $r(t)$ will be continuous functions and the improper integral $W(0)$ will exist. If $c(0) < \psi_1(k(0))$ then the phase path through $(k(0), c(0))$ will either lie in or enter and remain in (IV) and if $c(0) > \psi_1(k(0))$ then the phase path through $(k(0), c(0))$ will enter and remain in (II).

Similarly, if $k(0) < k^*$ then it is possible that $c(0)$ will be less than, equal to, or greater than $\psi_2(k(0))$. If $c(0) = \psi_2(k(0))$ then $(k(t), c(t))$ will converge asymptotically to (k^*, c^*) along the stable branch of the SP in (III). Again, the economy will converge asymptotically to the steady state growth path with capital-effective-labour ratio k^*. As required, (15.42) to (15.45) will again be satisfied by $c : [0, \infty) \to R_+$ and $k : [0, \infty) \to R$, $w(t)$ and $r(t)$ will be continuous functions and the improper integral $W(0)$ will exist. If $c(0) < \psi_2(k(0))$ then the phase path through $(k(0), c(0))$ will enter and remain in (IV) and if $c(0) > \psi_2(k(0))$ then the phase path through $(k(0), c(0))$ will either lie in or will enter and remain in (II).

Consider the phase paths in region (IV). Note that the equilibrium $(\bar{k}, 0)$ is a stable node, no two phase paths touch each other and $c = 0$ is itself a phase path along which (k, c) converges asymptotically to $(\bar{k}, 0)$. Therefore, along every phase path in region (IV), (k, c) converges asymptotically to $(\bar{k}, 0)$. However, since $f'(\bar{k}) < \delta + n + \mu$, therefore, if $c(0)$ is such that $k(t)$ converges asymptotically to $\bar{k}$ then $\lim_{t\to\infty} k(t) e^{-\int_0^t \{f'(k(v))-\delta-n-\mu\}dv} = \infty$ (condition (15.45) is violated) and the improper integral $W(0)$ does not exist. Hence, it follows that the value of $c(0)$ can never be such that the phase path passing through $(k(0), c(0))$ will either lie in or enter and remain in (IV).

Suppose $c(0)$ is such that the phase path through $(k(0), c(0))$ either lies in or enters and remains in (II). Then, for sufficiently large $t > 0$, $\dot{k}(t) < 0$ and $\ddot{k}(t) = \{f'(k(t)) - (\delta+n+\mu)\}\dot{k}(t) - \dot{c}(t) < 0$. This implies that there exists a value of $t > 0$ for which $k(t) = 0$. It follows that the value of $c(0)$ cannot be such that the phase path through $(k(0), c(0))$ either lies in or enters and remains in (II).

Hence, if $k(0) = k^*$ then the economy is always on a steady state growth path with $k(t) = k^*$ and $c(t) = c^*$ for all $t \geq 0$ and if $k(0) \neq k^*$ then $k(t)$ and $c(t)$ are both increasing or both decreasing functions of time and the economy converges asymptotically to a steady state growth path with $\lim_{t\to\infty} k(t) = k^*$ and $\lim_{t\to\infty} c(t) = c^*$.

Finally, note from (15.35) that for all $i \in \{1, 2, ..., N\}$,

$$\lim_{T\to\infty} \int_0^T \frac{\{c_i^*(t)e^{\mu t}\}^{1-\theta}}{1-\theta} e^{-\rho t} dt$$

$$= \frac{\{c_i^*(0)\}^{1-\theta}}{1-\theta} \lim_{T\to\infty} \int_0^T e^{\{(1/\theta)-1\}\int_0^t \{r(v)-\rho-n-\theta\mu\}dv} e^{-\{\rho-(1-\theta)\mu\}t} dt$$

$$= \frac{1}{1-\theta} \left\{\frac{c_i^*(0)}{c(0)}\right\}^{1-\theta} \lim_{T\to\infty} \int_0^T \{c(t)\}^{1-\theta} e^{-\{\rho-(1-\theta)\mu\}t} dt$$

From the above discussion we know that if $0 < k(0) \leq k^*$, then $c(t) \in [\psi_2(k(0)), c^*]$ for all $t \geq 0$ and if $k(0) > k^*$ then $c(t) \in [c^*, \psi_1(k(0))]$ for all $t \geq 0$. Remember that $\psi_2(k) > 0$ for all $k > 0$. Therefore, $\{c(t)\}^{1-\theta}$ is a positive-valued function which is bounded above. Given that $\rho > (1-\theta)\mu$, it then follows that $\int_0^T \{c(t)\}^{1-\theta} e^{-\{\rho-(1-\theta)\mu\}t} dt$ is an increasing function of T and is bounded above. Hence, the value of the objective functional is defined for household i's optimal consumption plan $c_i^* : [0, \infty) \to \Re_+$ according to the catching-up criterion. Now, note that $\int_0^T [\{c_i(t)\}^{1-\theta}/(1-\theta)]e^{-\{\rho-(1-\theta)\mu\}t} dt$ is a monotonic function of T for every admissible consumption plan $c_i : [0, \infty) \to R_+$ of household i. If there exists any admissible consumption plan c_i for which $\lim_{T\to\infty} \int_0^T [\{c_i(t)\}^{1-\theta}/(1-\theta)]e^{-\{\rho-(1-\theta)\mu\}t} dt = \infty$, c_i^* cannot be the optimal consumption plan for household i according to the catching-up criterion. It follows that for any admissible consumption plan c_i either the value of the objective functional is defined or $\lim_{T\to\infty} \int_0^T [\{c_i(t)\}^{1-\theta}/(1-\theta)]e^{-\{\rho-(1-\theta)\mu\}t} dt = -\infty$. Hence, for every household $i \in \{1, 2, ..., N\}$, the optimal consumption plan according to the catching-up criterion c_i^*, defined by (15.37), also maximizes the value of the objective functional on the set of admissible controls.

15.4 The Social Planner's Problem: An Alternative Approach

Consider again a single-good economy. The good can be consumed or stocks of the good can be used (along with homogenous labour) as capital in the production of the good itself. Suppose the total number of units of labour used for production in the economy in any given period is equal to the size of the population. 'Gross output' of the good in the economy in any period is defined as equal to the total number of new units of the good produced in the period plus capital at the beginning of the period less depreciation in the period. Let per capita gross output in any period $t \geq 0$ be a function $f(k(t))$ of the per capita capital stock $k(t)$ in the economy in period t. Assume that the function $f : R_+ \to R_+$ satisfies the following properties:

$$f(0) = 0; (\forall k)(k \in R_{++} \to f'(k) > 0 \wedge f''(k) < 0); \lim_{k \to 0} f'(k) = \infty; \lim_{k \to \infty} f'(k) = 0$$

Gross output in any period $t \geq 0$ can be utilized for consumption in the period or set aside as the stock of capital for the (next) period $t + 1$. If population grows at a constant compound rate of n per period then

$$c(t) + (1 + n)k(t + 1) \leq f(k(t))$$

where $c(t)$ denotes per capita consumption in period t. Obviously, in any period $t \geq 0$, $c(t) \geq 0$.

Given an initial capital stock per capita $k_0 > 0$, any feasible path of per capita capital stock $\{k(t)\}_{t=0}^{\infty}$ in the economy must be such that for all $t \geq 0$, $0 \leq (1 + n)k(t + 1) \leq f(k(t))$ and $k(0) = k_0$. Note that, given the properties of the intensive production function f, the output–capital ratio $f(k)/k$ is a strictly decreasing function of k for all $k > 0$, $\lim_{k \to 0} f(k)/k = \infty$ and $\lim_{k \to \infty} f(k)/k = 0$. Therefore, there exists a unique value of k, say $\bar{k}$, such that $f(k)/k \gtreqless (1 + n)$ according as $k \lesseqgtr \bar{k}$. If $k_0 \geq \bar{k}$ then along any feasible time-path for per capita capital stock, for all $t \geq 0$, $(1 + n)k(t + 1) \leq f(k(t)) < (1 + n)k(t)$ if $k(t) > \bar{k}$. This implies that if $k_0 \geq \bar{k}$, the per capita capital stock is decreasing so long as it is greater than $\bar{k}$ and must either converge asymptotically to $\bar{k}$ or become less than or equal to $\bar{k}$ in finite time. *In this discussion we will consider the case where $k_0 < \bar{k}$.* In this case, along any feasible time path for per capita capital stock, $k(t) \leq \bar{k}$ for all $t \geq 0$, because given that $f(k)$ is increasing in k there cannot exist $t \geq 0$ such that $f(k(t)) \geq (1 + n)k(t + 1) > (1 + n)\bar{k} = f(\bar{k})$.

Suppose social welfare in the economy is a function $\sum_{t=0}^{\infty} \beta^t u(c(t))$ of the time-path of per capita consumption $c : \mathbb{Z}_+ \to R_+$ in the economy, where $\beta \in (0, 1)$. Assume that the function $u : R_+ \to R$ satisfies the following conditions:

$$(\forall c)(c \in R_{++} \to u'(c) > 0 \wedge u''(c) < 0); \lim_{c \to 0} u'(c) = \infty.$$

In this section, we first prove that there exists a unique growth path which maximizes social welfare in the economy and then discuss the nature of this optimal growth path.

Note that, since $\lim_{c \to 0} u'(c) = \infty$ and $u'(c) > 0$ for all $c > 0$ it follows that along any growth path maximizing social welfare in the economy: $c(t) + (1+n)k(t+1) = f(k(t))$, for all $t \geq 0$.

Since per capita output in any period is determined completely by per capita capital stock in that period, an optimal growth path for output in the economy is associated with an optimal time-path for per capita capital stock. The latter can be thought of as an optimal plan for a benevolent social planner faced with the following decision-making problem:

$$\max_{k:\mathbb{Z}_+ \to [0,k]} \sum_{t=0}^{\infty} \beta^t u(f(k(t)) - (1+n)k(t+1))$$

$$\text{s.t. } k(t+1) \in \left[0, \frac{f(k(t))}{1+n}\right]$$

$$k(0) = k_0$$

We can compare the social planner's problem with Problem (D) in Chapter 14. In the social planner's problem, $\{x(t)\}_{t=0}^{\infty} = \{k(t)\}_{t=0}^{\infty}$, $X = \left[0, \bar{k}\right]$, $\Gamma(x_I) = [0, f(x_I)/(1+n)]$ for all $x_I \in \left[0, \bar{k}\right]$, $A = \{(x_I, x_o)|x_I \in \left[0, \bar{k}\right] \wedge x_o \in [0, f(x_I)/(1+n)]\}$ and $F(x_I, x_o) = u(f(x_I) - (1+n)x_o)$ for all $(x_I, x_o) \in A$.

$f(x_I)$ is a continuous function of x_I on R_+ and $(1+n)x_o$ is a continuous function of x_o on R. Therefore, $f(x_I) - (1+n)x_o$ is a continuous function on A with values in R_+. Given that u is a continuous function on R_+ it follows that $u(f(x_I) - (1+n)x_o)$ is a continuous function on A. Also, $k(t) \leq \bar{k}$ for all $t \geq 0$, along any feasible trajectory for per capita capital stock. Therefore, $u(f(x_I) - (1+n)x_o) \leq u(f(\bar{k}))$ for all $(x_I, x_o) \in A$. Thus, for the social planner's problem, the current net return function $F(x_I, x_o) = u(f(x_I) - (1+n)x_o)$ is a bounded and continuous function on A. Further, $\Gamma(x_I) = [0, f(x_I)/(1+n)]$ is a non-empty, closed, and bounded set in R for all $x_I \in \left[0, \bar{k}\right]$, and $A = \{(x_I, x_o)|x_I \in \left[0, \bar{k}\right] \wedge x_o \in [0, f(x_I)/(1+n)]\}$ is a closed and convex set in R^2. Therefore, a bounded and continuous value function $V : \left[0, \bar{k}\right] \to R$ exists for the social planner's problem satisfying the functional equation (14.27) in Chapter 14. That is,

$$V(x_I) = \max_{x_o \in \left[0, \frac{f(x_I)}{1+n}\right]} [u(f(x_I) - (1+n)x_o) + \beta V(x_o)], \text{ for all } x_I \in \left[0, \bar{k}\right]$$

Therefore, it follows that there exists an optimal plan for the social planner's problem and an optimal growth path for the economy.

Since f and u are strictly concave functions and u is a strictly increasing function, $F(x_I, x_o) = u(f(x_I) - (1+n)x_o)$ is a strictly concave function. Therefore, from our discussion of the chain rule in Chapter 14, it follows that V is a strictly concave function and there exists a continuous policy function $g : \left[0, \bar{k}\right] \to \left[0, \bar{k}\right]$ for the social planner's problem. Therefore,

$$V(x_I) = u(f(x_I) - (1+n)g(x_I)) + \beta V(g(x_I)), \text{ for all } x_I \in \left[0, \bar{k}\right]$$

Note that the existence of the policy function implies that there is a unique optimal plan for the social planner's problem. Let $\{k^*(t)\}_{t=0}^{\infty}$ be the optimal plan.

For the social planner's problem, $int\ A = \{(x_I, x_o)|x_I \in (0, \bar{k}) \wedge x_o \in (0, f(x_I)/(1+n))\}$. Since u and f are twice-differentiable functions on $\Re_{++}$ and $f(x_I) - (1+n)x_o$ is a continuous function on $int\ A$ with values in R_{++}, therefore, for all $(x_I, x_o) \in int\ A$, the function $F(x_I, x_o)$ has a partial derivative with respect to x_I given by

$$\frac{\partial F(x_I, x_o)}{\partial x_I} = u'(f(x_I) - (1+n)x_o).f'(x_I)$$

and the function $u'(f(x_I) - (1+n)x_o).f'(x_I)$ is continuous at all $(x_I, x_o) \in int\ A$. From our discussion in Chapter 14 it follows that V is a continuously differentiable function in $(0, \bar{k})$ if for all $x_I \in (0, \bar{k})$, $g(x_I) \in (0, f(x_I)/(1+n))$. Since $\lim_{c \to 0} u'(c) = \infty$ and $f(0) = 0$, therefore, for all $x_I \in (0, \bar{k}]$, the optimal policy $g(x_I) \in (0, f(x_I)/(1+n))$. Thus, for all $x_I \in (0, \bar{k})$,

$$V'(x_I) = u'(f(x_I) - (1+n)g(x_I)).f'(x_I) \tag{15.46}$$

Given that $g(x_I) = \arg\max_{x_o \in [0, f(\bar{k})/(1+n)]} [u(f(x_I) - (1+n)x_o) + \beta V(x_o)]$ and $g(x_I) \in (0, f(x_I)/(1+n))$ for all $x_I \in (0, \bar{k})$, it follows that for all $x_I \in (0, \bar{k})$,

$$-(1+n)u'(f(x_I) - (1+n)g(x_I)) + \beta V'(g(x_I)) = 0 \tag{15.47}$$

Since $k_0 > 0$, $g(k^*(t)) = k^*(t+1) \in (0, f(k^*(t)/(1+n))$, for all $t \geq 0$. Further, since $k_0 < \bar{k}$ and $\bar{k}(1+n) = f(\bar{k}) > f(k)$ for all $k < \bar{k}$, therefore, $k^*(t) \in (0, \bar{k})$ for all $t \geq 0$. This implies that for all $t \geq 0$,

$$V'(k^*(t)) = u'(f(k^*(t)) - (1+n)g(k^*(t))).f'(k^*(t)) \tag{15.48}$$

and

$$-(1+n)u'(f(k^*(t)) - (1+n)g(k^*(t))) + \beta V'(g(k^*(t))) = 0 \tag{15.49}$$

We next prove that g is a strictly increasing function on $(0, \bar{k})$. Suppose this is not true. Let $x_I' \in (0, \bar{k})$, $x_I'' \in (0, \bar{k})$, $x_I' > x_I''$ and $g(x_I') \leq g(x_I'')$. Then, since $f'(k) > 0$ for all $k > 0$, $f(x_I') > f(x_I'')$. Therefore, since $u''(c) < 0$ for all $c > 0$,

$$V'(g(x_I')) = \frac{1+n}{\beta}u'(f(x_I') - (1+n)g(x_I')) < \frac{1+n}{\beta}u'(f(x_I'') - (1+n)g(x_I'')) = V'(g(x_I''))$$

Since V is a strictly concave function it follows that $g(x_I') > g(x_I'')$. Therefore, there is a contradiction, and our supposition cannot be true. Therefore, g is a strictly increasing function on $(0, \bar{k})$.

Since, g is a strictly increasing function on $(0, \bar{k})$ and $k^*(t) \in (0, \bar{k})$ for all $t \geq 0$, therefore $\{k^*(t)\}_{t=0}^{\infty}$ is bounded and is a strictly increasing sequence, a constant sequence, or a strictly decreasing sequence according to whether $g(k_0) \gtrless k_0$. Since any bounded and monotonic sequence in E is a convergent sequence, let $k_\infty \in [0, \bar{k}]$ such that $\lim_{t \to \infty} k^*(t) = k_\infty$. Since g is a continuous function it follows that the sequence

$\{g(k^*(t))\}_{t=0}^{\infty}$ is such that $\lim_{t\to\infty} g(k^*(t)) = g(k_\infty)$. Given that $k^*(t+1) = g(k^*(t))$ for all $t \geq 0$, it follows that $g(k_\infty) = k_\infty$. Since $g(\bar{k}) < f(\bar{k})/(1+n) = \bar{k}$, therefore, $k_\infty < \bar{k}$.

We next prove that $k_\infty > 0$. Suppose $k_\infty = 0$. Then, $\{k^*(t)\}_{t=0}^{\infty}$ is a strictly decreasing sequence and $\lim_{t\to\infty} k^*(t) = 0$. Further, from (15.48) and (15.49) it follows that for all $t \geq 0$,

$$\frac{u'(f(k^*(t)) - (1+n)k^*(t+1))}{u'(f(k^*(t+1)) - (1+n)k^*(t+2))} = \frac{\beta f'(k^*(t+1))}{1+n}$$

This is the Euler equation for the social planner's problem. Since $\lim_{k\to 0} f'(k) = \infty$ it follows that for sufficiently large values of t, $f'(k^*(t+1)) > (1+n)/\beta$. Therefore, since $u''(c) < 0$ for all $c > 0$, $f(k^*(t)) - (1+n)k^*(t+1) < f(k^*(t+1)) - (1+n)k^*(t+2)$ for sufficiently large values of t. It follows that for sufficiently large values of t, $f(k^*(t)) - (1+n)k^*(t+1)$ is positive and increasing in t. But, since $\lim_{t\to\infty} k^*(t) = 0$ and $f(0) = 0$, therefore, $\lim_{t\to\infty}[f(k^*(t)) - (1+n)k^*(t+1)] = 0$. Therefore, our supposition that $k_\infty = 0$ cannot be true and $k_\infty \in (0, \bar{k})$.

Since $g(k_\infty) = k_\infty \in (0, \bar{k})$, from (15.46) and (15.47) it follows that

$$f'(k_\infty) = \frac{1+n}{\beta}$$

Therefore, we can conclude that if $k_0 = k_\infty$, where k_∞ is defined by the above equation, the optimal per capita capital stock in the economy remains unchanged over time and the optimal growth path is a steady state along which both output and capital stock grow at the same rate as the labour force. If $k_0 \neq k_\infty$ then the optimal trajectory for per capita capital stock is such that per capita capital stock converges monotonically to k_∞, per capita output converges monotonically to $f(k_\infty)$ and along the optimal growth path the economy tends asymptotically towards a steady state.

Further Readings for Section III

For dynamical systems we may refer to Hirsch and Smale (1974); Hurewicz (1958) for continuous time processes; and Sandefur (1990) for discrete dynamical systems for elementary treatments. For a more advanced treatment, see Guckenheimer and Holmes (1983). For applications of Predator–Prey Models in Economics, see Mukherji (1992) and Mukherji (2005).

For applications to economics, Lorenz (1993) and Stokey and Lucas (1989) would provide a discussion of many topics. The discrete price adjustment process analysed is based on Mukherji (1999).

For the theory of optimal control, additional details are available from: Seierstad and Sydsaeter (1987), Grass et al., (2008), Arrow (1968), Arrow and Kurz (1970), and Hartl et al. (1995). For examples relating to economic applications of Optimal Control Theory: Leonard and Van Long (1992), Chiang (1992), and Caputo (2005).

For an analysis of growth theory, and possible further extensions of the model discussed here, see Dasgupta (2010), Barro and Sala-i-Martin (2003), Guha (2008); for development along other interesting lines, see Uzawa (1968), Epstein and Hynes (1983), and Das (2003). For topics, which we have only provided a brief introduction, that is, discrete processes and dynamic programming, but which form an influential section of the literature, Grandmont (1989) and Stokey and Lucas (1989), respectively should provide additional material. For a more recent reference, the comprehensive treatment of dynamic programming in Chapter 2 (Majumdar et al. 2000) by Tapan Mitra should be considered.

KEY TERMS

Bifurcation	Scarf Example
Discrete Price Adjustment	Social Planner's Problem
Gross Substitutes	Stable Competitive Equilibrium
Ramsey-Cass-Koopmans Model	Weak Axiom of Revealed Preference

USEFUL WEB LINKS

http://hassler-j.iies.su.se/Courses/MacroII/Notes/book.pdf (accessed on 2 August 2010); http://www.ssc.upenn.edu/~rwright/courses/growth.pdf (accessed on 2 August 2010): social planner's problem

http://ocw.mit.edu/courses/economics/14-452-economic-growth-fall-2009/lecture-notes/MIT14_452F09_lec05_06.pdf (accessed on 2 August 2010);

http://www.econ.cam.ac.uk/faculty/cavalcanti/S200MLecture3_07.pdf: (accessed on 2 August 2010);

http://www.econ.cam.ac.uk/faculty/cavalcanti/S200MLecture4_07.pdf (accessed on 2 August 2010): Ramsay-Cass-Koopmans Model

Aliprantis, C.D. and S.K. Chakrabarti (1999) *Games and Decision Making*, Oxford University Press, New York.

Andronov, A.A., A.A. Vitt, and S.E. Khaikin (1966) *Theory of Oscillators* (translated from Russian by F. Immirzi; translation edited and abridged by W. Fishwick), Pergamon Press, Oxford.

Apostol, T. (1974) *Mathematical Analysis*, Second Edition, Addison Wesley, Reading, Mass.

Arrow, K.J. (1951) *Social Choice and Individual Values* (1963: Second Edition), Wiley, New York.

—— (1968) 'Applications of Control Theory to Economic Growth', in G.B. Dantzig and A.F. Veinott, Jr (eds), *Mathematics of the Decision Sciences*, American Mathematical Society, Providence, R.I.

—— (1970) *Essays in the Theory of Risk Bearing*, North-Holland, American Elsevier, Amsterdam.

Arrow K.J. and A.C. Enthoven (1961) 'Quasi-concave Programming', *Econometrica*, 29, 779–800.

Arrow, K.J., L. Hurwicz, and H. Uzawa (1961) Constraint Qualifications in Maximization Problems, *Naval Research Logistics Quarterly*, VIII, 175–91.

Arrow K.J. and M. Kurz (1970) *Public Investment, The Rate of Return and Optimal Fiscal Policy*, Johns Hopkins Press, Baltimore.

Barro R.J. and X. Sala-i-Martin (2003) *Economic Growth*, MIT Press, Cambridge, Mass.

David, Blackwell (1965) 'Discounted Dynamic Programming', *Annals of Mathematical Statistics*, 36(1), 226–35.

Cass, D. (1965) 'Optimum Growth in an Aggregative Model of Capital Accumulation', *Review of Economic Studies*, 32, 233–40.

Cass, D. and J. Stiglitz (1970) 'The Structure of Investor Preference and Asset Returns and Separability in Portfolio Allocation: A Contribution to the Pure Theory of Mutual Funds', *Journal of Economic Theory*, 2, 122–60.

Caputo, M.R. (2005) *Foundations of Dynamic Economic Analysis*, Cambridge University Press, Cambridge.

Chiang, A.C. (1992) *Elements of Dynamic Optimization*, McGrawHill, New York.

Copi I.M. and C. Cohen, (1997), *Introduction to Logic*, 9th Edition, Prentice Hall, New Delhi.

Day, R.H. (1983) 'The Emergence of Chaos from Classical Growth', *Quarterly Journal of Economics*, 98, 201–13.

Das, M. (2003) 'Optimal Growth with Decreasing Marginal Impatience', *Journal of Economic Dynamics and Control*, 27(10), 1881–98.

Dasgupta, D. (2010) *Modern Growth Theory*, Oxford University Press, New Delhi.

Debreu, G. (1954) 'Representation of a Preference Ordering by a Numerical Function', in R.M. Thrall, C.H. Coombs, and R.L. Davis (eds), *Decision Processes*, Wiley, New York.

—— (1959) *The Theory of Value: An Axiomatic Analysis of Economic Equilibrium*, Wiley, New York.

Dorfman, R., P.A. Samuelson, and R. Solow (1958) *Linear Programming and Economic Analysis*, McGraw Hill, New York.

Epstein, L.G. and J.A. Hynes (1983) 'The Rate of Time Preference and Dynamic Economic Analysis', *Journal of Political Economy*, 91(4), 611–35.

Fudenberg, D. and J. Tirole (1991) *Game Theory*, MIT Press, Cambridge, Mass.

Gale, D., (1960), *Theory of Linear Economic Models*, McGraw Hill, New York.

Gibbons, R., (1992), *A Primer on Game Theory*, Harvester Wheatsheaf, New York.

Grandmont, J-M. (1988) 'Non-linear Difference Equations, Bifurcations, and Chaos: An Introduction', CEPREMAP Discussion Paper No. 8811, Paris. Later version published in *Research in Economics*, 62, 122–77.

Grass, D., J.P. Caulkins, G. Feichtinger, G. Tragler, and D.A. Behrens (2008) *Optimal Control of Nonlinear Processes: With Applications in Drugs, Corruption, and Terror*, Springer, Berlin.

Guckenheimer, J. and P. Holmes (1983) *Nonlinear Oscillations, Dynamical Systems, and Bifurcations of Vector Fields*, Springer-Verlag, Berlin.

Guha, S. (2008) 'Dynamics of the Consumption–Capital Ratio, the Saving Rate and the Wealth Distribution in the Neoclassical Growth Model', *Macroeconomic Dynamics*, 12(4), 481–502.

Halkin, H. (1974) 'Necessary Conditions for Optimal Control Problems with Infinite Horizons', *Econometrica*, 42, 267–72.

Harsanyi, J.C. (1967–8) 'Games with Incomplete Information Played by Bayesian Players', *Management Science*, 14, 159–82, 320–34, 486–502.

Hartl, R.F., S.P. Sethi, and R.G. Vickson (1995) 'A Survey of the Maximum Principle for Optimal Control Problems with State Constraints', *SIAM Review*, 37(2), 181–218.

Herstein, I.N. and J. Milnor (1953) 'An Axiomatic Approach to Measurable Utility', *Econometrica*, 21, 291–7.

Hildenbrand, W. (1983a) 'On the 'Law of Demand', *Econometrica*, 51, 997–1019.

—— (1983b) 'Introduction', *Mathematical Economics, Twenty Papers of Gerard Debreu*, Cambridge University Press, Cambridge.

Hirsch, M.W. and S. Smale, (1974), *Differential Equations, Dynamical Systems, and Linear Algebra*, Academic Press, New York.

Huang, C-F. and R.H. Litzenberger (1988) *Foundations for Financial Economics*, Prentice Hall, N.J.

Hurewicz, W. (1958) *Lectures on Ordinary Differential Equations*, MIT Press, Cambridge, Mass.

Ito, T. (1978) 'A Note on the Positivity Constraint in Olech's Theorem', *Journal of Economic Theory*, 17, 312–18.

Kuhn, H.W. and A.W. Tucker (1950) 'Non-linear Programming', in J. Neyman (ed.), *Proceedings of the Second Berkeley Symposium on Mathematical Statistics and Probability*, Berkeley, 481–92.

Koopmans, T.C. (1957) *Three Essays on the State of Economic Science*, McGraw Hill, New York.

—— (1965) 'On the Concept of Optimal Economic Growth', *Pontificae Academiae Scientiarum Scripta Varia*, 28, 225–300.

Lauwerier, H.A. (1986) 'One Dimensional Iterative Maps', in A.V. Holden (ed.), *Chaos*, Manchester University Press, Manchester, 39–57.

Leonard, D. and N. Van Long (1992) *Optimal Control Theory and Static Optimization in Economics*, Cambridge University Press, Cambridge.

Lorenz, H.W. (1993) *Nonlinear Dynamical Economics and Chaotic Motion*, 2nd Edition, Springer-Verlag, Berlin.

Mas-Colell, A., M.D. Whinston, and J. R. Green (1995) *Microeconomic Theory*, Oxford University Press, New York.

Majumdar, M., T. Mitra, and K. Nishimura (2000) *Optimization and Chaos*, Springer, Berlin.

Mangasarian, O.L. (1966) 'Sufficient Conditions for the Optimal Control of Nonlinear Systems', *SIAM Journal of Control*, 4, 139–52.

McKenzie, L.W. (1960) 'Stability of Equilibrium and the Value of Positive Excess Demand', *Econometrica*, 28, 606–17.

—— (1981) 'The Classical Theorem on the Existence of the Competitive Equilibrium', *Econometrica*, 49, 819–42.

—— (2002) *Classical General Equilibrium Theory*, MIT Press, Cambridge, Mass.

Michel, P. (1982) 'On the Transversality Condition in Infinite Horizon Optimal Problems,' *Econometrica*, 50(4), 975–85.

Mukherji, A. (1977) 'The Existence of Choice Functions', *Econometrica*, 45(4), 889–94.

—— (1985) 'On Some Implications of the Separating Hyperplane Theorem,' *Keio Economic Studies*, XXII (2), 57–71.

—— (1989) 'Quasi-Concave Optimization: Sufficient Conditions for a Maximum', *Economic Letters*, 30, 341–3.

—— (1999) 'A Simple Example of Complex Dynamics', *Economic Theory*, 14(3), 741–9.

—— (2005) 'The Possibility of Cyclical Behaviour in a Class of Dynamic Models', *American Journal of Applied Sciences*, Special Issue, 27–38.

Mukherji, B. (1992) 'Bifurcations in the Theory of Economic Growth', in J. Halevi, D. Laibman, and E.J. Nell (eds), *Beyond the Steady State*, Macmillan, New York.

Mangasarian, O.L. (1966) 'Sufficient Conditions for the Optimal Control of Nonlinear Systems', *SIAM Journal of Control*, 4, 139–52.

—— (1994) *Nonlinear Programming*, Classics in Applied Mathematics Edition, Volume 10, SIAM Publishers, Philadelphia.

Olech, C. (1963) 'On the Global Stability of an Autonomous System on the Plane', *Contributions to Differential Equations*, 1, 389–400.

Pratt, J.W. (1964) 'Risk Aversion in the Small and in the Large', *Econometrica*, 32(1–2), 122–36.

Ramsey, F.P. (1928) 'A Mathematical Theory of Saving', *Economic Journal*, 38, 543–59.

Rao, C.R. (1974) *Linear Statistical Inference and Its Applications*, Wiley Eastern Private Limited, New Delhi.

Russak, I. Bert (1970) 'On General Problems with Bounded State Variables', *Journal of Optimization Theory and Applications*, 6(6), 424–52.

Russell, B. (1931) *The Scientific Outlook*, Allen and Unwin, London.

Sandefur, J.T. (1990) *Discrete Dynamical Systems, Theory and Applications*, Oxford University Press, New York.

Scarf, H. (1960) 'Some Examples of Global Instability of the Competitive Equilibrium', *International Economic Review*, 1, 157–72.

Seierstad, A. (1984) 'Sufficient Conditions in Free Final Time Optimal Control Problems', Memorandum from Department of Economics, University of Oslo, 15 January, cited in Seierstad and Sydsaeter (1987).

Seierstad, A. and K. Sydsaeter (1977) 'Sufficient Conditions in Optimal Control Theory', *International Economic Review*, 18(2), 367–91.

Seierstad, A. and K. Sydsaeter (1987) *Optimal Control Theory with Economic Applications*, North-Holland, Amsterdam.

Sen, A. (1970) *Collective Choice and Social Welfare*, Holden-Day, San Francisco.

Starmer, C. (2000) 'Developments of Non-expected Utility Theory: The Hunt for a Descriptive Theory of Choice under Risk', *Journal of Economic Literature*, XXXVIII, 332–82.

Stokey, N.L. and R.E. Lucas, Jr with E.C. Prescott (1989) *Recursive Methods in Economic Dynamics*, Harvard University Press, Cambridge, Mass.

Suppes, P. (1964) *Introduction to Logic*, 7th Printing, Van Nostrand, New Jersey.

Takayama, A. (1985) *Mathematical Economics* (Second Edition), Cambridge University Press, Cambridge, UK.

Uzawa, H. (1961) 'The Stability of Dynamic Processes', *Econometrica*, 29, 617–31.

—— (1968) 'Time Preference, the Consumption Function, and Optimum Asset Holdings', in J.N. Wolfe (ed.), *Value, Capital, and Growth: Papers in Honour of Sir John Hicks*, Edinburgh University Press, Edinburgh.

Index